CROP MANAGEMENT AND IMPROVEMENT

CROP MANAGEMENT AND IMPROVEMENT

Editor

Stafford Cowden

scitus academics

Crop Management and Improvement
Edited by **Stafford Cowden**

This Book is Printed on Acid-Free Paper. ♾

ISBN: 9781681179094
Library of Congress Control Number: 2017948937

SCITUS Academics LLC,
616, Corporate Way, Suite 2, 4766,
Valley Cottage, NY 10989

www.scitusacademics.com

Notice

Printed in United States of America.

Preface

Improvement of agricultural methods and productivity are required to reduce the need to cut down additional forest land. Tropical forests cannot be maintained unless agricultural productivity is greatly improved. However, to feed the projected population of the mid-21st century even at present levels, not to mention a level approaching that of developed countries, agricultural efficiencies would have to be far greater than is currently the case in most countries. We need increases in agricultural productivity of between 1.8% and 3% per year for many years. The demand for food, feed, and feedstocks for bioenergy and biofactory plants will increase proportionally due to population growth, prosperity, and bioeconomic growth. Securing food supply and meeting demand for biomass will involve many biological and agro-ecological aspects such as genetic plant improvement, sustainable land use, water-saving irrigation, and integrated nutrient management as well as control of pests, diseases and weeds. It will be necessary to raise biomass production and economic yield per unit of land –not only under optimum growing conditions, but even more under conditions constrained by climate, water availability, and soil quality. In the future we will face greater complexity. Meeting food security and biomass feedstocks will involve many biophysical and ecological aspects such as genetic plant improvement, sustainable land use, water saving irrigation, integrated nutrient management, and control of pests, diseases and weeds. Furthermore, socio-economic factors and consumer behavior (change of diets, fast versus slow food) are already playing a major role in a more urbanized world. Within 20 years about 70% of the world population will live in cities, which will depend more for food security on global trade than on local or regional production capacity. For the most important commodities prices on the world markets will become more important. Therefore, with the growing urbanization the availability of land is not a regional or even national issue but has to be addressed at a global scale. Land use change through population growth, agricultural intensification and urbanization has also transformed natural ecosystems locally, regionally, and globally. Thus, more emphasis is needed on sustainable use of land, taking into account ecosystem services and prevention from polluting emissions to the environment.

'Crop Improvement and Management' covers problems of significant field crops which can have either genetic or agronomic solutions to increase productivity, sustainability and utilization of major crops and cropping systems. The book focuses on challenges, progress and prospects of crop production. It contains of wide assortment of topics including latest agronomics practices for different crops to enhance productivity, mitigate the challenges imposed by climate change, improve water use efficiency, factors controlling dormancy, optimum use of fertilizers etc. This book will help researchers and students all over the world to attain new and interesting results in the field of alternative crops and cropping systems.

Table of Contents

CHAPTER 1

Integrated crop management practices for maximizing grain yield of double-season rice crop

Depeng Wang, Jianliang Huang, Lixiao Nie, Fei Wang, Xiaoxia Ling, Kehui Cui, Yong Li & Shaobing Peng

National Key Laboratory of Crop Genetic Improvement, MOA Key Laboratory of Crop Ecophysiology and FarmingSystem in the Middle Reaches of the Yangtze River, College of Plant Science and Technology, Huazhong AgriculturalUniversity, Wuhan, Hubei 430070, China.

AUTHOR CONTRIBUTION

D.W. was responsible for all of the experimental work and data analyses, and took joint responsibility with S.P. for writing the manuscript and drawing the figures. S.P. was the principal investigator and laboratory leader, and responsible for the experimental design and reviewing final manuscript. J.H., L.N., F.W., X.L., K.C. and Y.L. contributed to experimental work and data analyses.

ABSTRACT

Information on maximum grain yield and its attributes are limited for double-season rice crop grown under the subtropical environment. This study was conducted to examine key characteristics associated with high yielding double-season rice crop through a comparison between an integrated crop management (ICM) and farmers' practice (FP). Field experiments were conducted in the early and late seasons in the subtropical environment of Wuxue County, Hubei Province, China in 2013 and 2014. On average, grain yield in ICM was 13.5% higher than that in FP. A maximum grain yield of 9.40 and 10.53 t ha^{-1} was achieved under ICM in the early- and late-season rice, respectively. Yield improvement of double-season rice with ICM was achieved with the combined effects of increased plant density and optimized nutrient management. Yield gain of ICM resulted from a combination of increases in sink size due to more panicle number per unit area and biomass production, further supported by the increased leaf area index, leaf area duration, radiation use efficiency, crop growth rate, and total nitrogen uptake compared with FP. Further enhancement in the yield potential of double-season rice should focus on increasing crop growth rate and biomass production through improved and integrated crop management practices.

1. INTRODUCTION

Rice is the staple food for more than half of the world's population and for more than 65% of the China's population[1,2]. Increasing world rice production in a sustainable manner is vital for ensuring global food security3. Global crop production can be increased by expanding the area of croplands, increasing crop yield, and increasing multiple cropping index4. Cropland expansion is not feasible because of urbanization and environmental concerns such as biodiversity loss and greenhouse gas emission4. It is essential to maintain the increase of rice yield at an annual rate of 1.5%[5] and at the same time to increase the harvest frequency of existing croplands4 in order to keep pace with the food demand of the growing human population.

Grain yield can be increased by breeding new rice varieties with greater yield potential and by improving crop and resource management to enhance actual farm yields[6,7]. Optimum crop management especially nutrient management has proven to be highly effective in improving rice grain yield[7,8]. Other management practices such as planting methods and plant density, quality of seeds and seedlings, and irrigation regime can also affect grain yield to some extend[9,10,11]. Qin *et al.* argued that testing single component of management practices independently may not capture the impact a holistic package would have on enhancing rice grain yield[12]. Ladha *et al.* stated that closing the yield gap is becoming increasingly difficult to achieve by using a component technology in isolation[13]. A more integrated approach involving nutrients, water, and other agronomic management factors will allow the maximization of rice grain yield. Furthermore, simultaneously applying a number of the best compatible individual technologies could maximize overall benefits to farmers. Depending on the need and profitability of new technologies, farmers generally integrate new technologies with existing farmers' practice (FP), which has been referred to as integrated crop management (ICM) or best management practices[13]. Several recent studies have reported greater yield improvement with ICM compared with individual crop production factor[10,12,14].

In the subtropical climates, rice can be grown up to two times per year on the same field. In the subtropical environment of Hubei province in China, for example, double-season rice cropping is usually practiced with an early-season crop from April to July and a late-season crop from July to October[15]. The wide adoption of double-season rice systems in both China and elsewhere in Asia increases multiple cropping index and thus contributes substantially to global rice supply[4]. However, the area of double cropping rice has decreased substantially in the last decade in China due to the dramatic increase in labor cost and low grain yield[15,16].

Grain yield of single-season rice crop is higher than that of double-season rice crop[17]. Within the double-season rice cropping system, the early-season rice has lower grain yield than the late-season rice[12,15]. The relatively lower yield under early season mainly resulted from slower crop growth during the vegetative phase, which was caused by lower temperature. Reduction in grain filling period due to higher temperature was also responsible for lower grain yield in the early-season rice[12]. Wu *et al.* demonstrated that grain yield of double-season rice can be increased with improved nitrogen (N) management and proper plant density, especially for the early-season rice[15]. It is necessary to determine if ICM can further increase grain yield of double-season rice crop.

Grain yield, radiation use efficiency (RUE), and N use efficiency (NUE) under various crop management practices have been intensively studied for single-season rice crop in China[18,19,20]. However, relatively little is known about yield performance, yield attributes, and resource use efficiency of double-season rice crop under ICM. Objectives of this study were to (i) compare grain yield and RUE between ICM and FP, (ii) determine maximum grain yield of double-season rice crop in central China, and (iii) identify the traits for improving yield potential of double-season rice.

2. RESULTS

2.1 Climatic condition

There was relatively small difference in seasonal average daily minimum and maximum temperatures between the early- and late-season rice (Table 1). However, temperature displayed an increasing trend in the early season, but a decreasing trend in the late season from transplanting to maturity. There was also small difference in seasonal average daily minimum and maximum temperatures between 2013 and 2014. However, higher average temperature was observed in 2013 than in 2014 in the early-season rice from flowering to maturity and in the late-season rice from transplanting to panicle initiation. The opposite was true in the late-season rice from flowering to maturity. Average temperature from panicle initiation to flowering was relatively stable across the two seasons and the two years. There was no clear difference in average daily solar radiation between the early and late seasons. Growing period from flowering to maturity generally had lower average daily solar radiation than other growing periods. Seasonal average daily solar radiation in 2013 was higher than that in 2014 (Table 1).

2.2 Crop growth and development

Relatively small difference in duration from transplanting to flowering was observed across seasons and years (Table 2). The early-season rice had 7 to 9 d longer duration in the seedbed than the late-season rice, whereas the late-season rice had 14–21 d longer duration in the ripening phase (from flowering to maturity) than the early-season rice. Overall, the growth duration of the late season was longer than that of the early season, especially in the main field from transplanting to maturity. Longer growth duration of the late season compared with the early season was mainly due to the difference in the duration of the ripening phase (Table 2).

2.3 Grain yield and its attributes

Crop management treatments had a significant effect on grain yield in both seasons in the two years (Fig. 1). On average, grain yield in ICM was 12.8% and 14.1% higher than that in FP in the early and late seasons, respectively. Grain yields of ICM and FP were 1.60–3.45 t ha^{-1} higher than that of zero-N control (N0). The late-season rice produced 40.3% higher grain yield than the early-season rice in N0, but only 16.9–18.3% higher grain yield than the early-season rice in FP and ICM (Fig. 1). There was a small and inconsistent difference in grain yield between 2013 and 2014.

Table 1: Climate conditions by crop growth stages for the early- and late-season rice in 2013 and 2014.

Year	Season	Min T[a]	Max T	RAD
		Transplanting to Panicle initiation		
2013	Early	17.8	26.2	13.9
	Late	26.8	35.4	20.6
2014	Early	16.3	25.8	14.7
	Late	24.1	31.3	13.7
		Panicle initiation to Flowering		
2013	Early	22.0	29.4	15.8
	Late	22.4	30.4	15.1
2014	Early	22.1	29.8	14.9
	Late	22.4	30.8	13.2
		Flowering to Maturity		
2013	Early	25.3	32.0	15.0
	Late	13.9	25.4	11.4
2014	Early	23.7	29.9	12.4
	Late	16.9	26.8	12.2
		Transplanting to Maturity		
2013	Early	21.9	29.3	15.0
	Late	19.7	29.5	14.9
2014	Early	20.9	28.5	13.7
	Late	20.1	29.0	12.9

a Average daily minimum temperature (Min T, °C), average daily maximum temperature (Max T, °C), and average daily solar radiation (RAD, MJ m^{-2} day^{-1}) for each growing period.

Table 2: Growth duration (days) for the early- and late-season rice in 2013 and 2014. aSW, TR, PI, FL, and MA are sowing, transplanting, panicle initiation, flowering, and maturity, respectively.

Year	Season	SW to TR[a]	TR to PI	PI to FL	FL to MA	SW to MA	TR to MA
2013	Early	45	24	31	27	127	82
	Late	36	28	29	48	141	105
2014	Early	43	25	28	32	128	85
	Late	36	23	26	46	131	95

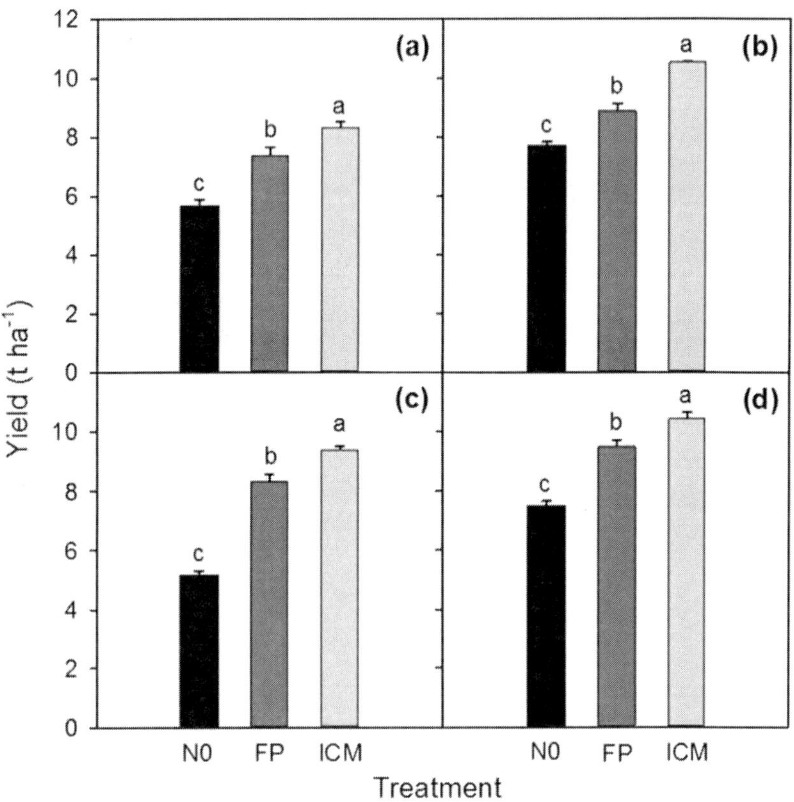

Figure 1 Grain yield in early (a) and late (b) seasons in 2013, and in early (c) and late (d) seasons in 2014. Different lowercase letters denote statistical differences between treatments of each season according to LSD test (0.05). Error bars represent ± 1 s.e. (n = 4, standard error of four replications).

Table 3: Yield components for the early- and late-season rice in 2013 and 2014.

Year	Season	Treat.	Panicles m^{-2}	Spikelets panicle^{-1}	Spikelets m^{-2} ($\times 10^3$)	Grain filling (%)	1000-grain weight (g)
2013	Early	N0[a]	255.3^{c}	108.6^{b}	27.7^{c}	81.2^{a}	23.3^{a}
		FP	308.5^{b}	138.7^{a}	42.8^{b}	70.0^{b}	22.2^{c}
		ICM	498.8^{a}	102.0^{b}	50.8^{a}	65.1^{c}	22.7^{b}
		Mean	354.2	116.4	40.4	72.1	22.7
	Late	N0	241.2^{c}	166.8^{a}	40.2^{c}	82.3^{a}	22.2^{a}
		FP	289.5^{b}	164.0^{a}	47.5^{b}	76.7^{b}	22.4^{a}
		ICM	395.7^{a}	142.4^{b}	56.3^{a}	78.3^{b}	22.4^{a}
		Mean	308.8	157.8	48.0	79.1	22.3
2014	Early	N0	229.8^{c}	88.6^{c}	20.3^{c}	92.2^{a}	25.2^{a}
		FP	341.6^{b}	120.3^{a}	41.1^{b}	81.9^{b}	24.1^{b}
		ICM	397.2^{a}	114.3^{b}	45.4^{a}	82.2^{b}	24.1^{b}
		Mean	322.8	107.7	35.6	85.4	24.5
	Late	N0	205.5^{c}	166.5^{a}	34.2^{c}	84.2^{a}	22.8^{b}
		FP	244.8^{b}	173.1^{a}	42.3^{b}	84.0^{a}	22.9^{b}
		ICM	377.4^{a}	134.3^{b}	50.7^{a}	80.7^{b}	23.2^{a}
		Mean	275.9	158.0	42.4	83.0	22.9

Higher grain yield of ICM over FP was mainly attributed to higher spikelets per m^2 (i.e. sink size), which was caused by the difference in panicles per m^2 between the two treatments (Table 3). Sink size of ICM was 10.5–18.7% and 18.5–19.9% higher than that of FP in early and late seasons, respectively. At the same time, ICM had 16.3–61.7% and 36.7–54.2% more panicles per m^2 than FP in early and late seasons, respectively (Table 3). Higher grain yield of late over early season was also due to higher sink size. The difference between the two seasons in sink size was attributed to the difference in panicle size (i.e. spikelets per panicle) instead of panicle number. The sink size of the late-season rice was 19.0% higher than that of the early-season rice, and the panicle size of the late-season rice was 35.6–46.7% higher than that of the early-season rice (Table 3). Grain filling percentage and 1000-grain weight were not responsible for the yield differences between ICM and FP or between the two seasons (Table 3). Average across seasons and years, daily grain yield of ICM and FP was 105.6 and 93.1 kg ha^{-1} d^{-1}, respectively (Table 4). There was no consistent difference in daily grain yield between the early and late seasons. Daily grain yield was higher in 2014 than in 2013, except for N0 in the early season (Table 4).

Yield difference between ICM and FP was due to the difference in aboveground total dry weight (TDW) rather than in harvest index (HI) (Tables 4 and 5). The TDW of ICM at maturity was 13.9–38.9% higher than that of FP (Table 5). From tillering to flowering, the late-season rice exhibited larger difference between ICM and FP in TDW than the early-season rice (Fig. 2a–d). Across the entire growing season, ICM had consistently higher crop growth rate (CGR) than FP in three out of the four field experiments (Fig. 2e–h). Yield advantage of the late season over the early season was due to both TDW and HI in 2013, but was due to TDW alone in 2014 (Tables 4 and 5). Overall, the early-season rice in 2013 had the lowest grain filling percentage and HI among the four field experiments, which was due to high temperature stress during flowering (daily maximum temperature of 36.2–37.0 °C on June 17–19).

2.4 Physiological characteristics

Maximum leaf area index (LAI) and leaf area duration (LAD) of ICM were significantly higher than those of FP (Table 4). The ICM treatment had higher LAI than FP throughout the growing season except for the vegetative stage in the early-season rice in 2014 (Supplementary Fig. 1). Higher LAI was responsible higher CGR and TDW in ICM compared with FP. The ICM treatment also had higher stems per m^2 than FP throughout the growing season except for the vegetative stage in the early-season rice in 2014 (Supplementary Fig. 2). The maximum stems per m^2 of ICM were higher than that of FP (Table 4).

Table 4: Maximum leaf area index, leaf area duration, maximum stem number, harvest index, and daily grain yield for the early- and late-season rice in 2013 and 2014. Within a column for each season and year,means followed by the same letters are not significantly different according to LSD (0.05). a Leaf area durationwas calculated from transplanting to maturity. bN0, FP, and ICM are zero-N, farmers' practice, and integrated crop management, respectively.

Year	Season	Treat.	Maximum leaf area index	Leaf area duration[a] (m^2 d m^{-2})	Maximum stem number m^{-2}	Harvest index (%)	Daily grain yield (kg ha^{-1} d^{-1})
2013	Early	N0[b]	2.71^{c}	134.3^{c}	303.9^{c}	46.0^{b}	69.0^{c}
		FP	4.56^{b}	210.9^{b}	364.7^{b}	48.2^{a}	89.9^{b}
		ICM	7.91^{a}	378.8^{a}	544.8^{a}	39.2^{c}	101.5^{a}
		Mean	5.06	241.4	404.4	44.5	86.8
	Late	N0	3.91^{c}	198.4^{c}	386.4^{b}	54.1^{a}	73.2^{c}
		FP	5.00^{b}	293.3^{b}	413.4^{b}	53.1^{a,b}	84.6^{b}
		ICM	8.00^{a}	468.2^{a}	691.2^{a}	50.1^{b}	100.3^{a}
		Mean	5.64	320.0	497.0	52.4	86.0
2014	Early	N0	2.32^{c}	117.6^{c}	299.4^{b}	55.4^{a}	60.7^{c}
		FP	5.52^{b}	291.3^{b}	466.3^{a}	55.3^{a}	98.1^{b}
		ICM	6.82^{a}	362.5^{a}	517.2^{a}	53.5^{b}	110.6^{a}
		Mean	4.89	257.1	427.6	54.8	89.8
	Late	N0	3.57^{c}	199.9^{c}	279.8^{c}	55.8^{a}	78.8^{c}
		FP	5.46^{b}	329.9^{b}	364.2^{b}	55.0^{a}	99.9^{b}
		ICM	8.11^{a}	556.9^{a}	530.5^{a}	51.4^{b}	109.8^{a}
		Mean	5.71	362.2	391.5	54.0	96.2

Table 5: Seasonal solar radiation utilization for the early- and late-season rice in 2013 and 2014. Within a column for each season and year, means followed by the same letters are not significantly different according to LSD (0.05). aIncident radiation, percent of intercepted radiation, intercepted radiation, aboveground total dry weight, and radiation use efficiency were calculated from transplanting to maturity. bN0, FP, and ICM are zero-N, farmers' practice, and integrated crop management, respectively.

Year	Season	Treat.	Incident radiation[a] (MJ m^{-2})	Intercepted percent (%)	Intercepted radiation (MJ m^{-2})	Total dry weight (g m^{-2})	Radiation use efficiency (g MJ^{-1})
2013	Early	N0[b]	1230.2	57.8^{c}	710.6^{c}	1136.0^{c}	1.60^{b}
		FP	1230.2	71.3^{b}	877.4^{b}	1376.1^{b}	1.57^{b}
		ICM	1230.2	73.4^{a}	902.6^{a}	1911.5^{a}	2.12^{a}
		Mean	1230.2	67.5	830.2	1474.5	1.76
	Late	N0	1564.4	72.7^{c}	1136.5^{c}	1354.4^{c}	1.19^{b}
		FP	1564.4	80.3^{b}	1255.3^{b}	1537.0^{b}	1.22^{b}
		ICM	1564.4	84.7^{a}	1325.8^{a}	1976.3^{a}	1.49^{a}
		Mean	1564.4	79.2	1239.2	1622.6	1.30
2014	Early	N0	1206.2	49.1^{b}	591.6^{b}	865.4^{c}	1.47^{c}
		FP	1206.2	70.1^{a}	845.2^{a}	1494.8^{b}	1.77^{b}
		ICM	1206.2	70.1^{a}	845.5^{a}	1702.2^{a}	2.01^{a}
		Mean	1206.2	63.1	760.8	1354.1	1.75
	Late	N0	1221.3	75.2^{c}	918.3^{c}	1185.0^{c}	1.29^{c}
		FP	1221.3	82.1^{b}	1003.1^{b}	1488.8^{b}	1.48^{b}
		ICM	1221.3	86.4^{a}	1055.1^{a}	1860.7^{a}	1.76^{a}
		Mean	1221.3	81.2	992.2	1511.5	1.51

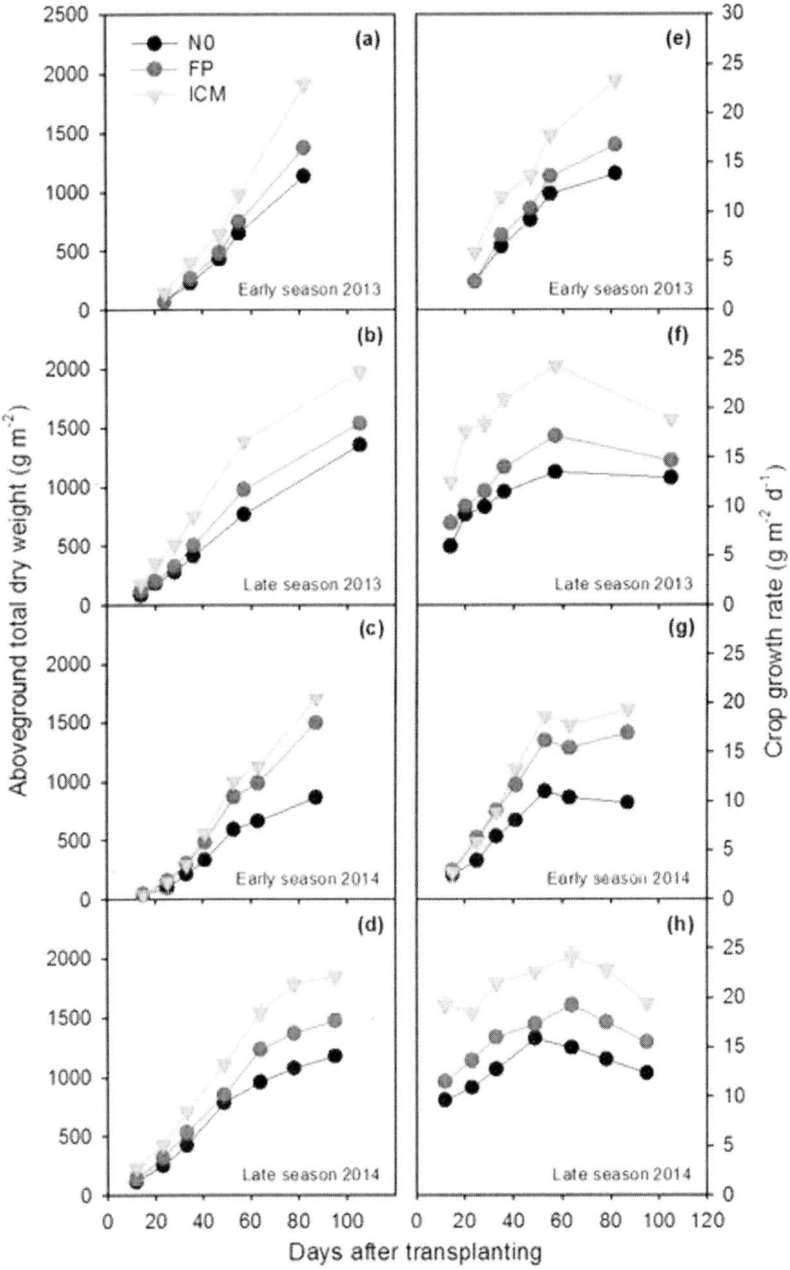

Figure 2 Aboveground total dry weight (a–d) and crop growth rate (e–h) in the early and late seasons of 2013 and 2014. Error bars represent ± 1 s.e. (n = 4, standard error of four replications).

Table 6: Nitrogen uptake and utilization for the early- and late-season rice in 2013 and 2014.

Year	Season	Treat.	N uptake^{a} (kg ha^{-1})	NHI (%)	NUE$_g$ (kg kg^{-1})	AE (kg kg^{-1})	RE (%)	PE (kg kg^{-1})	PFP (kg kg^{-1})
2013	Early	N0^{b}	109.4^{c}	57.0^{a}	47.8^{a}	—	—	—	—
		FP	172.2^{b}	57.8^{a}	38.5^{b}	7.2^{a}	32.2^{b}	22.3^{a}	34.0^{a}
		ICM	237.9^{a}	46.8^{b}	31.5^{c}	9.2^{a}	52.5^{a}	17.4^{b}	30.6^{b}
		Mean	173.2	53.8	39.3	8.2	42.4	19.9	32.3
	Late	N0	114.9^{c}	61.2^{a}	63.7^{a}	—	—	—	—
		FP	145.6^{b}	64.3^{a}	56.1^{b}	4.4^{b}	15.7^{b}	26.2^{a}	41.9^{a}
		ICM	212.2^{a}	54.5^{b}	46.6^{c}	9.9^{a}	37.4^{a}	26.4^{a}	38.1^{b}
		Mean	157.6	60.0	55.5	7.1	26.6	26.3	40.0
2014	Early	N0	89.6^{c}	62.1^{b}	53.5^{a}	—	—	—	—
		FP	176.4^{b}	63.2^{a}	46.9^{b}	17.8^{a}	44.5^{a}	39.9^{a}	42.4^{a}
		ICM	205.7^{a}	59.1^{c}	44.2^{c}	17.6^{a}	47.4^{a}	37.1^{b}	37.2^{b}
		Mean	157.3	61.5	48.2	17.7	45.9	38.5	39.8
	Late	N0	128.4^{c}	67.1^{a}	51.4^{a}	—	—	—	—
		FP	162.8^{b}	69.1^{a}	50.4^{a}	8.1^{b}	17.7^{b}	49.4^{a}	42.0^{a}
		ICM	208.3^{a}	55.1^{b}	45.9^{b}	11.3^{a}	30.7^{a}	36.9^{b}	36.8^{b}
		Mean	166.5	63.8	49.2	9.7	24.2	43.1	39.4

Higher panicles per m^2 of ICM was attributed to higher stem number, which was due to higher plant density with narrower hill spacing and more seedlings per hill compared with FP.

The RUE of ICM was 13.6–35.0% higher than that of FP (Table 5). The differences between ICM and FP in light interception percentage and intercepted radiation were relatively small compared with the difference in TDW between the two treatments. Higher RUE was observed in the early-season rice than in the late-season rice. There was no consistent difference in RUE between 2013 and 2014 (Table 5).

2.5 Nitrogen uptake and use efficiency

The ICM treatment has higher N uptake at maturity but lower N harvest index (NHI), N use efficiency for grain production, and partial factor productivity of applied fertilizer N than FP (Table 6). Agronomic N use efficiency was higher in ICM than in FP in the late-season rice, but no difference in the early-season rice. There was no consistent difference between ICM and FP in N recovery efficiency and physiological N use efficiency (Table 6), because no significant differences between ICM and FP were found in RE in the early season of 2014, and in PE in the late season of 2013. Overall, ICM tended to reduce NUE due to higher rate of N fertilizer application compared with FP. The early-season rice demonstrated higher N recovery efficiency than the late-season rice, but the reverse was true for NHI, N use efficiency for grain production, and physiological N use efficiency although the differences were relatively small in 2014 compared with in 2013 (Table 6). There was no consistent difference between the two seasons in other NUE-related traits.

Within a column for each season and year, means followed by the same letters are not significantly different according to LSD (0.05). [a] Nitrogen uptake at maturity, nitrogen harvest index (NHI), nitrogen use efficiency for grain production (NUE_g), agronomic nitrogen use efficiency (AE), nitrogen recovery efficiency (RE), physiological nitrogen use efficiency (PE), partial factor productivity of applied fertilizer nitrogen (PFP). bN0, FP, and ICM are zero-N, farmers' practice, and integrated crop management, respectively.

3. DISCUSSION

On average, ICM produced the grain yield of 9.67 t ha^{-1} compared to 8.52 t ha^{-1} from FP, resulting in a 13.5% increase in grain yield over FP. Significant increase in grain yield by ICM over FP was also reported in single-season rice in China[10,11,21-23], in double-season rice in China[12], and in double-season rice in India[14] and Bangladesh[9]. The higher grain yield of ICM was attributed to the increase in sink size, which was caused by more panicles per unit area compared with FP. It is interesting to observe that yield enhancement through improved crop management is generally realized through the increase in panicle number, while yield increase by varietal improvement is generally resulted from the large panicle size.

Higher biomass production instead of HI was responsible for higher grain yield of ICM over FP. The ICM had higher CGR than FP throughout the entire growing season in three out of four field experiments, which was associated with higher LAI, LAD, and RUE of ICM. Similar results were reported by Xue *et al.* and Qin *et al.* who attributed the yield gain of ICM to increased LAI and high radiation interception and RUE[12,21]. Notably, the RUE of ICM reached 2.01–2.12 g MJ^{-1} in the early-season rice and 1.49–1.76 g MJ^{-1} in the late-season rice in our study. These RUE values were similar to the potential values determined under high-yielding environment in previous studies[19,24,25].

Increased plant density with narrower hill spacing and more seedlings per hill in ICM contributed to higher stem number per unit area and higher CGR during the vegetative phase compared with FP. Higher stem number per unit area was the prerequisite for higher panicle number at maturity in ICM. Peltonen-Sainio stated that improved early season growth capacity supported good establishment for high interception of solar radiation, which, in turn, determines total plant biomass and grain yield[26].

Improvement in nutrient management in ICM with increased rates of N, P, and K application, and with more times of N and K application supported higher CGR throughout the growing season and higher N uptake and RUE as well compared with FP. Improved nutrient management in ICM with delayed N application was responsible for slower leaf senescence during ripening phase, as evidenced by higher ratio of flag leaf SPAD reading at maturity to that at flowering in ICM than FP (Supplementary Fig. 3). Slower leaf senescence of ICM could ensure the maintenance of higher LAI, CGR, RUE, and N uptake after flowering. Sui *et al.* also reported that N application at later reproductive growth stages had a benefit for grain yield, which might prevent and slow down leaf senescence, resulting in high photosynthetic activity[22]. Although the increased rates of N, P, and K application ensured that nutrients did not limit crop growth and yield formation in ICM, decline in nutrient use efficiency in ICM compared with FP would increase the risk of nutrient losses and cause environmental concerns.

The late-season rice produced 2.18 t ha^{-1} higher grain yield than the early-season rice in N0. The seasonal yield difference was reduced to 1.33–1.62 t ha^{-1} in FP and ICM. Yield difference was more than 1 t ha^{-1}between the two seasons in several provinces in central China15. As reported by Qin *et al.*[12], sink size due to the different panicle size was mainly responsible for the yield difference between the two seasons. The lower grain yield in the early-season rice compared to the late-season rice could be partially attributed to varietal difference and different climatic conditions between the two seasons[12]. There was no doubt that lower temperature of the early-season rice reduced CGR during the vegetative phase, while higher temperature shortened the ripening phase by 14–21 d compared with the late-season rice. One strategy to overcome the limitation of low temperature on early vegetative growth in the early-season rice is to increase the rate of basal N application. However, high rate of N application at the early growth stage when the plant's N uptake ability is still low could maximize the risk of N losses and reduce NUE. Both limited biomass accumulation during vegetative stage and shortened grain filling duration during gain development stage were detrimental to yield formation of the early-season rice. In addition, extremely high temperature may occur during flowering period in the early-season rice, which could induce spikelet sterility and reduce grain filling percentage and HI, and consequently lead to lower grain yield. This had happened in the early-season rice in 2013, as evidenced by lower grain filling percentage and HI compared with the other three field experiments (Tables 3 and 4). It appeared that the reduction in grain filling percentage and HI due to high temperature stress in 2013 early-season rice was more severe in ICM than in FP andN0, suggesting that caution should be taken when high nutrient input is used in ICM to enhance rice yield potential in high temperature-prone season or area.

Double-season rice generally has lower grain yield than single-season rice although its annual grain yield (i.e. summation of grain yield during both early and late seasons) is higher than the single-season rice[12,15,17.] Wu *et al.* stated that the attainable yield under double rice-cropping system is characterized by relatively lower grain yield of 5.46 t ha^{-1} in the early-season crop and 7.69 t ha^{-1} in the late-season crop[15]. Using data from on-farm experiments conducted in China's major rice-producing regions from 2000 to 2013, Xu *et al.* reported average grain yields of 6.5, 8.0, and 6.9 t ha^{-1} for the early-, middle-, and late-season rice, respectively[17.] Under the best crop management treatment, Qin *et al.* was able to achieve 8.3 and 9.5 t ha^{-1} grain yield in the early- and late-season rice, respectively[12]. Similarly, a grain yield of 9.5 t ha^{-1} was produced by the hybrid cultivar Liangyou-287 in the early-season rice27and by T-you 207 in the late season rice[28.] In our study, ICM achieved a maximum grain yield of 9.40 t ha^{-1} with hybrid cultivar Liangyou 287 in the early-season rice in 2014 and 10.53 t ha^{-1} with hybrid cultivar Tianyouhuazhan in the late-season rice in 2013. More importantly, daily grain yield in the main field of ICM was more than 100 kg ha^{-1} d^{-1} for both the early- and late-season rice crops. One of the criteria for "super" rice varieties in China is to produce 100 kg ha^{-1} d^{-1} in the main field excluding the period in the seedbed[29]. This is a plausible criterion because it eliminates the approach of improving yield potential by increasing crop growth duration so that cropping intensity could be maintained in the cropping system[30]. Daily grain yield is also an important criterion for judging the productivity of double-season rice crop due to limitation in total growth duration under subtropical conditions.

To achieve 9.0–10.5 t ha^{-1} grain yield in double-season rice, the following traits and their corresponding values should be considered: >45,000 spikelets m^{-2}, >80% grain filling, >50% in HI, >1,700 g TDW m^{-2}, >18 g m^{-2} d^{-1} in seasonal mean CGR, >7 in maximum LAI, >500 m^{-2} in maximum stem number, >70% in seasonal mean LI, >1.5 g MJ^{-1} in RUE, >200 kg N ha^{-1} in total N uptake, and >100 in kg ha^{-1} d^{-1} in daily grain yield. As suggested by Sui *et al.*, it is difficult to increase rice yield potential by improving a single trait of yield component[22]. For example, increase in grain yield not only needs to enlarge sink size by increasing the number of panicles but also requires adjustment of other yield formation processes. The ICM was very effective in breaking the negative relationship among the yield-related traits and achieving an overall improvement in grain yield[22].

In general, implementation of ICM involves in increased inputs in labor and resources[9]. Labor-demanding practices are less attractive to farmers as wages and the opportunity cost of labor are increasing with the progress in economic development[31]. Rice farmers in China are reluctant to invest more resources in the rice production because of lower rice prices[32]. The future research on ICM should consider the inclusion of labor-saving technologies, efficient nutrient management, and simplified crop management practices.

4. CONCLUSIONS

Yield improvement of double-season rice with ICM was achieved with the combined effects of increased plant density and optimized nutrient management. A maximum grain yield of 9.40 and 10.53 t ha^{-1} was achieved under ICM in the early- and late-season rice, respectively, indicating the potential to further increase the grain yield of double-season rice following a holistic and integrated agronomic approach[12]. Yield gain of ICM resulted from a combination of increases in sink size due to more panicle number per unit area and biomass production, further supported by the increased LAI, LAD, RUE, CGR, and total N uptake compared with FP. Further enhancement in the yield potential of double-season rice should focus on increasing CGR and biomass production through improved and integrated crop management practices. There was a tendency that nitrogen use efficiency declined under ICM due to higher rate of nitrogen fertilizer application compared with FP. Therefore, future study should consider more efficient nutrient management in ICM.

5. MATERIALS AND METHODS

5.1 Experiment design and plant materials

Experiments were conducted in 2013 and 2014 in a farmer's field at Zhougan Village (29°51′N, 115°33′E, 51 m altitude), Dajin Township, Wuxue County, Hubei Province, China. In each year, rice was grown in a double-season cropping system with an early-season rice from March to July and a late-season rice from June to October or November. Detailed dates of sowing, transplanting, and maturity were given in Supplementary Table 1. The soil has the following properties: pH 5.1, 29.7 g kg^{-1} organic matter, 2.7 g kg^{-1} total nitrogen (N), 38.3 mg kg^{-1} Olsen phosphorus (P), and 301.8 mg kg^{-1} exchangeable potassium (K). The soil test was based on samples taken from the upper 20 cm of the soil before the application of basal fertilizers in 2013.

In each experiment, crop management treatments were arranged in a complete randomized block design with four replicates. Crop management treatments included N0, FP, and ICM. The differences in crop management practices among the three treatments were summarized in Supplementary Table 2. For the ICM treatment, we increased (1) the rates of N, P, and K application, (2) the times of N and K application, and (3) plant density with narrower hill spacing and more seedlings per hill compared with FP. High nutrient input was used to ensure that the yield potential of double-season rice was not limited by nutrient supply.

For FP, N as ammonium bicarbonate was applied at basal while N as urea was applied at midtillering. For both N0 and FP, P as superphosphate was applied at basal while K as KCl was applied at midtillering. For ICM, basal N application was consist of 84.0 kg N ha^{-1}as compound fertilizer and 119.7 kg N ha^{-1} as ammonium bicarbonate in the early season, and 78.8 kg N ha^{-1} as compound fertilizer and 53.8 kg N ha^{-1} as ammonium bicarbonate in the late season. Urea was used for topdressing in ICM in both seasons. Phosphorus was applied only at basal for ICM with 36.7 kg P ha^{-1} as compound fertilizer and 36.7 kg P ha^{-1} as superphosphate in the early season, and 34.4 kg P ha^{-1} as compound fertilizer and 16.5 kg P ha^{-1} as superphosphate in the late season. Potassium for ICM was applied at basal, midtillering, and panicle initiation. At basal, 69.7 and 65.3 kg K ha^{-1} as compound fertilizer was applied in the early and late seasons, respectively. Potassium chloride was used for K topdressing with 37.4 kg K ha^{-1} at midtillering and 74.7 kg K ha^{-1} at panicle initiation in the early season, and 87.4 kg K ha^{-1} at midtillering and 52.3 kg K ha^{-1} at panicle initiation in the late season. Zinc as zinc sulfate heptahydrate was applied only at basal for all treatments in both seasons.

The N0 treatment was embedded in the FP treatment. Plot sizes were 30, 90, and 120 m^2 for N0, FP, and ICM, respectively. Varieties were Liangyou 287 for the early-season rice and Tianyouhuazhan for the late-season rice. Both varieties are F1 hybrids and widely grown for double-season rice crop in central China. Pre-germinated seeds were sown in nursery bed to produce uniform seedlings. Forty-three- to 45-day-old seedlings were manually transplanted for the early season, while 36-day-old seedlings were manually transplanted for the late season. A water depth of 5 to 10 cm was maintained until 7 days before maturity when the fields were drained. Weeds, insects, and diseases were controlled as required to avoid yield loss.

5.2 Measurements

5.2.1 Plant sampling

Sowing, transplanting, panicle initiation, flowering, and maturity dates were recorded for determining crop growth duration. Twelve hills were sampled from each plot with an interval of 7–15 days during the growing season to measure stem number, leaf area index (LAI), and aboveground total dry weight (TDW). Plants were separated into green leaves and stems. Green leaf area was measured with a leaf area meter (LI-3000, LI-COR Inc., Lincoln, NE, USA) and expressed as LAI. Leaf area duration (LAD) was calculated as the product of LAI and duration in days. The dry weight of leaves and stems were determined after oven-drying at 70 °C to constant weight. Crop growth rate was calculated as the ratio of increases in TDW to the duration of growing period. Flag leaf SPAD value (Chlorophyll meter SPAD-502, Minolta, Ramsey, NJ) was measured at flowering and maturity stages. The ratio of SPAD at maturity to SPAD at flowering was used to quantify the rate of flag leaf senescence.

5.2.2 Yield and yield components

At maturity, 12 hills were taken diagonally from a 5 m^2 area in each plot where grain yield was determined to measure LAI, TDW, harvest index (HI), and yield components. Panicles of each hill were counted to determine the panicle number per m^2. Plants were separated into green leaves, stems, and panicles. Panicles of all 12 hills were hand threshed and filled spikelets were separated from unfilled spikelets by submerging them in tap water. Three subsamples with each of 30 g filled spikelets and 10 g unfilled spikelets were taken to determine the number of spikelets. Dry weights of green leaves, stems, rachis, and filled and unfilled spikelets were measured after oven drying at 70 °C to constant weight. Spikelets per panicle, grain filling percentage (100 × filled spikelets/total spikelets), and HI (100 × filled spikelet weight/TDW) were calculated. Grain yield was determined from a 5 m^2area in each plot and adjusted to the standard moisture content of 0.14 g H_2O g^{-1} fresh weight. Grain moisture content was measured with a digital moisture tester (DMC-700, Seedburo, Chicago, IL, USA). Daily grain yield was calculated as the ratio of grain yield to growth duration from transplanting to maturity.

5.2.3 Nitrogen uptake and use efficiency

Tissue N concentration was measured at maturity. After dry weight measurement, different organs were ground using a mixer mill homogenizer (MM400, Retsch, Germany). Approximately 5.0 mg was used to measure N concentration using an NC analyzer (IsoPrime100 IRMS, Isoprime Ltd, UK). Nitrogen uptake at maturity, N harvest index (NHI), N use efficiency for grain production, agronomic N use efficiency, N recovery efficiency, physiological N use efficiency, partial factor productivity of applied fertilizer N were calculated in both seasons according to Peng *et al.*[33].

5.2.4 Radiation interception and use efficiency

In both seasons, climate data (daily solar radiation, minimum temperature, and maximum temperature) were collected from the weather station located within 2 km from the experimental site. A data logger (CR800, Campbell Scientific Inc., Logan, Utah, USA) was used as the measurement and control module. A silicon pyranometer (LI-200, LI-COR Inc., Lincoln, NE, USA) and temperature/RH probe (HMP45C, Vaisala Inc., Helsinki, Finland) were used to measure total solar radiation and temperature, respectively. Daily solar radiation from transplanting to maturity was used to determine seasonal accumulated radiation. Canopy light interception was measured between 1100 and 1300 h with an interval of 7–15 days during the growing season with a line ceptometer (AccuPAR LP-80, Decagon Devices Inc., Pullman, WA, USA). In each plot, light intensity inside the canopy was measured by placing the light bar in the middle of two rows and at approximately 5 cm above the water surface. Light intensity above the canopy was recorded immediately after the light measurement inside the canopy. Light interception was calculated as the percentage of light intercepted by the canopy [100 × (light intensity above canopy-light intensity below canopy)/light intensity above canopy]. Estimates of seasonal LI were made by linear interpolation of instantaneous measurements of LI with respect to days after transplanting according to Muurinen and Peltonen-Sainio34. Seasonal intercepted radiation during the entire growing season from transplanting to maturity was the product of seasonal LI and accumulated solar radiation during this growing period. Radiation use efficiency was calculated as the ratio of biomass production to seasonal intercepted radiation from transplanting to maturity.

5.3 Data analysis

Data were analyzed following analysis of variance[35] and means of crop management treatments were compared based on the least significant difference test (LSD) at the 0.05 probability.

ACKNOWLEDGEMENTS

This work was supported by the Program of Introducing Talents of Discipline to Universities in China (the 111Project no. B14032), National High Technology Research and Development Program of China (the 863 Project no. 2014AA10A605), Special Fund for Agro-scientific Research in the Public Interest of China from the Ministry of Agriculture (No. 201203096), and Major International Joint Research Project of China National Science Foundation (No. 31361140368).

REFERENCES

1. Maclean, J. L., Dawe, D. C., Hardy, B. & Hettel, G. P. *Rice Almanac*, 3rd ed. International Rice Research Institute, Los Baños,Philippines. 1–253 (2002).
2. Zhang, X., Wang, D., Fang, F., Zhen, Y. & Liao, X. Food safety and rice production in China. *Res. Agric. Modernization* 26, 85–88 (In Chinese with English abstract) (2005).
3. Normile, D. Reinventing rice to feed the world. *Science* 321, 330–333 (2008).
4. Ray, D. K. & Foley, J. A. Increasing global crop harvest frequency: recent trends and future directions. *Environ. Res. Lett*, doi:10.1088/1748-9326/8/4/044041 (2013).
5. Aureus, M. *The Rise of Rice. Rice Today*, Jan-Mar. p. 16 (International Rice Research Institute, Los Baños, Philippines, 2011).Horie, T. *et al.* Can yields of lowland rice resume the increases that they showed in the 1980s? *Plant Prod. Sci.* 8, 259–274 (2005).
6. Peng, S. *et al.* Improving nitrogen fertilization in rice by site-specific N management. A review. *Agron. Sustain. Dev.* 30, 649–656(2010). Zorrilla, G. *et al.* Improving rice production systems in Latin America and the Caribbean. In *Eco-Efficiency: From Vision to Reality*.
7. pp. 161–170 (CIAT, Cali, Columbia, 2012).
8. Alam, M. M., Karim, Md. R. & Ladha, J. K. Integrating best management practices for rice with farmers' crop management techniques: A potential option for minimizing rice yield gap. *Field Crops Res.* 144, 62–68 (2013).

9. Peng, X. *et al.* Crop management for increasing rice yield and nitrogen use efficiency in northeast China. *Agron. J.* 107, 1682–1690(2015).
10. Chu, G., Wang, Z., Zhang, H., Yang, J. & Zhang, J. Agronomic and physiological performance of rice under integrative crop management. *Agron. J.* 108, 117–128 (2016).
11. Qin, J. *et al.* Integrated nutrient, water and other agronomic options to enhance rice grain yield and N use efficiency in double season rice crop. *Field Crops Res.* 148, 15–23 (2013).
12. Ladha, J. K. *et al.* Integrating crop and resource management technologies for enhanced productivity, profitability and sustainability of the rice-wheat system in South Asia. In *Integrated Crop and Resource Management in the Rice-Wheat System of* South Asia (edsLadha, J. K. *et al.*). pp. 69–108 (International Rice Research Institute, Los Baños, Philippines, 2009).
13. Regmi, A. P. *et al.* Improving food security through integrated crop and resource management in the rice-wheat system in Nepal. In *Integrated crop and resource management in the rice-wheat system of* South Asia (eds Ladha, J. K. *et al.*). pp. 69–108 (International Rice Research Institute, Los Baños, Philippines, 2009).
14. Wu, W. *et al.* Toward yield improvement of early-season rice: Other options under double rice-cropping system in central China.*Europ. J. Agron.* 45, 75–86 (2013).
15. Hu, Z. *et al.* Premature heading and yield losses caused by prolonged seedling age in double cropping rice. *Field Crops Res.* 183,147–155 (2015).
16. Xu, X. *et al.* Estimating nutrient uptake requirements for rice in China. *Field Crops Res.* 180, 37–45 (2015).
17. Ying, J. *et al.* Comparison of high-yield rice in tropical and subtropical environments: I. Determinants of grain and dry matter yields *Field Crops Res.* 57, 71–84 (1998).
18. Katsura, K. *et al.* The high yield of irrigated rice in Yunnan China: a cross-location analysis. *Field Crops Res.* 101, 1–11 (2008).
19. Zhang, Y. *et al.* Yield potential and radiation use efficiency of super hybrid rice grown under subtropical conditions. *Field Crops Res.*114, 91–98 (2009).
20. Xue, Y. *et al.* An improved crop management increases grain yield and nitrogen and water use efficiency in rice. *Crop Sci.* 53, 271–284 (2012).
21. Sui, B. *et al.* Optimizing nitrogen supply increases rice yield and nitrogen use efficiency by regulating yield formation factors. *Field Crops Res.* 150, 99–107 (2013).
22. Chen, X. *et al.* Producing more grain with lower environmental costs. *Nature* 514, 486–489 (2014).
23. Kiniry, J. R. *et al.* Radiation-use efficiency in biomass accumulation prior to grain-filling for five grain-crop species. *Field Crops Res.* 20, 51–64 (1989).
24. Xiong, J., Ding, C., Wei, G., Ding, Y. & Wang, S. Characteristic of dry-matter accumulation and nitrogen-uptake of super-highyielding early rice in China. *Agron. J.* 105, 1142–1150 (2013).
25. Peltonen-Sainio, P., Forsman, K. & Poutala, T. Crop management effects on pre- and post-anthesis changes in leaf area index and leaf area duration and their contribution to grain yield and yield components in spring cereals. *J. Agron. & Crop Sci.* 179, 47–61 (1997).
26. Zhou, Y., Ju, C., Xu, G., Xie, P. & Gao, M. Liangyou-287, a new two-line early cropping super hybrid rice combination with fine grain quality. *Hybrid rice* 23, 71–72 (In Chinese) (2008).
27. Shen, Y. Performance of fine quality hybrid rice T-you 207 in experimental production and demonstration at Ningqiang, Shaanxi and its high yielding cultural techniques. *Hybrid Rice* 20, 42–43 (In Chinese) (2005).
28. Yuan, L. Breeding of super hybrid rice. In *Rice Research for Food Security and Poverty Alleviation* (eds Peng, S. & Hardy, B.). pp. 143–149 (International Rice Research Institute, Los Baños, Philippines, 2001).
29. Peng, S., Khush, G. S., Virk, P., Tang, Q. & Zou, Y. Progress in ideotype breeding to increase rice yield potential. *Field Crops Res.* 108, 32–38 (2008).
30. Dawe, D. *et al.* Nutrient management in the rice soils of Asia and the potential of site-specific nutrient management. In *Increasing productivity of intensive rice systems through site-specific nutrient management* (eds Dobermann, A. *et al.*). pp. 337–358 (Science Publishers and IRRI, Enfield, NH, and Los Baños, Philippines, 2004).

31. Peng, S. Reflection on China's rice production strategies during the transition period. *Scientia Sinica Vitae* 44, 845–850 (In Chinese with English abstract) (2014).
32. Peng, S. Crop improvement for nitrogen use efficiency in irrigated lowland rice. In *The Molecular and Physiological Basis of Nutrient Use Efficiency in Crops* (eds. Hawkesford, M. J. & Barraclough, P.). pp. 211–225 (John Wiley & Sons Ltd, West Sussex, UK, 2011).
33. Muurinen, S. & Peltonen-Sainio, P. Radiation-use efficiency of modern and old spring cereal cultivars and its response to nitrogen in northern growing conditions. *Field Crops Res.* 96, 363–373 (2006).
34. SAS Institute. SAS Version 9.1.2. 2002–2003 (SAS Institute, Inc., Cary, NC, 2003).

CHAPTER 2

Re-orienting crop improvement for the changing climatic conditions of the 21st century

Chikelu Mba[1*], Elcio P Guimaraes[2] and Kakoli Ghosh[1]

[1]*Plant Genetic Resources and Seeds Team, Plant Production and Protection Division, Food and Agriculture Organization of the United Nations (FAO), Rome, Italy*

AUTHORS' CONTRIBUTIONS

CM conceived of the study, developed the outline for a publication, conducted literature reviews, coordinated inputs and drafted the manuscript. EPG conducted literature reviews and contributed to the design and draft of the manuscript. KG conducted literature reviews and contributed to the design and draft of the manuscript. All authors read and approved the manuscript.

ABSTRACT

A 70% increase in food production is required over the next four decades to feed an ever-increasing population. The inherent difficulties in achieving this unprecedented increase are exacerbated by the yield-depressing consequences of climate change and variations and by the pressures on food supply by other competing demographic and socioeconomic demands. With the dwindling or stagnant agricultural land and water resources, the sought-after increases will therefore be attained mainly through the enhancement of crop productivity under eco-efficient crop production systems. 'Smart' crop varieties that yield more with fewer inputs will be pivotal to success. Plant breeding must be re-oriented in order to generate these 'smart' crop varieties. This paper highlights some of the scientific and technological tools that ought to be the staple of all breeding programs. We also make the case that plant breeding must be enabled by adequate policies, including those that spur innovation and investments. To arrest and reverse the worrisome trend of declining capacities for crop improvement, a new generation of plant breeders must also be trained. Equally important, winning partnerships, including public-private sector synergies, are needed for 21st century plant breeding to bear fruits. We also urge the adoption of the continuum approach to the management of plant genetic resources for food and agriculture as means to improved cohesion of the components of its value chain. Compellingly also, the National Agricultural Research and Extension System of developing countries require comprehensive overhauling and strengthening as crop improvement and other interventions require a sustained platform to be effective. The development of a suite of actionable policy interventions to be packaged for assisting countries in developing result-oriented breeding programs is also called for.

Keywords: Plant genetic resources for food and agriculture, PGRFA, Plant breeding, Crop improvement, Climate change, Biotechnology, Marker-aided selection, Genetic transformation, Induced mutations, Phenomics

1. INTRODUCTION

Population growth rates globally have so outstripped the linear rate of increases in food production that the Food and Agriculture Organization of the United Nations (FAO) estimated that 70% more food [1] must be produced over the next four decades in order to nourish adequately a human population projected to exceed 9 billion by the year 2050. The odds for attaining such an unprecedented increase, which would require the raising of the historically linear increases in annual food production by 37% [2], is substantially lessened by the consequences of climate change and variations on crop production systems [3, 4].

Table 1 Some expected negative impacts of climate change on crop production by regions[a]

Asia

- Crop yields could decrease by up to 30% in Central and South Asia
- More than 28million hectares (ha) in arid and semi-arid regions of South and East Asia will require substantial (at least 10%) increases in irrigation for a 1 °C increase in temperature.

Africa

- One of the most vulnerable continents to climate change and climate variability
- With many semi-arid regions and projected increase of 5% to 8% by the 2080s, likely reduction in the length of growing seasons will render further large regions of marginal agriculture out of production
- Projected reductions in crop yields of up to 50% by 2020
- Fall in crop net revenues by up to 90% by 2100
- Population of 75 to 250 million people at risk of increased water stress by the 2020s and 350 to 600 million people by the 2050s

Australia and New Zealand

- Agricultural production may decline by 2030 over much of southern and eastern Australia, and over parts of eastern New Zealand, due to increased drought and fire
- Change land use in southern Australia, with cropping becoming non-viable at the dry margins
- Production of Australian temperate fruits and nuts will drop on account of reduced winter chill
- Geographical spread of a major horticultural pest, the Queensland fruit fly (*Bactrocera tryoni*), may spread to other areas including the currently quarantined fruit fly-free zone

Europe

- Crop productivity is likely to decrease along the Mediterranean and in south-eastern Europe
- Differences in water availability between regions are anticipated to increase
- Much of European flora is likely to become vulnerable, endangered or committed to extinction by the end of this century

North America

- Increased climate sensitivity is anticipated in the south-eastern USA and in the USA corn belt making yield unpredictable
- Yields and/or quality of crops currently near climate thresholds (for example, wine grapes in California) are likely to decrease
- Yields of cotton, soybeans, and barley are likely to change

Latin America

- Risk of extinctions of important species
- By the 2050s, 50% of agricultural lands in drier areas may be affected by desertification and salinization
- Generalized reductions in rice yields by the 2020s
- Reductions in land suitable for growing coffee in Brazil, and reductions in coffee production in Mexico
- The incidence of the coffee leaf miner (*Perileucoptera coffeella*) and the nematode *Meloidogyne incognita* are likely to increase in Brazil's coffee production area
- Risk of *Fusarium* head blight in wheat is very likely to increase in southern Brazil and in Uruguay

Small islands

- Subsistence and commercial agriculture on small islands will be adversely affected by climate change
- In mid- and high-latitude islands, higher temperatures and the retreat and loss of snow cover could enhance the spread of invasive species including alien microbes, fungi, plants, and animals

[a]Adapted from the Second Report on the State of the World's Plant Genetic Resources for Food and Agriculture [38].

1.1 The scope of the problem

The frequent occurrences of drought and floods, that invariably result in acute food shortages such as the very recent ones in the Horn of Africa [5], are symptomatic of the grave implications of extreme weather conditions for crop production and, hence, food security. Chatham House [6] had, relying on data provided by the United Nation's Intergovernmental Panel on Climate Change (IPCC), concluded that an additional 40 to 170 million more people will be undernourished as a direct consequence of climate change. Indeed, the overwhelming prognosis is that extreme weather events such as heavy precipitation, heat waves, and rising sea levels will occur in many parts of the world during the 21st century [7] with resulting floods, drought, and salinity as the most critical consequences. The strategies for devising solutions to these constraints will vary across geographical regions as the types and magnitudes of the problems will vary. For instance, though there is the consensus that rainfall is expected to increase globally overall, some places will actually be receiving less annual rainfalls while the seasonality of rains and hence the timing of the cultivation of crops will also change. More worrisome yet, the frequencies of occurrence and durations of the extreme weather events are also expected to increase. Table 1 summarizes some of the expected negative impacts on crop production by regions of the world.

This generational challenge of producing enough food for a rapidly growing population under extreme and changing weather conditions is further exacerbated by dwindling agricultural land and water resources. There are no more redundant water resources and arable lands to deploy in augmenting the already over-stretched ones in many parts of the world. Other noteworthy drivers for food insecurity include the competing demands on scarce, depleted, and over-used arable lands and scarce food stuff for productions of bioenergy and livestock feeds. Equally confounding current conventional efforts to increase crop production sustainably is the prohibitive economic and environmental costs of the deployment of further agricultural chemicals as means for boosting yields.

The most vulnerable segments of society will be in poor developing countries, particularly in South Asia and sub-Saharan Africa, as they will suffer the most consequences of these changes to their food production systems [8, 9, 10]. In fact, Ejeta [11] estimated yield decreases of 10% to 20% for Africa's most important food crops in the coming decades. Similarly, Tester and Langridge [2] inferred that the greatest demand for yield increases as population continues to increase will be in the developing countries of the world though Interestingly, Foresight [12] averred that the applications of already existing knowledge and technology could increase yields two- to three-fold in the medium and low income countries of the world.

Success in attaining the imperative of producing more food under worsening climatic conditions and with a severely constrained natural resources base hinges on enhanced efficiencies, that is achieving more yield per unit of input. This consideration informed the advocacy by Chatham House [6] for the eco-friendly 'knowledge-intensive' 21st century Green Revolution that will replicate the dramatic yield increases of its 20th century 'input intensive' precursor [11, 13]. The growing of diverse 'smart' crop varieties that are capable of producing 'more with less' is in accord with this 'greener' perspective and will be critically important to achieving the *sine qua non* of enhanced efficiencies. This will of course require the re-orientation of many aspects of crop production systems with plant breeding and the cultivation of the resulting high yielding, well-adapted, input use-efficient, and resilient crop varieties constituting a major component of the interventions. In line with this perspective, Beddington *et al.* [4] aptly surmised that the concomitant attainment of food security and environmental sustainability would require innovative interventions as main driver for change.

2. GENETIC GAINS TRANSLATE TO 'SMART' CROP VARIETIES

Crop yields represent the net result of the intricate interactions between two main critical determinants, of approximately equal contributory effects, namely, the inherent genetic constitution of the crops and agronomic management practices [14]. Indeed, over the past seven decades in the United States, the percentage contribution of genetic gains to total on-farm yield increases in maize ranged between 33% and 94% with an average of about 50% to 60% [15, 16, 17]. Genetic gains, accruable from harnessing the potentials coded into the genetic blueprints of plant genetic resources for food and agriculture (PGRFA), could therefore make significant contributions to attaining this required 70% increases in food production.

Instances of the dramatic effects of genetic gains on crop yields include the development and massive dissemination of high yielding and resilient cereal crop varieties around the world in the course of the aforementioned Green Revolution starting in the late 1960s. The consequent marked increases in food production in many food deficit countries was credited with saving billions of people from starvation especially in Asia [11, 13]. More recently, the introduction of high yielding rice varieties, the New Rice for Africa (NERICA), in sub-Saharan Africa has also been credited with substantial increases in the production of the crop in the region [18, 19, 20, 21].

Improved crop varieties, that possess superior agronomic and quality traits, are the direct outputs of plant breeding, described by the Columbia Encyclopedia as the science of altering the heritable patterns of plants to increase their value [22]. Foresight [12] had, in recommending the use of new scientific and technological tools to address the significant challenges of producing substantially more food with minimal environmental footprints, specifically identified 'plant breeding using conventional and new techniques to improve yields ... increase water, nutrient and other input efficiencies' as means to attaining this goal. The World Economic Forum [23] also situated the breeding of new crop varieties at the top of the agenda of its industry partners' coalition of global companies to address food insecurity. This paper contributes to the ongoing discussions on how plant breeding could be rendered more responsive to these challenges. We highlight some of the strategic policy, scientific, technological, and partnership interventions that can aid national programs, especially of developing countries, to have responsive result-oriented crop improvement activities.

3. PROFILE OF THE DESIRED 'SMART' CROP VARIETIES

FAO [24] posited that 'a genetically diverse portfolio of improved crop varieties, suited to a range of agroecosystems and farming practices, and resilient to climate change' is key to sustainable production intensification. In addition to high yields, the new elite varieties envisioned to address the bourgeoning drivers for food insecurity must be adapted to extreme weather conditions and the attendant continually evolving new strains and biotypes of pests and diseases. Extreme and changing patterns of drought and salinity are probably the most critical consequences of climate change and variations for which plant breeding must develop well-adapted varieties. Additionally, the 21st century plant breeding must cater to different prevailing farming systems and conditions - including rain-fed agriculture that accounts for a significant proportion of global food production in places where erratic rainfall patterns are expected. The new elite varieties must make more efficient use of inputs, and have improved nutritional qualities that meet the myriad dietary preferences of an increasingly more affluent, health-conscious, and generally more discerning consumer. Breeding objectives and strategies must also lead to those crop varieties that fit into ecosystem-based approaches such as conservation agriculture that emphasizes zero tillage. The breeding of multi-purpose crop varieties which biomass are severally suited for use as food, bioenergy substrates, livestock feeds, and fiber will contribute to assuaging the effects of the ever increasing competing demands from these industries on arable lands, water resources, and even foodstuff.

4. UNLOCKING THE INHERENT POTENTIALS OF PGRFA

Deliberate human interventions, including hybridizations and selection pressures, in the last 10,000 years have resulted in the domestication of wild ancestors into the hundreds of thousands of breeds of both plants and animals that now form the basis for food and agriculture [25, 26]. An unintended consequence of this human intervention in the otherwise natural process of evolution and speciation has been the narrowing of the genetic base of the plants cultivated for food [2]. The extremely narrow genetic base of crops, as evidenced in the similarities and shared close ancestries of cultivars, imperil food security grievously as a majority of the cultivars of the world's most important food crops would be vulnerable to the same stresses. In Russia, for instance, 96% of all winter wheat varieties are descendants of either one or both of two cultivars, Bezostaya 1 and Mironovskaya 808 [27]. This scenario evokes the specter of the potato blight and ensuing famine in Ireland in the mid-19th century and more recently in the summer of 1970, the major devastation of corn fields by a strain of *Helminthosporium maydis* in the middle and south central part of the United States. With climate change and variations, the threat of wide ranging major crop failures as a result of biotic and abiotic stresses is all too real. This threat can be mitigated by sourcing and/or inducing and deploying new allelic variations in plant breeding.

5. WIDENING THE SOURCES OF HERITABLE VARIATIONS

Scientists are mindful of the shortcomings in the genetic diversity - and hence, increased vulnerabilities - of crops. Wild relatives of crops, land races, and other non-adapted genetic materials, even if usually low yielding and harboring undesirable traits, should be used more routinely in genetic improvement as means to addressing this shortcoming [2, 25]. The investments of efforts in the use of such non-adapted materials in plant breeding have been quite rewarding. Instances include the use of genes located on a translocated chromosome arm of rye in the genetic improvement of wheat [28]. Gur and Zamir [29] also demonstrated that the introduction of genes from the wild relative of tomato, the drought-tolerant green-fruited *Solanum pennelli*, increased yields by up to 50%. Two centers of the Consultative Group on International Agricultural Research (CGIAR), the International Institute of Tropical Agriculture, Ibadan, Nigeria and the International Center for Tropical Agriculture, Cali, Colombia, have severally used wild relatives of cassava to enhance disease resistance, improve nutritional qualities and extend shelf life of the fresh roots of the crop [30, 31, 32, 33, 34]. The legendary contribution of the reduced height gene from the Japanese wheat variety, Norin 10, to the Green Revolution is widely chronicled and certainly, other efforts have yielded significant results as well.

In general, crop wild relatives (CWRs), underutilized crops, and neglected species, that are conserved *ex situ,*on-farm, and *in situ,* are veritable repositories of the beneficial heritable traits lost in the course of domestication [29], including those for adapting to climate change [35]; these can be assembled into the envisaged 'smart' crop varieties. McCouch [25] had aptly surmised that in crop improvement, 'the surest way to succeed in a reasonable amount of time is to have access to a large and diverse pool of genetic variation'. This imperative is at the core of the work of the International Treaty on Plant Genetic Resources for Food and Agriculture (the International Treaty) which aims at the conservation, access, and sustainable use of PGRFA [36, 37].

It is indeed paradoxical that PGRFA is the least tapped resource [38] in the quest for increased food production under worsening climate change and variations scenarios even though there is ample compelling evidence to the contrary. We recommend the harnessing of the widest possible spectrum of the inherent potentials of crops and their relatives as reversal to this trend of sub-optimal use of PGRFA in crop improvement. The accruable benefits from using these non-adapted materials certainly outweigh the additional efforts and costs in time and resources for breaking linkage drags and eliminating unwanted deleterious alleles - the main reason why breeders repeatedly and largely invariably always use the same set of 'safe bet' parents. A large scale global project aimed at collecting and using wild relatives of crops in plant breeding being implemented by the Global Crop Diversity Trust, for instance, is an example of internationally driven multi-stakeholder efforts to redress this shortcoming [39]. Pre-breeding, whereby germplasm curators and plant breeders work together to use heritable variations from non-traditional gene donors to produce populations of intermediate materials that can then be used in breeding, should be adopted universally in achieving this diversification of the genetic base of improved crop varieties. The e-learning course on pre-breeding [40, 41] developed by FAO and partners under the auspices of the Global Partnership Initiative for Plant Breeding Capacity Building (GIPB; [42]), is contributing to capacity development in this novel aspect of crop improvement. Pre-breeding facilitates the broadening of the genetic base of crops through the integration of new alleles of genes into elite novel crop varieties.

Through its Global System on PGRFA [43], FAO makes available relevant policy instruments, information systems, and other mechanisms that facilitate the conservation and sustainable use of PGRFA for food security. These include the World Information and Early Warning System (WIEWS; [44]) which provides online access to 19 databases and 13 organizations, instruments, and entities relevant to PGRFA and the World Information Sharing Mechanism on the implementation of the GPA [45] which provides access to PGRFA information of 71 countries, most of which also have their own portals. FAO's Global System for PGRFA also includes landmark publications such as the Second Report on the State of the World's PGRFA [38] which provides a periodic comprehensive report on not only the status of conservation and use of PGRFA worldwide but also the relevant emerging trends. Most recently in 2011, the Second Global Plan of Action for Plant Genetic Resources for Food and Agriculture (the Second GPA; [46]) was adopted by countries as a global framework to strengthen the capacities of countries in the conservation of crop diversity and the development and deployment of a genetically diverse portfolio of improved varieties with new traits that meet food and nutritional security needs (Table 2).

All these information repositories are aiding the access to, and use of, genetic variability even across national boundaries. They facilitate access to the 1,750 national, regional, and international genebanks around the world which collectively hold about 7.4 million accessions [38]. These genebanks have been particularly successful with the collection, characterization, evaluation, and conservation of crop germplasm. Complementing the roles of these *ex-situ* gene repositories are about 2,500 botanical gardens which provide refuge for innumerable CWRs *in-situ* and the Svalbard Global Seed Vault, Norway which holds over 400,000 duplicate copies of crop germplasm from around the world [38]. Continued support, through sustained funding and enabling policies, is important for these repositories to be able to avail access to the widest possible genetic variation for improving crops. A major critical weakness in the conservation of PGRFA is the absence of a concerted, possibly global mechanism that mirrors the management of *ex-situ* collections, for *in-situ* conservation. CWRs continue to be lost as their refuges are appropriated for agricultural production or development projects so time is of essence in this regard.

Table 2 Priority Activities of the Second Global Plan of Action on PGRFA[a]

Theme	Priority activity
In situ conservation and management	1. Surveying and inventorying plant genetic resources for food and agriculture
	2. Supporting on-farm management and improvement of plant genetic resources for food and agriculture
	3. Assisting farmers in disaster situations to restore crop systems
	4. Promoting *in situ* conservation and management of crop wild relatives and wild food plants
Ex situ conservation	5. Supporting targeted collecting of plant genetic resources for food and agriculture
	6. Sustaining and expanding *ex situ* conservation of germplasm
	7. Regenerating and multiplying *ex situ* accessions
Sustainable use	8. Expanding characterization, evaluation, and further development of specific subsets of collections to facilitate use
	9. Supporting plant breeding, genetic enhancement, and base-broadening efforts
	10. Promoting diversification of crop production and broadening crop diversity for sustainable agriculture
	11. Promoting development and commercialization of all varieties, primarily farmers' varieties/landraces and underutilized species
	12. Supporting seed production and distribution
Building sustainable institutional and human capacities	13. Building and strengthening national programmes
	14. Promoting and strengthening networks for plant genetic resources for food and agriculture
	15. Constructing and strengthening comprehensive information systems for plant genetic resources for food and agriculture
	16. Developing and strengthening systems for monitoring and safeguarding genetic diversity and minimizing genetic erosion of plant genetic resources for food and agriculture
	17. Building and strengthening human capacity
	18. Promoting and strengthening public awareness on the importance of plant genetic resources for food and agriculture

[a]Adapted from the Second Global Plan of Action for PGRFA, Food and Agriculture Organization of the United Nations, Rome, Italy [46].

6. INDUCED MUTATIONS

In situations where it is either impossible or impractical to source heritable variations from existing germplasm, the induction of allelic variations becomes an appealing option. Mutation, the heritable alteration to the genetic blueprint, has been the main driver for evolution and hence speciation and domestication of both crops and animals. Following the sublime discovery of X-rays and other forms of radiation in the early 20th century and the subsequent demonstration that these could alter the genetic material permanently, scientists have induced mutations in plants using both physical and chemical agents [47, 48, 49]. Induced mutation is hence an established crop improvement strategy and is credited with the development of over 3,200 officially released elite crop varieties and ornamental plants being cultivated all over the world [50].

The induction of mutation is a chance event so scientists traditionally enhance their chances of success at inducing useful mutation events by generating massive numbers of putative mutants that are then subsequently screened. This is expensive and time-consuming with the associated sheer drudgery cited as main reason for seeking other means for exploiting heritable variations in crops. Biotechnology applications are now being used to enhance the efficiency levels for producing and evaluating large populations. For instance, the high throughput reverse genetics technique, TILLING, short for Targeted Induced Local Lesions IN Genomes [51, 52, 53] permits the efficient screening of large populations of plants for specific mutation events [54, 55, 56, 57, 58, 59, 60, 61, 62, 63, 64]. The specificity, and hence efficiency, of TILLING - it identifies mutation events in predetermined genome regions - holds great promise for the use of induced mutations to broaden the genetic base of crops.

Cell and tissue biology techniques are also used to enhance the efficiency of mutation induction. For instance, with doubled haploidy [65, 66], homozygosity of the mutated segments of the genome is achieved rapidly while *in vitro* propagation techniques are used to dissociate chimeras quickly (to generate solid homohistonts) and to produce and manage large mutant populations in cost-, time-, and space-efficient manners [67]. The critical importance of other uses of cell biology techniques, for instance, in germplasm conservation, in overcoming hybridization barriers and in the rapid multiplication of disease-free planting materials makes it an indispensable tool in crop improvement in general.

7. A RE-INVIGORATED PLANT BREEDING FOR A CHANGING WORLD

Translating the combinations of the widest possible sources of heritable variations efficiently into crop varieties whose increased yields, improved nutritional quality attributes and enhanced adaptations to abiotic and biotic stresses exceed those of the prior gains of the 20th century Green Revolution cannot be attained with a business-as-usual mindset. The current yield-centric breeding practices, of oftentimes weak breeding programs, whose objectives are largely conceived solely by the plant breeders, must evolve into participatory, multidisciplinary, and demand-driven programs that, underpinned by nurturing policy environments, make use of the most suitable scientific and technological tools to harness the potentials of PGRFA. Plant-breeding activities must perforce be re-oriented in order to have a reasonable chance of succeeding in the development of the envisaged portfolio of 'smart' crop varieties. We discuss some of the specific attributes that must characterize the result-oriented crop improvement programs of the 21st century.

8. PARTICIPATORY PLANT BREEDING

Factoring in the perspectives of the growers and other stakeholders such as consumers, extensionists, vendors, industry, and rural cooperatives in the crop improvement endeavor of developing new varieties is known as Participatory Plant Breeding (PPB; [68]). The need for this paradigm in plant breeding is probably greatest in developing countries relative to the industrialized countries where market forces determine agricultural research and development (R&D) themes including plant-breeding objectives. By having farmers and other end-users involved in the development of varieties, feedback mechanisms are enhanced hence improving the relevance of the breeding activities to the needs of the growers. Farmers' participation in plant breeding can be categorized under the three stages of design, testing, and diffusion [69]. During the design stage, breeding goals are set and variability to be used created while at the testing stage, the breeding materials are evaluated and narrowed down to the few promising ones. The diffusion stage encompasses activities spanning varietal release, on-farm trials under farmer management and the identification of the mechanisms for the dissemination of the seeds and planting materials of the improved varieties.

Farmers, as the custodians of PGRFA, have over the several millennia of selecting from, improving, and exchanging local genetic diversity contributed immensely to the diversity of plants we grow. With the upsurge in the ready availability of modern crop varieties bred in research institutes, the roles of farmers in ensuring diversity and adding value to PGRFA have waned significantly. One effect of this shift is the precariously narrow genetic base of the modern crop varieties. The obvious threat that this poses to food security calls for the systematic re-integration of farmers' knowledge and perspectives in the developing of modern crop varieties. PPB is a veritable and validated means for ensuring this. The International Treaty, through its Article 9, also requires of contracting parties the safeguarding of farmer's rights to access and benefit from PGRFA. Those rights are not safeguarded when crop varieties that do not meet their food security and nutritional needs and/or do not enhance the resilience of their farming systems are all that are available to them.

In general, PPB facilitates the rapid and enthusiastic adoption of crop varieties [70]. The related Participatory Varietal Selection (PVS) is a means for involving these stakeholders in breeding when elite materials are already available to select from and is relatively more rapid and cost-effective than the more resource-intensive PPB [71]. Ashby [69] identified the impact pathways for PPB and PVS and concluded that their characteristic of producing more acceptable varieties and hence increasing adoption was the most compelling incentive for plant breeders to adopt this paradigm. Indeed, a CGIAR-wide review of plant breeding had recommended that PPB constitute 'an organic part of each center's breeding program' [72].

9. NOVEL PLANT-BREEDING TECHNIQUES

The incredible advances in biotechnology demonstrably hold great promise for crop improvement [73]. For instance, molecular breeding, the integration of molecular biology techniques in plant breeding [74], through enhanced efficiencies, has great potentials for changing permanently the science and art of plant breeding. Molecular breeding encompasses both the use of distinguishing molecular profiles to select breeding materials and the applications of recombinant deoxyribonucleic acid (DNA) methods, that is genetic transformation, to add value to PGRFA. There are also a number of other emerging molecular biology-based techniques that hold promise for enhancing the efficiency levels of plant breeding activities. We provide some overview of the use of these technologies and techniques in developing novel crop varieties.

10. MARKER-ASSISTED SELECTION

The increasingly available rapid, efficient, high throughput, and cost-effective molecular biology tools for identifying the sources, and tracing the inheritance, of desired traits are revolutionizing the management of PGRFA in general and plant breeding in particular. Advances in molecular biology, including the ever cheaper sequencing of whole genomes, have resulted in the availability of significant amounts of information on, and hence tools for assaying, the totality of an individual's genetic make-up, that is the genome; this is known as genomics. The related proteomics (the study of proteins) and metabolomics (the study of metabolites), made possible by an ever growing volume of publicly accessible DNA, gene, and protein sequence information, are also novel ways for investigating the heredity of traits. Equally significant, advances in bioinformatics and computational molecular biology which are facilitated greatly by the novel sophisticated and powerful information technology platforms for storing and analyzing the huge volumes of data generated through these molecular biology strategies, permit the making of valid inferences in the molecular characterization of germplasm, assessments of genetic diversity and for the selections of breeding materials.

The ability to use appropriate molecular approaches in identifying genome segments that discriminate between individuals (that is molecular markers) and to apply statistical algorithms in identifying precisely where these 'landmarks' are located on the genome has changed plant breeding permanently and will be key in developing the 'smart' crops of the 21st century. Molecular markers are now demonstrably the tools of choice for tracing the inheritance of target regions of genomes in breeding materials, a plant breeding methodology known as marker-assisted (or -aided) selection (MAS).

MAS entails the use of environment-neutral molecular markers to trace the inheritance of genes, and hence the trait(s) they control, in a breeding program with or without phenotypic selection [75]. The utility of MAS is greatest for genes whose effects are difficult, time-consuming, or otherwise expensive to evaluate in a population. This may be on account of the phenotypic effects being evident only at maturity, low heritabilities, the absence of the particular stress factor being bred for or as a result of confounding environmental influences on the trait.

The use of MAS is relatively straightforward in breeding for qualitative monogenic traits with clear-cut differences between phenotypes, such as disease resistance in plants, as the genetic mapping of the associated marker results in the mapping of the trait also and vice versa. For quantitative traits, the validation of the trait-marker association through large-scale field experimentations and statistical methods in order to more precisely identify the target genome segments, that is quantitative trait loci (QTL), is additionally required [76, 77]. In general, once the marker-trait association has been verifiably established, the transmission of trait genes from parent to offspring is monitored by querying segregating materials for closely linked markers using suitably designed marker-assisted backcrossing, for instance. The utility of MAS in breeding for polygenic traits can also be derived in gene pyramiding, that is the accumulation of two or more genes, say for disease and pest resistance, which seems feasible only with this method [2].

It has been demonstrated that consistently, MAS, either as a standalone strategy or in combination with phenotyping, significantly reduces the number of generations for evaluating segregating breeding materials and generally increases efficiency levels [2, 74, 75, 78- 93]. Indeed, it has been demonstrated that MAS permits a seven-fold increase in data handling and ultimately halves the time required for breeding a new crop variety [94]. Nonetheless, the cost-benefit analysis for adopting MAS relative to phenotypic selection is always a critical consideration that must be borne in mind in devising breeding strategies especially for developing countries.

Already routinely applied in the private sector breeding companies, such as the multinational companies, Monsanto [94]; Pioneer Hi-Bred [95] and Syngenta [96], MAS is yet to take hold in public crop improvement programs mostly on account of high set-up costs and intellectual property rights (IPR) restrictions. This implies that public sector plant breeding is clearly missing out on this singularly promising opportunity to innovate. Thro et al. [97] captured the immense expectations riding on the investments in plant genomics in relation to crop improvement in characterizing plant breeding as the 'translator' of knowledge into improved crop varieties. Public sector plant breeding is yet to assume this 'translator' role in the new dispensation of crop improvement that must be 'knowledge-intensive'.

An encouraging trend, though, is the progressive decline in the cost and the concomitant improvement in the high throughput applicability of molecular biology assays and equipment. It is logical to assume that at some point in the near future, set-up costs would be generally affordable and routine assays sufficiently efficient [98] as to permit wide adoption of MAS in the public sector. The continued successful use of MAS in the private sector is providing the much needed validation and proof of concept for this paradigm. This is critically important as capacity for this breeding methodology will be critical in handling the large populations of new breeding materials to be produced from pre-breeding activities using non-adapted genetic resources, for instance. The Integrated Breeding Platform (IBP) of the Generation Challenge Program of the CGIAR [99] is an example of multi-stakeholder efforts to extend the use of MAS to developing elite varieties of food security crops in developing countries.

11. GENETIC TRANSFORMATION

Recombinant DNA technology, involving the use of molecules containing DNA sequences derived from more than one source to create novel genetic variation, has become an important crop improvement option. This is known as genetic modification (or transformation) with the new variants referred to as transgenics or simply genetically modified organisms (GMOs). The procedures involve the incorporation of exogenous DNA or ribonucleic acid (RNA) sequences, using either biolistics or vectors, into the genome of the recipient organism which, as a result, expresses novel and agronomically useful traits. Though transgenic varieties of only four crops, maize, soybean, canola, and cotton, harboring two transformation events, that is herbicide tolerance and insect resistance or their combinations, have been grown commercially since the first approvals in 1996, James [100] estimated that there had been a 94-fold increase in hectarage in the 16 years of the commercialization of genetically modified (GM) crops (from 1.7 million hectares in 1996 to 160 million hectares in 2011). Grown in 29 countries (19 developing and 10 industrial), the author estimated the value of the GMO seed market at US$13.2 billion in 2011 while the produce for GM maize, soybean, and cotton were valued in excess of US$160 billion for the same year.

In spite of the low numbers of commercial GM crops and the transformation events that confer the modified agronomic traits, four and two, respectively, the development and deployment of GM crops signal a trend in crop improvement that can no longer be ignored. This is more so as approvals for the importation of GM crops and release to the environment had been approved in 31 other countries [100]. Tester and Langridge [2] pointed out that, though the major contributions to crop improvement for this decade will be non-GM, the production and evaluation of GM crops remained an actively researched theme with only political and bioethical considerations (both driven mostly by public negative perceptions for the technology) constituting the main hindrances to wider access to the technology by growers in more countries.

Technically, the drawbacks to more widespread development of GM varieties include the lack of efficient genotype-independent regeneration systems for most crops. Also, the lingering technical difficulties with the stacking of transformation events severely limits the utility of genetic transformation in breeding for polygenic straits such as resistance to the abiotic stresses, for example salinity and drought, being caused by climate change and variations. However, the successful stacking of genes conferring insect resistance and herbicide tolerance [100] is indicative of progress in addressing this constraint. Also, research efforts must target the increasing of the range of agronomic traits being improved through this method; the two transformation events in commercial varieties are simply inadequate for GM technology to become a dominant crop improvement method.

Probably the most limiting of all factors, however, is the associated intellectual property rights (IPR) protections that restrict access to the technology. Such IPR regimes have made GMOs remain the exclusive preserve of multinational plant breeding and seed companies in developed countries that effectively use patents to restrict access to several technologies relevant to the R&D efforts for the production of the transgenic crops. These constraints must be addressed in order that this technology be used fully in realizing its possible contributions to the development of the 'smart' crop varieties of this century. With GMO crops currently grown in developing countries, for example about 60 million hectares in South America in 2011 and with millions of small holder farmers cultivating transgenic cotton in both India and China [100, 101, 102], it is plausible to expect that the IPR regimes will be changing in the future. Another hindrance to wider adoption of the GM technology is the absence of biosafety regulatory frameworks as specified by the Cartagena Protocol on Biosafety to the Convention on Biological Diversity [103] in many countries.

Efforts to address the constraints that impede both the use of the GM technology in R&D and the cultivation of GMOs have been significant as well. For instance, the African Agricultural Technology Foundation (AATF; [104]), based in Nairobi, Kenya, is acquiring and deploying proprietary agricultural technologies in sub-Saharan Africa. In one instance, AATF obtained 'a royalty-free, nonexclusive license to Monsanto technology, a *Bacillus thuringiensis* (Bt) gene (cry-1Ab)' which is being used in the development of cowpea varieties with resistance to the cowpea pod borer [105]. Similarly, the US-based Public Sector Intellectual Property Resource for Agriculture (PIPRA; [106]), assists 'foundations, not-for-profit organizations, universities, international aid agencies, and governments' in dealing with IPR issues in order to enable access to proprietary technologies. Also, Cambia, an Australian private, non-profit research institute, publishes relevant patents, white papers, and provides tutorials as means 'to provide technical solutions that empower local innovators to develop new agricultural solutions' [107]. The activities of these organizations underscore the seriousness of the impediments that IPR protections pose for innovations in agriculture and the countervailing efforts to extend the reach of the technologies and applications especially into the public goods and commons R&D domains.

12. EMERGING BIOTECHNOLOGY TECHNIQUES OF RELEVANCE TO PLANT BREEDING

The integration of biotechnologies into crop improvement is a very dynamic field of endeavor that is changing continually. A snapshot of the status of emerging technologies is provided by Lusser et al. [108] in response to a request by the European Commission 'to provide information on the state of adoption and possible economic impact of new plant breeding techniques'. The authors identified eight new such techniques and concluded that the new varieties ensuing from these techniques might be released within 3 years. These new techniques and their features are:

- Zinc finger nuclease (ZFN): Single mutations or short indels are generated or new genes are introduced into pre-determined target sites of the genome
- Oligonucleotide directed mutagenesis (ODM): Targeted mutations of one or a few nucleotides are induced
- Cisgenesis and intragenesis: GMOs are produced by the insertion of hereditary materials derived from the species itself or from a cross-compatible species and are contiguous and unchanged (cisgenesis) or the inserted DNA may be a new combination of DNA fragments but must still be from the species itself or from a cross-compatible species
- RNA-dependent DNA methylation (RdDM): Still being refined, modified gene expressions are epigenetic with the new phenotypes inherited only over a few generations
- Grafting (on GM rootstock): Desired improvements are achieved by the grafting of non-transgenic scions onto GM rootstock
- Reverse breeding: A combination of recombinant DNA techniques and cell biology procedures is used to generate suitable transgene-free homozygous parental lines rapidly for reconstituting elite heterozygous genotypes
- Agro-infiltration: Used mostly in research settings, for example to study plant-pathogen interaction in living tissues, to select parental lines or to evaluate the efficacy of transgenes, a liquid suspension of *Agrobacterium* sp. containing the desired gene(s) is used to infiltrate plant tissues, mostly leaves, so that the genes are locally and transiently expressed at high levels
- Synthetic genomics: Large functional DNA molecules that are synthesized without any natural templates are used for constructing viable minimal genomes which can serve as platforms for the

biochemical production of chemicals such as biofuels and pharmaceuticals

Lusser et al. [108] concluded that ODM, cisgenesis/intragenesis, and agro-infiltration were the most commonly used techniques with the crops developed using them having reached the commercial development phase. On the other hand, the ZFN technology, RdDM, grafting on GM rootstocks, and reverse breeding were the less used techniques in breeding. The authors further projected that the first commercial products derived from these technologies that will be released for production would be herbicide resistant oilseed rape and maize using ODM and fungal resistant potatoes, drought tolerant maize, scab resistant apples, and potatoes with reduced amylose content developed using cisgenesis and/or intragenesis.

The clearly identified needs for the further fine-tuning of technical impediments to the routine adoptions and use of these new techniques notwithstanding, it would appear that policy regulations that are expensive to comply with and public perceptions, rather than the ability to innovate, are holding back the unleashing of the incredible advances of science and technology in crop improvement. Considering that Blakeney [109] opined that 'the right to patent agricultural innovations is increasingly located within a political context', it is plausible that the magnitude of the worsening threats to global food security may ultimately serve as the critical inducement for policy-makers, interest groups, and leaders of thought and industries to unravel the thorny issues that constrain the scope of the integration of biotechnology into crop improvement.

13. HIGH THROUGHPUT PHENOTYPIC EVALUATIONS

The selections of few promising individuals out of large populations of segregating materials can be a very daunting task. With MAS, the volume of assays that can be carried out and data points generated per unit time has increased substantially. For the workflow to be wholly efficient, the assessments of the phenotypes must also keep pace with high throughput molecular assays. Indeed, for molecular data used in breeding to be reliable, the corresponding phenotypic data for which inferences are made, must also be accurate [110]. Phenomics, the study of phenomes - the sum total of an individual's phenotype is the term that describes the novel high throughput measurements of the physical and chemical attributes of an organism. Somewhat imprecisely named in this seeming analogy to genomics, it is defined by Houle *et al.* [111] as 'the acquisition of high-dimensional phenotypic data on an organism-wide scale'. High throughput imaging of parts of a living plant, for example roots and leaves, using thermal infra-red, near infra-red, fluorescence, and even magnetic resonance imaging permit non-destructive physiological, morphological, and biochemical assays as means for dissecting complex traits such as drought and salinity tolerances into their component traits [112, 113]. Though significant technical challenges, such as data management, still require addressing, phenomics facilities are increasingly being set up with a number of them providing high throughput phenotyping services to requestors. These new facilities include the High Resolution Plant Phenomics Centre in Canberra and the Plant Accelerator in Adelaide, both in Australia [114]; LemnaTec in Wuerselen [115] and Jülich Plant Phenotyping Centre in Jülich [116] both in Germany; and Ecotron [117] and Ecophysiology Laboratory of Plant Under Environmental Stress (LEPSE; [118]) both in in Montpellier, France. In Canada, there is the The Biotron Experimental Climate Change Research Centre in London, Ontario [119]. The high set-up costs and technical know-how may impede the access of developing countries to such platforms for some considerable time.

14. OVERARCHING POLICY ENVIRONMENT FOR THE PGRFA MANAGEMENT CONTINUUM

The benefits of value addition to PGRFA, that is improved crop varieties that meet the needs of the growers, can be derived sustainably, especially for the most at-risk food insecure countries in the developing world, only with the comprehensive strengthening of, and forging of linkages between, the three components of the PGRFA value chain: (1) conservation; (2) plant breeding; and (3) the delivery of high quality seeds and planting materials to growers. This is the 'PGRFA continuum' [120], the seamless dovetailing of the three components, as distinct from targeting the strengthening of any of the three in isolation. Based on the cohesion in this value chain - that characterizes the activities of private sector commercial breeding companies and the PGRFA management of some emerging countries such as Brazil, China, and India [94] - it is logical to conclude that the real value of crop germplasm lies in its use in plant breeding. Pragmatically also, the efforts invested in breeding come to naught if there is no effective delivery system for the seeds and planting materials underscoring therefore the need to interlock all three components.

The successful implementation of the Second GPA [46] also envisages the adoption of this continuum approach. The 18 priority activities (Box 1) of the GPA provide a most practical template for countries for concerted interventions at the three components of the PGRFA value chain. These PAs are subdivided into four main themes: *in-situ* conservation and management; *ex-situ* conservation; sustainable use; and building sustainable institutional and human capacities.

The sustainable use of PGRFA encompasses activities relating to direct utilization of PGRFA by farmers and to their uses in crop improvement. The International Treaty, especially in its Article 6, equally requires of contracting parties not only to conserve their genetic resources but to use them (for value addition) and to deliver the improved varieties efficiently. FAO [121] opined that 'any weakness in this continuum truncates the value chain and effectively scuttles all the efforts to grow the most suitable crop varieties'. It is in this vein that FAO and partners are working with developing countries to articulate National PGRFA Strategies for institutionalizing the continuum approach to managing PGRFA [120]. The strategy identifies priority crops and relevant stakeholders; prescribes time-bound action plans across the continuum and enunciates governance mechanisms and means for monitoring implementation. Nurturing policy environments, especially those that enable countries adopt the continuum approach to the management of PGRFA, are critically important for reaping the most sustainable benefits from PGRFA, namely, the improved crop varieties. FAO's normative activities provide support for the implementations of the International Treaty and the Second GPA and for developing the necessary policies, and legislations as means for attaining this goal.

15. WINNING PARTNERSHIPS

The reorientation of crop improvement in order to be responsive to the drivers of food insecurity, especially in developing and emerging economies, will require a wider range of partnerships beyond the traditional National Agricultural Research and Extension Systems (NARES). FAO [38] reported the prevailing trend whereby the private sector (multinational and local commercial plant breeding and seed companies) is increasingly developing and deploying elite crop varieties especially in instances where markets, favorable policy regimes, and legal frameworks that spur investments are in place. In tandem, public investment in crop breeding programs is contracting implying therefore that the breeding and dissemination of elite varieties of crops that fall outside of the business remit of the private sector could, as is increasingly the case, be neglected to the detriment of food security. Equally important is the role of non-governmental organizations and myriad civil society actors in the provision of agricultural extension services in developing countries. These bourgeoning dynamics must influence the articulation of policies and the building of collaborations and wide-ranging partnerships. For such partnerships to succeed, local knowledge must be integrated just as relevant private and public sector entities including the NARES, centers of the CGIAR, and regional R&D networks are assembled. The safeguarding of intellectual property rights, including plant variety protection, and the respect of patents are means for attracting private sector investments. Public-private partnerships, for example the ongoing joint activities between Syngenta and public African NARES [122, 123], are particularly important for technology transfer, a critical vehicle for increasing the access of developing countries to novel biotechnologies that impact on crop improvement, for instance. On the other hand, public sector investments in food security must be ensured as the private sector, especially in developing countries, do not cater for all crops that are important for food security. Partnerships must also be cross-sectoral, for instance between ministries responsible for the environment, science and technology, commerce, education, and the ministry of agriculture. This ensures access to the full spectrum of PGRFA that may be needed for value addition while also ensuring a means for delivering the planting materials efficiently to the growers in gainful manners.

16. NATIONAL CAPACITIES FOR CROP IMPROVEMENT

The GIPB surveyed 81 countries for capacities in plant breeding and related biotechnologies [124] and subsequently conducted in-depth analysis of the plant-breeding and seed systems sectors of six of the countries: Ghana, Kenya, Malawi, Bangladesh, Thailand, and Uruguay [125]. The findings reflected the deduction by FAO [38] that, in general, the scope of funding, staffing and hence, activities per capita, of publicly-funded plant-breeding programs were either dwindling progressively or had stagnated over time. In Africa, instances of up to a 10-fold decrease in funding of plant breeding activities have occurred between 1985 and 2001 [126, 127]. The worrisome global trend of ageing and retiring plant breeders that were not being replaced by younger ones was captured in these surveys also; over 40% of plant breeders in the countries surveyed were aged 50 years and above. Indeed, to compound the problem, too few new plant breeders are being trained in universities in both developed and developing countries [127, 128, 129]. It would appear though that there was no perceptible downward trend in the award of plant breeding degrees in the USA between 1995 and 2000 [130] implying that this problem might either have been more acute in developing countries [128] or had assumed a global dimension only in the last decade. Currently, there is a general consensus however that the current capacity for plant breeding is inadequate to deal with the generational challenges of food insecurity with Knight [131] encapsulating the sense of despair in the somberly titled article, 'A Dying Breed'.

The training of future plant breeders is generally considered a major component of the preparedness for sustained food security and has been the subject of copious analyses and studies. For instance, the symposium 'Plant Breeding and the Public Sector: Who Will Train Plant Breeders in the U.S. and around the World?' held at Michigan State University in the US was aimed at charting a course for addressing this critical constraint through the devising of curricula, raising awareness, and fostering partnerships [126, 132-139]. The symposium concluded that future plant breeders, at PhD level, must in addition to possessing skills in the traditional disciplines of experimental design, applied statistics, Mendelian (transmission) genetics, population and quantitative genetics, and principles and practice of plant breeding also be trained in myriad areas ranging from subjects in the biological sciences including plant physiology, ecology, pathology, entomology, molecular biology, and genomics through business management to law, especially IPR [137]. More recently, Repinski *et al.* [129] in analyzing a very wide ranging Delphi study for articulating the curriculum of the future plant breeder came to the same conclusions regarding the need for broadening the scope of the curriculum to reflect the realities of modern breeding techniques and the fact that a significant number of plant breeders work in the private sector where legal and policy issues are critically important. Multidisciplinary teams, staffed by personnel with specialized skills in these areas, will compensate for the reality that no one plant breeder will be adept at sufficient levels of skill in all these disciplines.

Granted, most private sector plant breeders graduated from publicly-funded institutions ([132] estimated that most private sector breeders in the US attended publicly-funded land grant universities, for instance) but the public sector's role in the training of plant breeders is very critical and must be considered a contribution to public good [133] that cannot be ceded wholly to the private sector without compromising the future of plant breeding and hence food security. While the role of the private sector is also critical in this regard, in the provision of fellowships, for instance [139], it should not be expected to play the leading role as funding could not be guaranteed this way.

The centers of the CGIAR are also considered valuable partners in the training of plant breeders [133]. With improved funding, these centers, appropriately located in developing countries and working on food security crops, could provide the much-needed training facilities that many developing country governments cannot provide. The IBP, for instance, is spearheading the training of plant breeders from developing countries in molecular breeding techniques. The African Centre for Crop Improvement

(ACCI; [139]) at the University of Kwazulu-Natal, South Africa and the West Africa Centre for Crop Improvement (WACCI;) at the University of Ghana, Legon, Ghana, both funded under the auspices for the Alliance for a Green Revolution in Africa, are producing highly skilled plant breeders that are trained in Africa to work on African food security crops. Both universities partner with Cornell University, Ithaca, New York in the US in this endeavor. This is a very laudable model that is bridging the gap created by the continued inability of countries to establish and fund training facilities adequately.

17. CONCLUSIONS AND FUTURE PERSPECTIVES

There is a compelling urgency to institute measures that ensure that farmers worldwide, but especially the small-scale farmers that produce the majority of the food in food insecure countries, can grow the portfolio of suitable crop varieties that are amenable to the eco-efficient production systems of the sustainable crop production intensification (SCPI) paradigm needed to feed the world in the 21st century. The major hindrances to the attainment of SCPI include: inadequate investment; sub-optimal human resources; inability to innovate as evidenced in prevailing inadequate deployment of appropriate science and technology; weak institutions; sub-optimal R&D infrastructure; and poor policy regimes. Crop improvement, by fostering genetic gains that aid food production through enhanced productivities, is a very critical component of SCPI. We make the case therefore that plant breeding, by translating the potentials inherent in PGRFA into 'smart' crop varieties, can engender a most significant impetus for sustained food security even as human population increases and extremely inclement weather conditions constrain crop production. To achieve this, plant breeding must be re-oriented in a number of very critical ways.

17.1 Broadened genetic diversity of crops

Firstly, the extremely narrow genetic base of crops, which puts food security at risk, must be broadened at both the intra- and inter-specific levels. Conserved PGRFA, *ex-situ* and *in-situ,* and the heritable diversity available on-farm, including in landraces, must be explored to source the novel alleles that confer enhanced productivities. FAO through its Global PGRFA System, the International Treaty and the Global Crop Diversity Trust; the CGIAR centers, regional networks, and the NARES around the world must continue to invest considerable efforts to ensure that breeders have access to the genetic variations they require for their work. Some harmonization of the information dissemination mechanisms is called for to ensure enhanced efficiencies. International norms are now being leveraged to facilitate the sourcing of these much needed genetic variations even across national boundaries. Induced mutations, an established scientific method that has been used for almost one century to mimic nature, is increasingly important for inducing the unmasking of novel alleles of genes to which plant breeders do not otherwise have access. The current constraints to crop productivities deny humanity the limitless space and time for the natural process of spontaneous mutations to make these novel heritable variations available. Pre-breeding is critical in achieving this broadened genetic base of crops. The introduction of new genes and their variants into crops from novel sources will be critical to replicating the impacts of the Green Revolution as the current generational challenges demand.

17.2 Defining the breeding objectives

A second area for re-orienting plant breeding is in the 'what'. What should be the breeding objectives? Without de-emphasizing yield, resistances to biotic and abiotic stresses of import in climate change adaptation, enhanced nutritional quality traits, and the multipurpose use of crop biomass (including for bioenergy, livestock feed, and fiber) are key objectives. Also, the amenability to low-input eco-efficient farming systems will increasingly constitute standard breeding objectives. The enthusiastic adoption of NERICA in sub-Saharan Africa is an example of the efficacy of the alignment of breeding objectives to addressing the constraints posed by empirically determined drivers. In general, market forces which reflect end-user preferences will be the main driver in the definition of breeding objectives.

17.3 Innovating for result-oriented plant breeding

Thirdly, the 'how' of plant breeding will probably attract the most innovative interventions. How should crops be bred? Increased use of the immensely powerful biotechnologies that have revolutionized the biological sciences is imperative. Demonstrably, MAS, supported by the tools of genomics and the other -omics and information technology platforms, permits high throughput evaluations of breeding materials. Genetic transformation and the resulting GM crops are increasingly cultivated around the world; the technology holds promise and countries need capacity building in order to, at the minimum, make evidence-based decisions as to its adoption. Equally, the other emerging biotechnologies such as ZFN, ODM, transgenesis and cisgenesis, RdDM, grafting on GM stock, reverse breeding, agro-infiltration, and synthetic genomics, though requiring further refinements to varying degrees, will also become quite important in the very near future. Countries will increasingly require support in navigating the IPR regimes that govern access to these technologies and the regulatory issues pertaining to their adoptions. As massive numbers of new breeding materials are generated through pre-breeding, MAS must be complemented by phenomics in order that reliable predictions of the breeding values can be made. Private sector plant breeding and seed companies have taken the lead in leveraging these innovations in producing highly successful crop varieties and provide models for re-tooling the public sector crop improvement programs.

17.4 Policy and strategic interventions

A fourth consideration is the 'where' in the agricultural R&D environment for situating plant breeding. Certainly, an enabling environment is required for breeding to be relevant and, hence, thrive. The erstwhile piecemeal interventions at the three components of the PGRFA value chain, namely, conservation, breeding, and dissemination of seeds and planting materials is, simply, inadequate. A result-oriented plant breeding must have access to the widest possible source of heritable variations just as it needs an effective mechanism to deliver high quality seeds and planting materials to the growers. This is the PGRFA continuum that significantly enhances the ability of plant breeding to deliver need-based outputs. We posit that not only all three individual components but their intervening linkages must be strengthened in tandem. A National PGRFA Strategy helps to institutionalize this paradigm that demonstrably mirrors the operations of the highly successful private sector crop improvement multinationals.

17.5 Winning partnerships for the reinvigorated crop improvement

The 'who' of the 21st century plant breeding is the fifth critical consideration. Who are the main stakeholders in the crop improvement component of the PGRFA management continuum? The increasingly pivotal roles of the private sector must be factored into policy-making and in the development of strategies. The private sector is not only marketing seeds and planting materials but also breeding the new varieties; its continued participation in these activities must be encouraged especially where comparative advantages are demonstrated. Enabling policy, legal, and market environments that spur innovation and investments of capital are key to fostering the much needed public-private partnerships required for operating at scale. A healthy balance must be struck between IPR (and the innovations and investments that they encourage) and the imperative of contributing to public good. The roles of the International Convention for the Protection of New Varieties of Plants, that is UPOV, and various national, regional, and global industry interest groups will be critically important in this regard.

17.6 Capacity enhancements for the 21st century plant breeder

A sixth consideration is the 'by whom'. What is the profile of the 21st century plant breeder? In fact, the 'plant breeder' is the multidisciplinary team that makes use of the most appropriate scientific and technological tools in generating new crop varieties and the germplasm curators, farmers, and seed marketers that they work with. Technically, the multidisciplinary team driving a breeding program will include persons skilled in the traditional disciplines of plant breeding as well as those with in-depth knowledge of various ancillary biotechnological techniques. Skills in information technology, business management, law, and so on will also be required in such teams. Aside from private sector plant breeding and seed companies, such a suite of expertise does not exist in most public sector breeding concerns. The training of the future plant breeder, though mentioned often now, is still not receiving as much attention, in terms of funding, facilities, skilled trainers, and the number of available opportunities, which it deserves. Capacity building will require wide-ranging public-private partnerships in order that the curriculum being developed can be effective. The role of the CGIAR centers will remain critical. The regional training hubs, ACCI and WACCI, provide models worth emulating and scaling up. The highly successful land grant universities scheme of the United States demonstrates the lasting impacts that concerted investment of resources in training can have.

17.7 Strengthening the NARES

Finally, the re-oriented crop improvement programs require a sustaining platform, in this case, the NARES. As we have indicated, the continued decline in funding for agricultural R&D has led to weakened NARES; breeding programs are ill-staffed and poorly equipped while extension systems have become moribund in many developing countries. Equally disturbing is the dearth of reliable mechanisms for the dissemination of high quality seeds and planting materials of improved varieties. Indeed, while the work of the CGIAR centers in filling this gap cannot but be commended, the manifest over dependence of many NARES on these centers can only be injurious in the long run. For one thing, the mandates of these centers preclude work on many important food security crops. United in the recognition of the imperative for re-orienting agriculture, development organizations including FAO, the World Bank, the International Fund for Agricultural Development (IFAD), the CGIAR, and so on have severely recommitted their resolves to stamp out hunger. The strengthening of the NARES, the ultimate bulwark between hunger and the populace in many developing countries, must be at the top of the agenda. Bold initiatives underpinned by political will have strengthened and re-oriented agriculture in the past. For instance, the contributions of the land grant universities, including the extension services, to the food security of the US are legendary. Many national governments sadly lack the political will to strengthen their NARES as means for ending hunger. Support to national governments must therefore include mechanisms that contribute to fostering the nurturing policy environments for investments to bear fruit. In the final analysis, the ultimate responsibilities for crop improvement, just as in safeguarding food security in general, lies with national governments and by extension, their NARES. These responsibilities may be abdicated only at the peril of food security and at the certain risk of consequent instability and retarded development. The well-funded and adequately staffed Embrapa, Brazil's Agricultural Research Corporation, for instance, demonstrates very clearly the recent significant impacts that government policies can have on the viability of a country's agricultural R&D sector.

The coalescence of the consequences of climate change and variations with other critical demographic, economic, social, and industrial pressures pose unprecedented monumental risks to food security and people's general wellbeing. Unarguably, crop improvement and its outputs of 'smart' crop varieties can contribute to mitigating these threats. Multilateral organizations, civil society, and national governments must ride the momentum of the current reinvigorated attention to food security and strengthen capacities for crop improvement in innovative manners. Countries need assistance with suites of actionable policy interventions that leverage validated technologies and strategies in aid of result-oriented crop improvement. Such policy items or measures that countries can adopt in strengthening the three components of, and the linkages between, the PGRFA continuum in tandem are not readily available in forms amenable to ease of dissemination. The re-orienting of crop improvement would require the packaging of validated measures into a 'toolbox' to act as a one-stop shop for actionable intervention instruments. The work of the GIPB and similar multi-stakeholder platforms in articulating and assembling such tools serve as examples of multi-stakeholder efforts that deserve continued support especially in order to operate successfully at scale.

ACKNOWLEDGEMENTS

Generous funding provided by the Bill and Melinda Gates Foundation - through the Global System Project implemented by the Global Crop Diversity Trust - in support of the activities of the Global Partnership Initiative on Plant Breeding Capacity Building is gratefully acknowledged.

REFERENCES

1. Food and Agriculture Organization of the United Nations: How to Feed the World in 2050. 2009, FAO, Rome
2. Tester M, Langridge P: Breeding technologies to increase crop production in a changing world. Science. 2010, 327: 818-822.
3. Food and Agriculture Organization of the United Nations: Mitigation of Climate Change in Agriculture Series No. 1. FAO, Global Survey of Agricultural Mitigation Projects, Rome
4. Beddington J, Asaduzzaman M, Fernandez A, Clark M, Guillou M, Jahn M, Erda L, Mamo T, Van Bo N, Nobre CA, Scholes R, Sharma R, Wakhungu J: Achieving food security in the face of climate change: Summary for policy makers from the Commission on Sustainable Agriculture and Climate Change. Copenhagen, Denmark: CGIAR Research Program on Climate Change. 2011, Agriculture and Food Security (CCAFS), Copenhagen
5. International Food Policy Research Institute (IFPRI): Global Hunger Index: The Challenge of Hunger. Taming Price Spikes and Excessive Food Price Volatility. 2011, , Washington, DC
6. Evans A: The Feeding of the Nine Billion: Global Food Security for the 21st Century. 2009, Chatham House, London
7. Intergovernmental Panel on Climate Change (IPCC): Summary for Policymakers. Managing the Risks of Extreme Events and Disasters to Advance Climate Change Adaptation. Edited by: Field CB, Barros V, Stocker TF, Qin D, Dokken DJ, Ebi KL, Mastrandrea MD, Mach KJ, Plattner G-K, Allen SK, Tignor M, Midgley PM. 2012, Cambridge University Press, Cambridge
8. Nelson GC, Rosegrant MW, Koo J, Robertson R, Sulser T, Zhu T, Ringler C, Msangi S, Palazzo A, Batka M, Magalhaes M, Valmonte-Santos R, Ewing M, Lee D: Food Policy Report No. 21. Climate Change: Impact on Agriculture and Costs of Adaptation. 2009, 19-International Food Policy Research Institute, Washington DC, USA
9. Hertel TW, Burke MB, Lobell DB: The poverty implications of climate-induced crop yield changes by 2030. Global Environmental Change. 2010, 20: 577-585. 10.1016 /j.gloenvcha.2010.07.001.
10. Rosegrant MW: Impacts of climate change on food security and livelihoods. Food security and climate change in dry areas: proceedings of an International Conference, 1-4 February 2010, Amman, Jordan. Edited by: Solh M, Saxena MC. 2011, 24-26. International Center for Agricultural Research in the Dry Areas (ICARDA), Aleppo
11. Ejeta G: Revitalizing agricultural research for global food security. Food Sec. 2009, 1: 391-401. 10.1007/s12571-009-0045-8.
12. Foresight: The Future of Food and Farming: Final Project Report. 2011, The Government Office for Science, London
13. Ejeta G: African Green Revolution Needn't Be a Mirage. Science. 2010, 327: 831-832.
14. Fernandez-Cornejo J: Agriculture Information Bulletin No. 786. The Seed Industry in US Agriculture: An exploration of Data and Information on Crop Seed Markets, Regulation, Industry Structure, and Research and Development. 2004, 213-Economic Research Service, US Department of Agriculture, Washington DC, USA

15. Duvick DN: Genetic contributions to advances in yield of US maize. Maydica. 1992, 37: 69-79.
16. Duvick DN: Heterosis: Feeding People and Protecting Natural Resources. The Genetics and Exploitation of Heterosis in Crops. Edited by: Coors JG, Pandey S, Madison WI. 1999, American Society of Agronomy, Inc., Crop Science Society of America, Inc., Soil Science Society of America, Inc, MAdison, USA
17. Duvick DN: Genetic Progress in Yield of United States Maize (Zea mays L.). Maydica. 2005, 50: 193-202.
18. Dalton TJ, Guei RG: Productivity gains from rice genetic enhancements in West Africa: countries and ecologies. World Development. 2003, 31: 359-374. 10.1016/S0305-750X(02)00189-4.
19. Diagne A: Diffusion and adoption of NERICA rice varieties in Côte d'Ivoire. The Developing Economies. 2006, XLIV-2: 208-231.
20. Oikeh SO, Nwilene F, Diatta S, Osiname O, Touré A, Okeleye KA: Responses of upland NERICA rice to nitrogen and phosphorus in forest agroecosystems. Agron J. 2008, 100: 735-741. 10.2134/agronj2007.0212.
21. Wopereis MCS, Diagne A, Rodenburg J, Sié M, Somado EA: Why NERICA is a successful innovation for African farmers. Outlook on Agriculture. 2008, 37: 169-176. 10.5367/000000008785915502.
22. The Columbia Encyclopedia: Sixth ednth editionNew York, NY: ColumbiaUniversity Press; 2008http://www.encyclopedia.com/topic/plant_breeding.aspx
23. World Economic Forum: Realizing a New Vision for Agriculture: A roadmap for stakeholders. 2010, World Economic Forum, Geneva
24. Food and Agriculture Organization of the United Nations: Save and Grow - A Policy Maker's Guide to the Sustainable Intensification of Smallholder Crop Production. 2011, FAO, Rome
25. McCouch S: Diversifying selection in plant breeding. PLoS Biol. 2004, 2: 1507-1512.
26. Waines GJ, Ehdaie B: Domestication and crop physiology: roots of green-revolution wheat. Annals of Botany. 2007, 100: 991-998.
27. Martynov SP, Dobrotvorskaya TV: Genealogical analysis of diversity of Russian winter wheat cultivars (Triticum aestivum L.). Genetic Resources and Crop Evolution. 2006, 53: 386-386.
28. Rabinovich SV: Importance of wheat-rye translocations for breeding modern cultivars of Triticum aestivum L. Euphytica. 1998, 100: 323-340. 10.1023/A:1018361819215.
29. Gur A, Zamir D: Unused natural variation can lift yield barriers in plant breeding. PLoS Biol. 2004, 2: 1610-1615.
30. Hahn SK, Terry ER, Leuschner K: Breeding cassava for resistance to cassava mosaic disease. Euphytica. 1980, 29: 673-683. 10.1007/BF00023215.
31. Akano AO, Dixon AGO, Mba C, Barrera E, Fregene M: Genetic mapping of a dominant gene conferring resistance to the cassava mosaic disease (CMD). Theor Appl Genet. 2002, 105: 521-525.
32. Fregene M, Matsumura H, Akano A, Dixon A, Terauchi R: Serial analysis of gene expression (SAGE) of host-plant resistance to the cassava mosaic disease (CMD). Plant Molecular Biology. 2004, 56: 563-571.
33. Chavez AL, Sanchez T, Jaramillo G, Bedoya JM, Echeverry J, Bolanos EA, Ceballos H, Iglesias CA: Variation of quality traits in cassava roots evaluated in landraces and improved clones. Euphytica. 2005, 143: 125-133. 10.1007/s10681-005-3057-2.
34. Blair MW, Fregene MA, Beebe SE, Ceballos H: Marker-assisted selection in common beans and cassava. Marker-assisted selection: Current status and future perspectives in crops, livestock, forestry and fish. Edited by: Guimaraes EP, Ruane J, Scherf BD, Sonnino A, Dargie JD. 2007, 81-115. FAO, Rome
35. Maxted N: Ford-Lloyd BV, Kell SP, Iriondo JM, Dulloo ME, Turok J (eds): Crop Wild Relative Conservation and Use. 2008, CABI Publishing, Wallingford
36. The International Treaty on Plant Genetic Resources for Food and Agriculture, http://www.planttreaty.org/

37. Food and Agriculture Organization of the United Nations (FAO): The International Treaty on Plant Genetic Resources for Food and Agriculture. 2009, FAO, Rome
38. Food and Agriculture Organization of the United Nations: Second Report on the State of the World's Plant Genetic Resources for Food and Agriculture. 2010, FAO, Rome
39. The Global Crop Diversity Trust,http://www.croptrust.org/main/
40. Pre-breeding for Effective Use of Plant Genetic Resources.http://km.fao.org/gipb/,
41. Food and Agriculture Organization of the United Nations: Pre-breeding for effective use of plant genetic resources: an e-Learning Course. 2011, FAO, Rome
42. Global Partnership Initiative for Plant Breeding Capacity Building (GIPB),http://km.fao.org/gipb/
43. The FAO Global System on Plant Genetic Resources for Food and Agriculture, http://www.fao.org/agriculture/crops/core-themes/theme/seeds-pgr/gpa/ gsystem/es/
44. The World Information and Early Warning System of Plant Genetic Resources for Food and Agriculture (WIEWS),http://apps3.fao.org/wiews/wiews.jsp
45. The World Information Sharing Mechanism on the Implementation of the Global Plan of Action for the Conservation and Sustainable Use of Plant Genetic Resources for Food and Agriculture.http://www.pgrfa.org/gpa/selectcountry.jspx,
46. Food and Agriculture Organization of the United Nations: Second Global Plan of Action on Plant Genetic Resources for Food and Agriculture. 2011, FAO, Rome
47. Maluszynski M, Nichterlein K, Van Zanten L, Ahloowalia BS: Officially Released Mutant Varieties - The FAO/IAEA Database. 2000, 84-International Atomic Energy Agency, Vienna
48. Ahloowalia BS, Maluszynski M, Nichterlein K: Global impact of mutation-derived varieties. Euphytica. 2004, 135: 187-204.
49. Shu QY: Proceedings of International Symposium on Induced Mutations in Plants: Induced Plant Mutations in the Genomics Era: 11-15 August 2008; Vienna, Austria. 2009, FAO, Rome
50. The Joint FAO/IAEA Mutant Varieties and Genetic Stocks Database,http://mvgs.iaea.org/AboutMutantVarieties.aspx
51. McCallum CM, Comai L, Greene EA, Henikoff S: Targeting induced local lesions IN genomes (TILLING) for plant functional genomics. Plant Physiol. 2000, 123: 439-442.
52. Colbert T, Till BJ, Tompa R, Reynolds S, Steine MN, Yeung AT, McCallum CM, Comai L, Henikoff S: High-throughput screening for induced point mutations. Plant Physiol. 2001, 126: 480-484.
53. Greene EA, Codomo CA, Taylor NE, Henikoff JG, Till BJ, Reynolds SH, Enns LC, Burtner C, Johnson JE, Odden AR, Comai L, Henikoff S: Spectrum of chemically induced mutations from a large-scale reverse-genetic screen in Arabidopsis. Genetics. 2003, 164: 731-740.
54. Caldwell DG, McCallum N, Shaw P, Muehlbauer GJ, Marshall DF, Waugh R: A structured mutant population for forward and reverse genetics in barley (Hordeum vulgare L.). Plant J. 2004, 40: 143-150.
55. Till BJ, Reynolds SH, Weil C, Springer N, Burtner C, Young K, Bowers E, Codomo CA, Enns LC, Odden AR, Greene EA, Comai L, Henikoff S: Discovery of induced point mutations in maize genes by TILLING. BMC Plant Biol. 2004, 4: 12-
56. Triques K, Sturbois B, Gallais S, Dalmais M, Chauvin S, Clepet C, Aubourg S, Rameau C, Caboche M, Bendahmane A: Characterization of Arabidopsis thaliana mismatch specific endonucleases: application to mutation discovery by TILLING in pea. Plant J. 2007, 51: 1116-25.
57. Till BJ, Cooper J, Tai TH, Colowit P, Greene EA, Henikoff S, Comai L: Discovery of chemically induced mutations in rice by TILLING. BMC Plant Biol. 2007, 7: 19-
58. Slade AJ, Fuerstenberg SI, Loeffler D, Steine MN, Facciotti D: A reverse genetic, nontransgenic approach to wheat crop improvement by TILLING. Nat Biotechnol. 2005, 23: 75-81.
59. Sato Y, Shirasawa K, Takahashi Y, Nishimura M, Nishio T: Mutant selection from progeny of gamma-ray-irradiated rice by DNA heteroduplex cleavage using Brassica petiole extract. Breeding Science. 2006, 56: 179-183. 10.1270/jsbbs.56.179.

60. Cooper JL, Till BJ, Laport RG, Darlow MC, Kleffner JM, Jamai A, El-Mellouki T, Liu S, Ritchie R, Nielsen N, Bilyeu KD, Meksem K, Comai L, Henikoff S: TILLING to detect induced mutations in soybean. BMC Plant Biology. 2008, 8: 9-
61. Suzuki T, Eiguchi M, Kumamaru T, Satoh H, Matsusaka H, Moriguchi K, Nagato Y, Kurata N: MNU-induced mutant pools and high performance TILLING enable finding of any gene mutation in rice. Mol Genet Genomics. 2008, 279: 213-223.
62. Tadele Z, Mba C, Till BJ: TILLING for mutations in model plants and crops. Molecular Techniques in Crop Improvement. Edited by: Jain SM, Brar DS. 2009, 307-322. Springer Publishing, Inc, Dordrecht, 2
63. Till BJ, Afza R, Bado S, Huynh OA, Jankowicz-Cieslak J, Matijevic M, Mba C: Global TILLING Projects. Proceedings of International Symposium on Induced Mutations in Plants: Induced Plant Mutations in the Genomics Era: 11-15 August 2008. Edited by: Shu QY. 2009, 237-239. FAO, Rome
64. Forster BP, Heberle-Bors E, Kasha KJ, Touraev A: The resurgence of haploids in higher plants. Trends Plant Sci. 2007, 12: 368-375.
65. Szarejko I, Forster BP: Doubled haploidy and induced mutation. Euphytica. 2007, 158: 359-370. 10.1007/s10681-006-9241-1.
66. Mba C, Afza R, Jankowicz-Cieslak J, Bado S, Matijevic M, Huynh A, Till BJ: Enhancing genetic diversity through induced mutagenesis in vegetatively propagated plants. Proceedings of International Symposium on Induced Mutations in Plants: Induced Plant Mutations in the Genomics Era: 11-15 August 2008. Edited by: Shu QY. 2009, 293-296. FAO, Rome
67. Sperling L, Ashby JA, Smith ME, Weltzien E, McGuire S: A framework for analyzing participatory plant breeding approaches and results. Euphytica. 2001, 122: 439-450. 10.1023/A:1017505323730.
68. Ashby JA: The Impact of Participatory Plant Breeding. Plant Breeding and Farmer Participation. Edited by: Ceccarelli S, Guimaraes EP, Weltzien E. 2009, 649-671. FAO, Rome
69. Efisue A, Tongoona P, Derera J, Langyintuo A, Laing M, Ubi B: Farmers' perceptions on rice varieties in Sikasso region of Mali and their implications for rice breeding. J Agronomy & Crop Science. 2008, 194: 393-400.
70. Witcombe JR, Joshi A, Joshi KD, Sthapit BR: Farmer participatory crop improvement. I. Varietal selection and breeding methods and their impact on biodiversity. Experimental Agriculture. 1996, 32: 445-460. 10.1017/S0014479700001526.
71. Technical Advisory Committee of the Consultative Group on International Agricultural Research (TAC):Systemwide Review of Breeding Methodologies in the CGIAR. 2001, Science Council of the CGIAR
72. Suslow TV, Thomas B, Bradford KJ: Biotechnology provides new tools for plant breeding. 2002, 19-Seed Biotechnology Center, University of California Davis, Davis, CA
73. Moose SP, Mumm RH: Molecular plant breeding as the foundation for 21st century crop improvement. Plant Physiology. 2008, 147: 969-977.
74. Stam P, : Eucarpia Leafy Vegetables 2003: Proceedings of the EUCARPIA Meeting on Leafy Vegetables Genetics and Breeding: 19-21 March 2003. Edited by: Hintum TJL, Lebeda A, Pink D, Schut JW. 2003, 117-124. Centre for Genetic Resources (CGN), Wageningen
75. Burstin J, Charcosset A: Relationship between phenotypic and marker distances: theoretical and experimental investigations. Heredity. 1997, 79: 477-483. 10.1038/hdy.1997.187.
76. Spelman RJ, Bovenhuis H: Moving from QTL experimental results to the utilization of QTL in breeding programmes. Animal Genetics. 1998, 29: 77-84.
77. Hospital F, Moreau L, Lacourde F, Charcosset A, Gallais A: More on the efficiency of marker-assisted selection. Theor Appl Genet. 1997, 95: 1181-1189. 10.1007/s001220050679.
78. Hospital F, Charcosset A: Marker-assisted introgression of quantitative loci. Genetics. 1997, 147: 1469-1485.
79. Tanksley SD, McCouch SR: Seed banks and molecular maps: Unlocking genetic potential from the wild. Science. 1997, 277: 1063-1066.

80. Bernacchi D, Beck-Bunn T, Emmatty D, Inai S: Advanced backcross QTL analysis of tomato II. Evaluation of near-isogenic lines carrying single-donor introgressions for desireable wild QTL alleles derived from Lycopersicon hirsutum and L. pimpinellifolium. Theor Appl Genet. 1998, 97: 170-180. 10.1007/s001220050882.
81. Bernacchi D, Beck-Bunn T, Eshed Y, Lopez J, Petiard V, Uhlig J, Zamir D, Tanksley S: Advanced backcross QTL analysis in tomato I. Identification of QTLs for traits of agronomic importance from Lycopersicon hirsutum. Theor Appl Genet. 1998, 97: 381-397. 10.1007/s001220050908.
82. Stuber CW, Polacco M, Senior ML: Synergy of Empirical breeding, marker-assisted selection, and genomics to increase crop yield potential. Crop Sci. 1999, 39: 1571-1583. 10.2135/cropsci1999.3961571x.
83. Moreau L, Lemarie S, Charcosset A, Gallais A: Economic efficiency of one cycle of marker-assisted selection. Crop Sci. 2000, 40: 329-337. 10.2135/cropsci2000.402329x.
84. Meuwissen THE, Hayes BJ, Goddard ME: Prediction of total genetic value using genome-wide dense marker maps. Genetics. 2001, 157: 1819-1829.
85. Charcosset A, Moreau L: Use of molecular markers for the development of new cultivars and the evaluation of genetic diversity. Euphytica. 2004, 137: 81-94.
86. Lecomte L, Duffé P, Buret M, Servin B, Hospital F, Causse M: Marker-assisted introgression of five QTLs controlling fruit quality traits into three tomato lines revealed interactions between QTLs and genetic backgrounds. Theor Appl Genet. 2004, 109: 658-668.
87. Collard BCY, Jahufer MZZ, Brouwer JB, Pang ECK: An introduction to markers, Quantitative trait loci (QTL) mapping and marker-assisted selection for crop improvement: The basic concepts. Euphytica. 2005, 142: 169-196. 10.1007/s10681-005-1681-5.
88. Tuberosa R, Salvi S: Genomics-based approaches to improve drought tolerance of crops. TRENDS in Plant Science. 2006, 11: 405-412.
89. Brady SM, Provart NJ: Extreme breeding: Leveraging genomics for crop improvement. J Sci Food Agric. 2007, 87: 925-929. 10.1002/jsfa.2763.
90. Ribaut J-M, Ragot M: Marker-assisted selection to improve drought adaptation in maize: the backcross approach, perspectives, limitations, and alternatives. J Exp Bot. 2007, 58: 351-360.
91. Collard BCY, Mackill DJ: Marker-assisted selection: an approach for precision plant breeding in the twenty-first century. Phil Trans R Soc B. 2008, 363: 557-572.
92. Choudhary K, Choudhary OP, Shekhawat NS: Marker assisted selection: a novel approach for crop improvement. American-Eurasian Journal of Agronomy. 2008, 1: 26-30.
93. Eathington SR, Crosbie TM, Reiter RS, Bull JK: Molecular markers in a commercial breeding program. Crop Sci. 2007, (Suppl 3): S154-S163.
94. Pioneer Hi-Bred.http://www.pioneer.com/home/site/about/research/plant-breeding/
95. Syngenta,http://www.syngentafoundation.org/index.cfm?pageID
96. Thro AM, Parrott W, Udall JA, Beavis WD: Proceedings of the Symposium on Genomics and Plant Breeding: The Experience of the Initiative for Future Agricultural and Food Systems: 13 November 2002. 2004, 1893-1919. Crop Science Society of America; 2004, Crop Sci, Indianapolis, IN, USA
97. Ikeda N, Bautista NS, Yamada T, Kamijima O, Ishii T: Ultra-simple DNA extraction method for marker-assisted selection using microsatellite markers in rice. Plant Molecular Biology Reporter. 2001, 19: 27-32. 10.1007/BF02824075.
98. The Integrated Breeding Platform of the Generation Challenge Program of the Consultative Group on International Agricultural Research.http://www.generationcp.org/ibp,
99. James C: Executive Summary Global Status of Commercialized Biotech/GM Crops: 2011. ISAAA Brief 2011. 2011, 30-ISAAA, Ithaca, NY
100. Glover D: Monsanto and smallholder farmers: A case-study on corporate accountability. IDS Working Paper 277. 2007, 53-University of Sussex, UK, Institute of Development Studies, Brighton

101. Glover D: Made by Monsanto: The corporate shaping of GM crops as a technology for the poor. STEPS Working Paper 11. 2008, 45-STEPS Centre, Brighton
102. Convention on Biological Diversity (CBD): Cartagena Protocol on Biosafety to the Convention on Biological Diversity. 1992, Secretariat of the Convention on Biological Diversity, Montreal
103. The African Agricultural Technology Foundation,http://www.aatf-africa.org/
104. Boadi RY, Bokanga M, : The African Agricultural Technology Foundation Approach to IP Management. Intellectual Property Management in Health and Agricultural Innovation: A Handbook of Best Practices. Edited by: Krattiger A, Mahoney RT, Nelsen L, Thomson JA, Bennett AB, Satyanarayana K, Graff GD, Fernandez C, Kowalski SP. Davis, CA, USA: PIPRA; Fiocruz, Rio de Janeiro, Brazil: Oswaldo Cruz Foundation; 2007:1765-1774, MIHR, Ithaca, NY, USA
105. Atkinson RC, Beachy RN, Conway G, Cordova FA, Fox MA, Holbrook KA, Klessig DF, McCormick RL, McPherson PM, Rawlings HR, Rapson R, Vanderhoef LN, Wiley JD, Young CE: Public sector collaboration for agricultural IP management. Science. 2003, 301: 174-175.
106. Delmer DP, Nottenburg C, Graff GD, Bennett AB: Intellectual property resources for international development in agriculture. Plant Physiology. 2003, 133: 1666-1670.
107. Lusser M, Parisi C, Plan D, Rodríguez-Cerezo E: New plant breeding techniques. 2011, Publications Office of the European Union, State-of-the-art and prospects for commercial development. Luxembourg
108. Blakeney M: Patenting of plant varieties and plant breeding methods. J Exp Bot. 2012, 63: 1069-1074.
109. Xu Y, Crouch JH: Marker-assisted selection in plant breeding: from publications to practice. Crop Sci. 2008, 48: 391-407. 10.2135/cropsci2007.04.0191.
110. Houle D, Govindaraju DR, Omholt S: Phenomics: the next challenge. Nature Reviews Genetics. 2010, 11: 855-866.
111. Finkel E: With 'Phenomics', plant scientists hope to shift breeding into overdrive. Science. 2009, 325: 380-381.
112. The Australian Plant Phenomics Facility,http://www.plantphenomics.org.au/
113. Lemna Tec,http://www.lemnatec.com/
114. Jülich Plant Phenotyping Centre,http://www2.fz-juelich.de/icg/icg-3/jppc
115. Ecotron.http://www.ecotron.cnrs.fr/,
116. The Ecophysiology Laboratory of Plant under Environmental Stress:,http:// www1.montpellier.inra.fr/ibip/lepse/english/
117. The Biotron Experimental Climate Change Research Centre,http://www.thebiotron.ca/
118. Mba C, Guimaraes EP, Guei GR, Hershey C, Paganini M, Pick B, Ghosh K: Mainstreaming the continuum approach to the management of plant genetic resources for food and agriculture through national strategy. Plant Genetic Resources. 2011, 10.1017/S1479262111000943.
119. Food and Agriculture Organization of the United Nations: Crop Improvement for the 21st Century. 2012, FAO; Information Document for 23rd Session of the Committee on Agriculture, Rome
120. Anthony VM, Ferroni M: Agricultural biotechnology and smallholder farmers in developing countries. Curr Opin Biotechnol. 2011, 23: 1-8.
121. Ferroni M, Castle P: Public-private partnerships and sustainable agricultural development. Sustainability. 2011, 3: 1064-1073. 10.3390/su3071064.
122. Plant Breeding and Related Biotechnology Capacity Assessment.http://gipb.fao.org/Web-FAO-PBBC/index.cfm?where,
123. Plant Breeding Studies.http://km.fao.org/gipb/capacity-assessments/country-case-studies/en/,
124. Guimaraes EP, Kueneman E, Carena MJ: Assessment of national plant breeding and biotechnology capacity in Africa and recommendations for future capacity building. HortScience. 2006, 41: 50-52.
125. Bliss FA: Education and preparation of plant breeders for careers in global crop improvement. Crop Sci. 2007, 47: S250-S261.

126. Gepts P, Hancock J: The future of plant breeding. Crop Sci. 2006, 46: 1630-1634. 10.2135/cropsci2005-12-0497op.
127. Repinski SL, Hayes KN, Miller JK, Trexler CJ, Bliss FA: Plant Breeding Graduate Education: Opinions about Critical Knowledge, Experience, and Skill Requirements from Public and Private Stakeholders Worldwide. Crop Sci. 2011, 2011: 2325-2336.
128. Guner N, Wehner TC: Survey of U.S. land-grant universities for training of plant breeding students. Crop Sci. 2003, 43: 1938-1944. 10.2135/cropsci2003.1938.
129. Knight J: A Dying Breed. Nature. 2003, 421: 568-570.
130. Hancock JF: Introduction to the symposium. HortScience. 2006, 41: 28-29.
131. Morris M, Edmeades G, Pehu E: The global need for plant breeding capacity: What roles for the public and private sectors?. HortScience. 2006, 41: 30-39.
132. Baenziger PS: Plant breeding training in the U.S. HortScience. 2006, 41: 40-44.
133. Bliss F: Plant breeding in the U.S. private sector. HortScience. 2006, 41: 45-47.
134. Khush GS: Impact of international centers on plant breeding training. HortScience. 2006, 41: 48-49.
135. Ransom C, Particka C, Ando K, Olmstead J: Report of breakout group 1. What kind of training do plant breeders need, and how can we most effectively provide that training?. HortScience. 2006, 41: 53-54.
136. Weebadde C, Mensah C: Report of breakout group 2. How will we provide improved varieties of specialty, minor and subsistence crops in the future?. HortScience. 2006, 41: 55.
137. Terpstra K, Oraby H, Vallejo V: Report of breakout group 3. How can the public and private sectors most effectively partner to train new generations of plant breeders?. HortScience. 2006, 41: 56.
138. The African Centre for Crop Improvement,http://www.acci.org.za/main.asp?nav
139. The West Africa Centre for Crop Improvement,http://www.wacci.edu.gh/index.html

CHAPTER 3

Genomics reveals new landscapes for crop improvement

Michael W Bevan* and Cristobal Uauy

ABSTRACT

The sequencing of large and complex genomes of crop species, facilitated by new sequencing technologies and bioinformatic approaches, has provided new opportunities for crop improvement. Current challenges include understanding how genetic variation translates into phenotypic performance in the field.

Keywords: Genomics crop, improvement, genomic selection, systems breeding

1. BACKGROUND

Genomics, the analysis of an organism's complete DNA sequence, has been one of the most transformative influences on biological studies. The genome sequences of organisms are fundamentally important for understanding the functions of individual genes and their networks, for defining evolutionary relationships and processes, and for revealing previously unknown regulatory mechanisms that coordinate the activities of genes. These genomics-based approaches are having a profound influence on both human disease diagnostics and treatment [1] and, equally importantly, on the improvement of crops for food and fuel production. In this review, we summarize progress in sequencing crop genomes, identify remaining technical challenges, and describe how genomics-based applications can aid crop improvement. We then assess the impact of genomics on plant breeding and crop improvement, showing how it is accelerating the improvement of staple and 'orphan' crops, and facilitating the utilization of untapped allelic variation. Finally, we speculate about the future impacts of genomics on plant biology and crop improvement by developing the concept of systems breeding, which integrates information on gene function, genome states, and regulatory networks across populations and species to create a predictive framework for estimating the contributions of genetic and epigenetic variation to phenotypes and field performance.

2. PROGRESS IN CROP GENOME SEQUENCING AND ANALYSIS

Advances in sequencing crop genomes have mirrored the development of sequencing technologies (Table 1). Until 2010, Sanger sequencing of bacterial artificial chromosome (BAC)-based physical maps was the predominant approach used to access crop genomes such as rice, poplar and maize [1, 2, 3]. The rice genome comprises complete sequences of individual BACs assembled into physical maps that are anchored to genetic maps, whereas for maize, the sequences of individual BACs were not completely finished. For poplar, grapevine, sorghum and soybean [2, 4, 5, 6], whole genome shotgun (WGS) reads of libraries of randomly sheared fragments of different sizes and of BAC end sequences (BES) were assembled with powerful assembly algorithms such as ARACHNE [7, 8]. The trade-offs that shaped genome-sequencing strategies in the era before next-generation sequencing became available involved coverage, time and expense. Physical maps of BACs provide a good template for completing gaps and errors, but genome coverage of physical maps can be non-representative due to cloning bias. In addition, intensive hand-crafting is required to assess physical map integrity and to close gaps; this effort scales directly with genome size and complexity.

The sorghum genome [1, 2, 3, 5] was the first crop genome to be sequenced completely by the exclusive use of WGS sequence assemblies, which were then assessed for integrity using high-density genetic maps and physical maps. This pioneering analysis showed that scaffolds of Sanger sequence assemblies accurately span extensive repetitive DNA tracts and extend into telomeric and centromeric regions. The larger soybean genome was subsequently sequenced to similar high standards.

Table 1 Progress in crop genome sequencing

Species (common name)	Genome size	Ploidy	Sequence strategy	Publication date	Assembly features	Reference
Oryza sativa (rice)	389 Mb	2n = 2x = 24	BAC physical map, Sanger sequencing	Aug 2005	Essentially complete chromosome arm coverage	[2]
Populus trichocarpa (black cottonwood)	550 Mb	2n = 2x = 38	BAC physical map, WGS, Sanger sequencing	Sep 2006	2,447 [c]scaffolds containing 410 Mb, 82% of sequence genetically anchored	[3]
Vitus vinifera (pinot noir grape)	475 Mb	2n = 2x = 36	WGS, Sanger sequencing	Sep 2007	3,514 [c]supercontigs containing 487 Mb, 69% of sequence genetically anchored	[5]
Sorghum bicolor (sorghum)	700 Mb	2n = 2x = 20	WGS, Sanger sequencing	Jan 2009	229 scaffolds containing 97% of the genome, 88% of sequence genetically anchored	[6]
Zea mays (maize)	2,300 Mb	2n = 2x = 20, one [a]WGD allotetraploid	BAC physical map, BAC sequence 4-6 x deep	Nov 2009	2,048 Mb in 125,325 [b]contigs forming 61,161 scaffolds	[4]
Glycine max (soybean)	1,115 Mb	Two WGD 2n = 2x = 40 allopolyploid	WGS, Sanger sequencing	Jan 2010	397 scaffolds containing 85% of the genome, 98% of sequence genetically anchored	[7]
Malus x domestica (apple)	750 Mb	One WGD 2n = 2x = 34	WGS, Sanger, Roche 454	Oct 2010	1,629 [c]metacontigs containing 80% of the genome, 71% of sequence genetically anchored	[10]
Theobroma cacao (cacao)	430 Mb	2n = 2x = 20	WGS, Sanger, Illumina, Roche 454	Dec 2010	524 scaffolds containing 80% of the genome, 67% of sequence genetically anchored	[11]
Fragaria vesca (woodland strawberry)	240 Mb	2n = 2x = 14	WGS, Roche 454, Illumina, SOLiD	Dec 2010	272 scaffolds containing 95% of the genome, 94% of sequence genetically anchored	[13]

Phoenix dactylifera (date palm)	658 Mb	2n = 2x = 36	WGS, Illumina	June 2011	57,277 scaffolds containing 60% of the genome	[12]
Solanum tuberosum (potato)	844 Mb	2n = 4x = 48	Double monoploid DM and diploid RH, WGS, Illumina, Roche 454	July 2011	443 superscaffolds containing 78% of the genome, 86% of the assembly genetically anchored	[14]
Brassica rapa (Chinese cabbage)	485 Mb	Three WGD 2n = 2x = 20	WGS, Illumina, BAC end Sanger sequencing	Aug 2011	288 Mb in scaffolds, 90% of the assembly genetically anchored	[15]
Medicago truncatula (alfalfa relative)	375 Mb	WGD 2n = 2x = 16	BAC physical map, Sanger, Illumina	Dec 2011	8 pseudomolecules containing 70% of the genome, 100% in optical map	[16]
Manihot esculenta (cassava)	770 Mb	2n = 2x = 36	WGS, Roche 454, BAC end Sanger sequencing	Jan 2012	12,977 scaffolds containing 80% of the genome	[19]
Cajanus cajan (pigeonpea)	833 Mb	2n = 2x = 22	WGS, Illumina	Jan 2012	137,542 scaffolds containing 73% of the genome	[20]
Setaria italic (foxtail millet)	500 Mb	2n = 2x = 18	WGS, Sanger, Illumina, BAC end sequence	May 2012	597 scaffolds containing 80% of the genome, 99% of the assembly genetically anchored	[21]
Solanum lycopersicum (tomato)	900 Mb	2n = 2x = 24	WGS, Roche 454, Illumina and SOLiD, BAC end Sanger sequencing	May 2012	91 scaffolds containing 85% of the genome, 99% of the assembly genetically anchored	[17]

Cucumis melo (melon)	312 Mb	Three WGD $2n = 2x = 24$	WGS, Roche 454, BAC end sequencing	July 2012	1,584 scaffolds containing 83% of the genome, 88% of the assembly genetically anchored	[22]
Musa acuminate (Cavendish banana)	523 Mb	$2n = 2x = 22$	WGS, Roche 454, Sanger, Illumina	Aug 2012	24,425 contigs containing 90% of the genome, 70% of the assembly genetically anchored	[33]
Citrus sinensis (Valencia sweet orange)	367 Mb	$2n = 2x = 18$	Dihaploid WGS, Illumina	Jan 2013	4,811 scaffolds containing 82% of the genome, 73% of the assembly genetically anchored	[23]
Gossypium raimondii (D genome cotton)	880 Mb	$2n = 2x = 26$	WGS, Illumina	Aug 2012	4,715 scaffolds containing 85% of the genome, 73% of the assembly genetically anchored	[24]
Hordeum vulgare (barley)	5,100 Mb	$2n = 2x = 14$	WGS, Illumina, BAC physical map, BAC sequence (Roche 454, Illumina)	Nov 2012	Physical map (4.98 Gb), BAC sequence (1.13 Gb), WGS assemblies (1.9 Gb); integrated by physical map and syntenic order	[26]
Triticum aestivum (bread wheat)	17,000 Mb	$2n = 6x = 42$ allopolyploid	WGS, Roche 454	Nov 2012	Orthologous group assembly, 437 Mb	[27]
Gossypium raimondii (D genome cotton)	880 Mb	$2n = 2x = 26$	WGS, Sanger, Roche 454, Illumina	Dec 2012	1,084 scaffolds containing 86% of the genome, 98% anchored and oriented to genetic map	[25]
G. hirsutum (upland cotton)		AtDt allopolyploid	Illumina		82x coverage	
Cicer arietinum (chickpea)	738 Mb	$2n = 2x = 16$	WGS, Illumina BAC end sequence	Jan 2013	7,163 scaffolds containing 64% of the genome	[31]

Phylostachys heterocycla (bamboo)	2 Gb	2n = 2x = 48	WGS, Illumina BAC end sequence	Apr 2013	80% of the 2.05 Gb assembly maps to 5,499 scaffolds of less than 62 kb	[34]
Picea abies (Norway spruce)	20,000 Mb	2n = 2x = 24	fosmid pools with both haploid (megagametophyte) and diploid WGS	May 2013	Merged assembly 12.0 Gb, with 4.3 Gb in ≥10 kb scaffolds	[42]
Pinus taeda (Loblolly pine)	24,000 Mb	2n = 2x = 24	WGS single haploid megagametophyte assembly	In progress		
Miscanthus sp. (elephant grass)	1,500 Mb	One WGD, diploid progenitors 2n = 2x = 38	WGS	In progress		
Elais guineensis Elais oleifera (oil palm)	1,890 Mb	2n = 2x = 32 commercial F_1 hybrids	WGS, BAC physical maps	In progress		
Saccharum officinarum x *S. spontaneum* (sugar cane)	>15,000 Mb	Diploid progenitors x = 10; 2n = 80; x = 8; 2n = 40-128	WGS	In progress		

[a]WGD alloploids have a whole-genome duplication in recent lineage. [b]A contig is an unambiguous linear assembly of sequences with no physical gaps in coverage, but which can contain errors. [c]The terms supercontig, scaffold or metacontig are used interchangeably to describe a set of contigs that are linked by a known physical distance but that contain sequence gaps. These scaffolds are usually created using mate-pair reads and BAC end sequences. [d]Pseudomolecule is a term applied to a chromosome-scale assembly of contigs and scaffolds that is anchored to a long-range framework using genetic markers and other chromosome features, including cytogenetic features and deletions.

The soybean genome is thought to be pseudo-diploid, derived from the diploidization of an allopolyploid in the past 50 million years [2, 4, 5, 6, 9], and this project successfully demonstrated that WGS assemblies are not confounded by large-scale genome duplication events.

By 2010 to 2011, a mixture of sequencing technologies, all using WGS assembly methods, were being successfully applied to trees (apple, cacao and date palm), fruit (strawberry), vegetables (potato and Chinese cabbage) and forage crops (alfalfa relative) [10, 11, 12, 13, 14, 15, 16]. The *Medicago* and tomato [17] projects, which were initiated in the BAC-based Sanger sequencing era, were completed using next-generation sequencing. The contiguity of assemblies varied according to genome composition and size, with very high contiguity being achieved in potato and alfalfa by alignment to BAC sequences. The *Brassica* genomes are among the most challenging to sequence with respect to achieving large-scale assemblies because they have undergone three recent whole-genome duplications followed by partial diploidization [18]. Polyploidy has a centrally important role in plant genome evolution and in the formation of important crop genomes. Figure 1 illustrates three examples of polyploidy and how these events contribute to crop genetic diversity in different ways. In *Brassica* species, polyploidy has led to extensive structural heterogeneity and gene copy number variation when compared with their close relative *Arabidopsis*. The *Brassica rapa* genome sequence remains fragmentary, but alignments of *Brassica* chromosome segments to the *Arabidopsis* genome are exceptionally useful for advanced genetic analysis [19].

In its early stages, crop genomics relied on many small-scale science laboratories joining forces to generate the sequence data. However, this has changed radically with the emergence and leadership of large-scale genome sequencing centers, which focused their expertise and resources on important crop genomes. Two examples are the Joint Genome Institute (JGI) in the USA and the Beijing Genome Institute (BGI, Shenzen) in China, both of which provide exceptional expertise, capacity and levels of engagement with researchers. These centers and others are currently sequencing the genomes of many crucially important food and fuel crops, and are working in collaboration with science groups worldwide on improving our functional understanding of these genomes.

Since 2012, analyses of the sequences of 12 crop genomes have been published, accounting for nearly half of the total published (Table 1) [20, 21, 22, 23, 24, 25, 26, 27, 28]. This explosion of data has been driven by cheaper and more effective sequencing technologies (primarily the Illumina [29] and Roche 454 [30] methods) coupled with increasingly sophisticated sequence and assembly strategies [31], which are generally delivered by large genome centers. Access to these technologies makes even a reasonably large crop genome project affordable and feasible within the period of a single research grant, and is having a major influence on strategies in crop genomics.

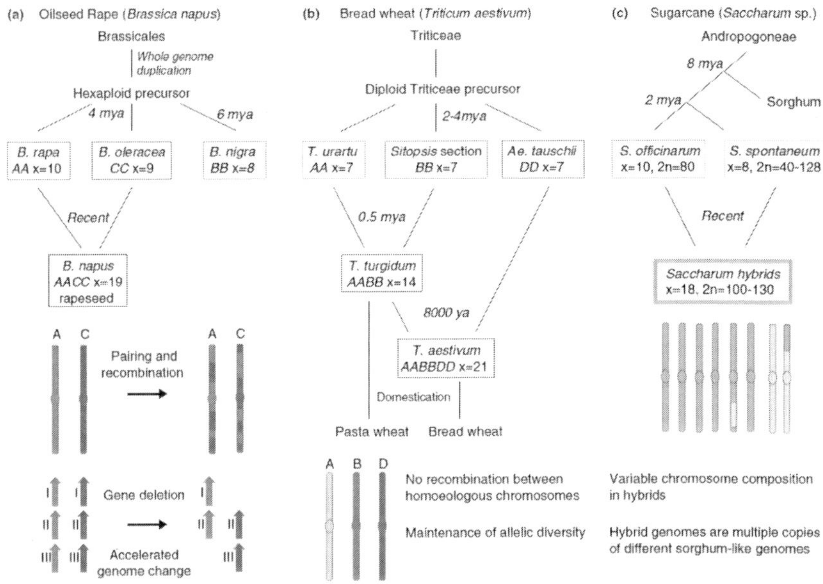

Figure 1 DIVERSE OUTCOMES OF POLYPLOIDY IN CROP SPECIES. Three examples of the consequences of allopolyploidy (in which hybrids have sets of chromosomes derived from different species) in important crop species are shown. **(A)**Oilseed rape (canola) is derived from a recent hybridization of *Brassica rapa* (Chinese cabbage, turnip) and *Brassica oleraceae* (broccoli, cauliflower, cabbage). The progenitor of these *Brassica* species was hexaploid (compared to *Arabidopsis*) after two rounds of whole-genome duplication. Extensive gene loss, possibly via deletion mechanisms [18], has occurred in these species. Upon hybridization to form allotetraploid *Brassica napus*, gene loss is accelerated, producing novel patterns of allelic diversity [19]. **(B)** Bread wheat is an allohexaploid derived from the relatively recent hybridization of allotetraploid durum (pasta) wheat and wild goat grass, *Aeglilops tauschii*. The *Ph1* locus in the B genome [37] prevents pairing between the A, B and D genomes, leading to diploid meiosis and genome stability. This maintains the extensive genetic diversity from the three progenitor Triticeae genomes that underpins wheat crop productivity. **(C)** Sugarcane (*Saccharum* sp.) is a complex and unstable polyploid that is cultivated by cuttings. Hybrids between *S. officinarum*, which has high sugar content, and *S. spontaneum*, a vigorous wild relative, have variable chromosomal content from each parent. The genomes are closely related to the ancestral diploid *Sorghum* [42].

For example, the role of multi-partner coordination has changed from raising and coordinating research funding and managing the distribution of research activities to a focus on data analyses, distribution and applications. These changes will further accelerate and greatly diversify the range of plant species and varieties sequenced.

The date palm genome [12] was sequenced using just paired end reads and remains fragmentary. Although this could be partly attributed to repeat composition, it is clear that the use of mate-pair libraries of different lengths, which provide accurately spaced pairs of sequence reads, substantially improves contiguity across medium-sized genomes of up to 1,000 Mb, as can be seen for citrus, diploid cotton, pigeonpea, chickpea and banana [21, 24, 25, 32, 33]. Contig and scaffold sizes were further increased in chickpea and pigeonpea by incorporating BES generated by Sanger sequence that have much longer read lengths paired over a 100 kb span. Increased lengths of Illumina reads, of up to 250 bases, are now available to users and should further improve contiguity. Using new assembly algorithms, the large genome of bamboo, a plant of major industrial and ecological significance, has recently been published [34].

Table 1 shows progress in sequencing two much larger Triticeae genomes, those of diploid barley (5,100 Mb) [27] and hexaploid bread wheat (17,000 Mb) [28]. Both the exceptional scale and high repeat content (approximately 80%) of these genomes provide significant challenges to straightforward WGS sequencing and assembly, with genes being separated by hundreds of kb of repeats such as nested retroelements [35]. In barley, a physical map of 67,000 BAC clones with a cumulative length of 4.98 Gb provided 304,523 BES reads as a framework for integration of 50X Illumina paired end and 2.5 kb mate pair reads. Contig median size was just 1.5 kb because the repeat content collapsed longer assemblies. Sequence assemblies were integrated with genetic and physical maps, and genic assemblies were assigned to chromosome arms. The chromosomal order of barley genes was then interpolated using synteny across multiple sequenced grass genomes and by ordering the genes according to the genetic or physical maps [36] (Figure 2).

The bread wheat genome is a recent hexaploid composed of three related genomes (A, B and D), each the size of the barley genome, which do not pair and recombine, leading to their independent maintenance [37] (Figure 1). The challenge for wheat WGS strategies was to provide independent assemblies covering and representing genes from each homoeologous genome. The two closest diploid progenitors of the A and D genomes were sequenced to identify polymorphisms that could be used to assess WGS gene assemblies. Low coverage (5X) Roche 454 sequence was generated, and orthologous gene sequences from multiple grasses were used to guide assemblies.

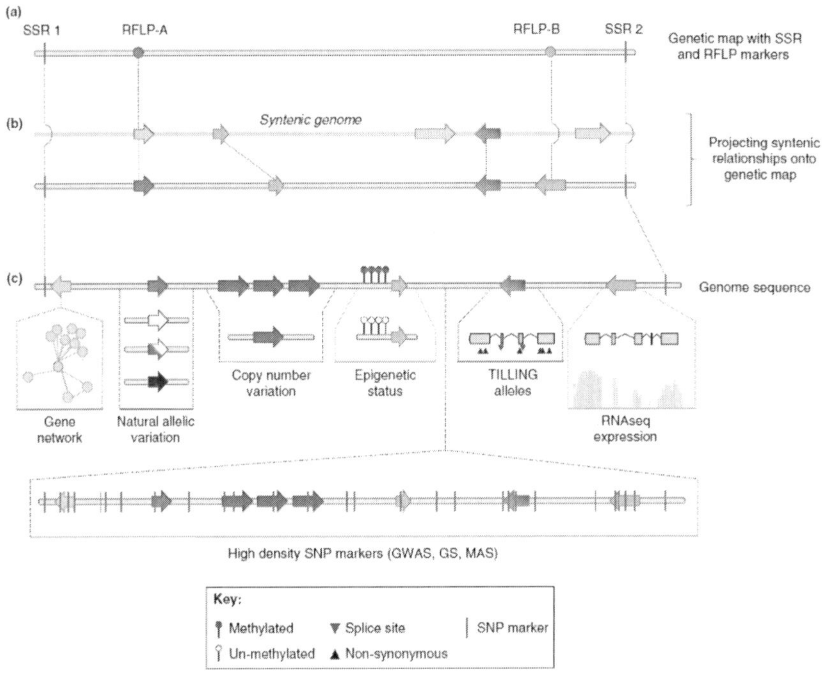

Figure 2 THE IMPACT OF WHOLE GENOME SEQUENCING ON BREEDING. **(A)** Initial genetic maps consisted of few and sparse markers, many of which were anonymous markers (simple sequence repeats (SSR)) or markers based on restriction fragment length polymorphisms (RFLP). For example, if a phenotype of interest was affected by genetic variation within the SSR1-SSR2 interval, the complete region would be selected with little information about its gene content or allelic variation. **(B)** Whole genome sequencing of a closely related species enabled projection of gene content onto the target genetic map. This allowed breeders to postulate the presence of specific genes on the basis of conserved gene order across species (synteny), although this varies between species and regions. **(C)**Complete genome sequence in the target species provides breeders with an unprecedented wealth of information that allows them to access and identify variation that is useful for crop improvement. In addition to providing immediate access to gene content, putative gene function and precise genomic positions, the whole genome sequence facilitates the identification of both natural and induced (by TILLING) variation in germplasm collections and copy number variation between varieties. Promoter sequences allow epigenetic states to be surveyed, and expression levels can be monitored in different tissues or environments and in specific genetic backgrounds using RNAseq or microarrays. Integration of these layers of information can create gene networks, from which epistasis and target pathways can be identified. Furthermore, re-sequencing of varieties

identifies a high density of SNP markers across genomic intervals, which enable genome-wide association studies (GWAS), genomic selection (GS) and more defined marker-assisted selection (MAS) strategies.

Approximately 94,000 genes were assembled and positively assigned to the A and D genomes using genome-specific single nucleotide polymorphisms (SNPs), with the remaining assemblies tentatively assigned to the B genome. Wheat gene assemblies, which are fragmentary compared to barley gene assemblies, were assigned to chromosomes using high-density genetic maps and conserved gene order.

The current wheat and barley gene-based assemblies are suitable for developing genetic markers [38] and for creating genetic maps for map-based cloning and marker-assisted breeding. To increase the gene coverage and contiguity of the barley genome, BACs in the physical map are being multiplex-sequenced using Illumina methods. This will result in chromosome assemblies with fewer gaps and more precisely ordered genes. This should establish barley as the pre-eminent genomic template and genetic reference for the Triticeae. On-going efforts in sequencing the bread wheat genome include sequencing purified flow-sorted chromosome arms to increase gene coverage and the complete assignment of homoeologous genes to the A, B, or D genome [39]. Constructing physical maps of BAC libraries made from purified chromosomes is also underway, with the chromosome 3B physical map [40] and BAC sequencing completed. Given sufficient funding and time, this strategy will provide the necessary high-quality reference genome. Since homoeologous genes can now be assembled and assigned to their genome, WGS can be used to improve the contiguity of wheat gene sequences by using long mate pair spans, in non overlapping increments up to 40 kb using fosmid vectors [41], coupled to longer read lengths. New template preparation methods, such as Illumina Moleculo, which breaks assemblies down into separate 10 kb units, could be used to span large repeat units and to facilitate accurate assemblies covering large tracts of repeats. Although a colossal amount of sequencing is required, a whole-genome strategy for wheat, supplemented by the flow-sorted chromosome arm data, has the potential to provide users with a high-quality draft sequence relatively quickly and cheaply.

Several industrially important species, such as the conifers Norway spruce (*Picea abies*) [42] and loblolly pine (*Pinus taeda*), have very large genomes (approximately 20,000 to 24,000 Mb, respectively). They are being sequenced using WGS strategies involving fosmid pool sequencing and Illumina long-mate pair methods [43]. These tree species have particular characters that facilitate their genome analysis, including the absence of whole-genome duplication in their ancestry, relatively inactive retroelements and the presence of a large multicellular haploid gametophyte, the sequence of which does not exhibit heterozygosity.

Sugar cane, another important crop plant, is a hybrid between *Saccharum officinarum* and *Saccharumspontaneum*. These species are closely related to sorghum [44] and have haploid contents of 8 and 10 base chromosomes, respectively. Both *S. officinarum* and *S. spontaneum* have a monoploid genome size close to that of sorghum (760 Mb), but they are highly autopolyploid (2n = 80 and 2n = 40-128, respectively), resulting in a genome size of >15 Gb for hybrid sugar cane. Commercial cultivars are derived by backcrossing hybrids to *S. officinarum*, resulting in lines that have different chromosome contributions from each parental species [45]. The highly variable and heterozygous composition of commercial sugarcane genomes is a major challenge to genome sequencing. The sequencing of progenitor genomes, using WGS strategies and sorghum genes as templates, could create high-stringency orthologous gene assemblies. As in the analysis of the wheat draft genome, this strategy would generate information on ortholog copy number and identify sequence polymorphisms that could be used to genetically map desirable traits in the two progenitor species. Upon the development of commercial hybrids from sequenced progenitors, re-sequencing could identify desired genotypes and gene copy numbers.

A similar approach could be used for the biomass crop *Miscanthus* x *giganteus*, a sterile triploid derived from *Miscanthus sinensis* and tetraploid *Miscanthus sacchariflorus*. A recent genetic analysis has shown that *M. sinensis* has recently undergone whole-genome duplication [46] and a single dysploid chromosome fusion [47], neither of which occurred in the closely related sorghum genome [48]. The WGS strategy developed for wheat could be also applied to *M. sinensis* and its hybrids to determine gene copy numbers and to identify genetic variation in homoeologous gene copies.

3. ACCESSING AND MEASURING SEQUENCE VARIATION AND THE EPIGENOME

It is reasonable to predict that within the next two years useful genome sequences will be available to support the genetic improvement of most of the important food and fuel crops. Crop improvement will depend, however, on the identification of useful genetic variation and its utilization by breeding and transformation. Such variation can be identified at a genome scale by comparison of multiple sequence reads to a single 'reference'. For example, in rice, low-coverage sequence of 1,083 *Oryza sativa* and 466 *Oryza rufipogon* (the progenitor species of cultivated rice) accessions [49] provided deep insights into the domestication of rice and the geographical distribution of variation, while providing material for quantitative trait loci (QTL) and genome-wide association studies (GWAS) [50]. The gene spaces of maize and wheat varieties are being re-sequenced using sequence capture methods that are based on the solution hybridization of sheared genomic DNA with biotinylated long overlapping oligos designed from gene sequences [51, 52]. The captured DNA is highly enriched in genic sequences, and its deep sequencing can distinguish closely related genes, including wheat homoeologs [53]. These approaches will facilitate the high-throughput sequencing of the gene space of multiple lines of crops, even those with very large polyploid genomes. These methods offer the ability to sequence rapidly the genomes or gene space of multiple accessions, wild relatives and even new species, which will undoubtedly accelerate the incorporation of unexplored and underutilized genetic variation into crops worldwide [54]. DNA sequence variation remains a primary focus, but extensive evidence from several crop species [55, 56] suggests that epigenetic changes are responsible for a range of stably heritable traits, and that epigenetic variation can be both induced and selected for during domestication [57]. The methylation status of captured DNA can be measured using bisulfite treatment followed by deep sequencing in a method called reduced representation bisulfite sequencing (RRBS) [58]. These important technological advances in sequence template preparation will permit the exceptionally detailed and cost-effective definition of variation in the sequences and epigenomes of multiple lines or species of crops, independently of their genome size and polyploid status [59].

4. APPLYING NEXT-GENERATION GENOMICS TO CROP IMPROVEMENT

Accessing genome-wide sequence variation by re-sequencing significantly improves the availability of information that can be used to develop markers, thereby enhancing the genetic mapping of agronomic traits. For example, in wheat, fewer than 500 SNP markers were available in 2008 [60] with that number increasing to 1,536 in 2010 [61], 10,000 in 2011 and over 90,000 in 2012 [38]. This relatively high-density SNP information is proving extremely useful across different systems, including QTL mapping in bi-parental crosses and recombinant inbred lines, GWAS, and mapping QTL in advanced inter-cross lines such as those in multi parent advanced generation inter-cross (MAGIC) [62] and nested association mapping (NAM) [63] populations. These approaches generally identify loci and causal genes for traits with relatively large phenotypic effects. The genomic segments that contain desired allelic variation can then be bred and combined in a single genetic background using markers to track the segments through marker-assisted selection (MAS).

Many important agricultural traits such as yield, however, result from relatively small effects across multiple loci. This implies that these loci might not be optimally identified through QTL or GWAS approaches and that their pyramiding through MAS will be inefficient. Therefore, breeders have begun to address these problems by developing a knowledge base of associations of polymorphic markers with phenotypes in breeding populations [64, 65]. These associations are used to develop a breeding model in which the frequency of desired marker alleles is optimized, thereby maximizing the estimated breeding value [66, 67]. Multiple cycles of selection are used to accumulate favorable alleles that are associated with desired phenotypes, although no causal relationship between a specific gene and a phenotype is established. This approach, termed genomic selection (GS) is incorporated into industrial-scale breeding processes that require very cheap high-throughput marker assays [68]. Next-generation sequencing of parental lines is influencing GS in several ways: by continuing to identify polymorphisms throughout the genome in both genic and inter-genic regions; by providing estimates of gene expression levels; and by providing information on the epigenetic states of genes (Figure 2). The first removes any limitations on marker density, while the latter two features are 'genomic features' that will surely have predictive power for complex traits. Speculatively, the encyclopedia of DNA elements (ENCODE) concept [69] of total genomic knowledge could eventually be incorporated into models for predicting performance from genomic information revealed by next-generation sequencing.

Breeding uses natural allelic variation to improve crop performance. Sequence variation can be experimentally enhanced using, for example, ethyl methanesulphonate (EMS) to alkylate bases. TILLING (targeted induced local lesions in genomes) [70] is then used to screen for base changes in genes of interest to assess gene function and to create advantageous alleles for breeding. It is now feasible to use genome capture to sequence an entire mutant population, even in complex polyploid genomes such as wheat [52]. Here, polyploidy provides an advantage by buffering the influence of otherwise deleterious mutation loads.

Genetic manipulation using the *Agrobacterium tumefaciens*-mediated transfer of genes from any other organism is a mature technology that has been adapted for use in many of the crop species listed in Table 1. The precise modification of gene sequences using zinc-finger nucleases (ZFN) that can be engineered to recognize specific DNA sequences has been applied to a target locus in maize [71]. More recently a new type of precision tool for genome engineering has been developed from the prokaryotic clustered regularly interspaced short palindromic repeats (CRISPR) Cas9 immune system [72, 73]. The Cas9 nuclease is guided to specific target sequences for cleavage by an RNA molecule. Several types of genome editing are possible, such as the simultaneous editing of multiple sites, inducing deletions, and inserting new sequences by nick-mediated repair mechanisms.

5. GENOMIC FEATURES FOR FUTURE BREEDING

Genomics has radically altered the scope of genetics by providing a landscape of ordered genes and their epigenetic states, access to an enormous range of genetic variation, and the potential to measure gene expression directly with high precision and accuracy (Figure 2). This not only has important practical advantages for breeding but also facilitates systematic comparison of gene functions across sequenced genomes, bringing the wealth of knowledge of gene function and networks obtained in experimental species directly into the ambit of crop improvement. Given a suitable cyber-infrastructure, the integration of biological knowledge and models of networks across species, in a two-way flow from crops to experimental species and back again, will begin to generate new layers of knowledge that can be used for crop improvement. One layer is provided by ENCODE-level analyses [69]; although yet to start in plants, these analyses can guide the interpretation of gene function and variation, thus providing new information to inform the prediction of phenotype from genotype. Another information layer is provided by the systems-level integration of gene function into networks, such as those controlling flowering time in response to day-length and over-wintering (Figure 2). These networks have been identified in *Arabidopsis* and rice, with allelic variation in key 'hubs' strongly influencing network outputs. Evolutionary processes, such as gene duplication, and the possible footprints of domestication can be mapped to networks such as those controlling flowering time [74, 75]. Such 'systems breeding' approaches can use diverse genomic information to increase the precision with which phenotype can be predicted from genotype, thereby accelerating crop improvement and helping to address food security.

ACKNOWLEDGEMENTS

This work was supported by grants BB/J004588/1, BB/J004596/1 and BB/J003557/1 from the UK Biotechnology and Biological Sciences Research Council (BBSRC) and the John Innes Foundation.

REFERENCES

1. Sequencing Project IRG: The map-based sequence of the rice genome. *Nature* 2005, 436:793-800.
2. Tuskan GA, DiFazio S, Jansson S, Bohlmann J, Grigoriev I, Hellsten U, Putnam N, Ralph S, Rombauts S, Salamov A, Schein J, Sterck L, Aerts A, Bhalerao RR, Bhalerao RP, Blaudez D, Boerjan W, Brun A, Brunner A, Busov V, Campbell M, Carlson J, Chalot M, Chapman J, Chen GL, Cooper D, Coutinho PM, Couturier J, Covert S, Cronk Q, *et al.*: The genome of black cottonwood, *Populus trichocarpa* (Torr. & Gray). *Science* 2006, 313:1596-1604.
3. Schnable PS, Ware D, Fulton RS, Stein JC, Wei F, Pasternak S, Liang C, Zhang J, Fulton L, Graves TA, Minx P, Reily AD, Courtney L, Kruchowski SS, Tomlinson C, Strong C, Delehaunty K, Fronick C, Courtney B, Rock SM, Belter E, Du F, Kim K, Abbott RM, Cotton M, Levy A, Marchetto P, Ochoa K, Jackson SM, Gillam B, *et al.*: The B73 maize genome: complexity, diversity, and dynamics. *Science* 2009, 326:1112-1115.
4. Jaillon O, Aury J-M, Noel B, Policriti A, Clepet C, Casagrande A, Choisne N, Aubourg S, Vitulo N, Jubin C, Vezzi A, Legeai F, Hugueney P, Dasilva C, Horner D, Mica E, Jublot D, Poulain J, Bruyère C, Billault A, Segurens B, Gouyvenoux M, Ugarte E, Cattonaro F, Anthouard V, Vico V, Del Fabbro C, Alaux M, Di Gaspero G, Dumas V, *et al.*: The grapevine genome sequence suggests ancestral hexaploidization in major angiosperm phyla. *Nature* 2007, 449:463-467.
5. Paterson AH, Bowers JE, Bruggmann R, Dubchak I, Grimwood J, Gundlach H, Haberer G, Hellsten U, Mitros T, Poliakov A, Schmutz J, Spannagl M, Tang H, Wang X, Wicker T, Bharti AK, Chapman J, Feltus FA, Gowik U, Grigoriev IV, Lyons E, Maher CA, Martis M, Narechania A, Otillar RP, Penning BW, Salamov AA, Wang Y, Zhang L, Carpita NC, *et al.*: The sorghum bicolor genome and the diversifi cation of grasses. *Nature* 2009, 457:551-556.
6. Schmutz J, Cannon SB, Schlueter J, Ma J, Mitros T, Nelson W, Hyten DL, Song Q, Thelen JJ, Cheng J, Xu D, Hellsten U, May GD, Yu Y, Sakurai T, Umezawa T, Bhattacharyya MK, Sandhu D, Valliyodan B, Lindquist E, Peto M, Grant D, Shu S, Goodstein D, Barry K, Futrell-Griggs M, Abernathy B, Du J, Tian Z, Zhu L, *et al.*: Genome sequence of the palaeopolyploid soybean. *Nature* 2010, 463:178-183.
7. Hood L, Heath JR, Phelps ME, Lin B: Systems biology and new technologies enable predictive and preventative medicine. *Science* 2004, 306:640-643.
8. Batzoglou S: ARACHNE: a whole-genome shotgun assembler. *Genome Res* 2002, 12:177-189.
9. Gill N, Findley S, Walling JG, Hans C, Ma J, Doyle J, Stacey G, Jackson SA: Molecular and chromosomal evidence for allopolyploidy in soybean. *Plant Physiol* 2009, 151:1167-1174.
10. Velasco R, Zharkikh A, Aff ourtit J, Dhingra A, Cestaro A, Kalyanaraman A, Fontana P, Bhatnagar SK, Troggio M, Pruss D, Salvi S, Pindo M, Baldi P, Castelletti S, Cavaiuolo M, Coppola G, Costa F, Cova V, Dal Ri A, Goremykin V, Komjanc M, Longhi S, Magnago P, Malacarne G, Malnoy M, Micheletti D, Moretto M, Perazzolli M, Si-Ammour A, Vezzulli S, *et al.*: The genome of the domesticated apple (*Malus* × *domestica* Borkh.). *Nat Genet* 2010, 42:833-839.
11. Argout X, Salse J, Aury J-M, Guiltinan MJ, Droc G, Gouzy J, Allegre M, Chaparro C, Legavre T, Maximova SN, Abrouk M, Murat F, Fouet O, Poulain J, Ruiz M, Roguet Y, Rodier-Goud M, Barbosa-Neto JF, Sabot F, Kudrna D, Ammiraju JSS, Schuster SC, Carlson JE, Sallet E, Schiex T, Dievart A, Kramer M, Gelley L, Shi Z, Bérard A, *et al.*: The genome of *Theobroma cacao*. *Nat Genet* 2010, 43:101-108.
12. Al-Dous EK, George B, Al-Mahmoud ME, Al-Jaber MY, Wang H, Salameh YM, Al-Azwani EK, Chaluvadi S, Pontaroli AC, DeBarry J, Arondel V, Ohlrogge J, Saie IJ, Suliman-Elmeer KM, Bennetzen JL, Kruegger RR, Malek JA: *De novo* genome sequencing and comparative genomics of date palm (*Phoenix dactylifera*). *Nat Biotechnol* 2011, 29:521-527.
13. Shulaev V, Sargent DJ, Crowhurst RN, Mockler TC, Folkerts O, Delcher AL, Jaiswal P, Mockaitis K, Liston A, Mane SP, Burns P, Davis TM, Slovin JP, Bassil N, Hellens RP, Evans C, Harkins T, Kodira C, Desany B, Crasta OR, Jensen RV, Allan AC, Michael TP, Setubal JC,

Celton J-M, Rees DJG, Williams KP, Holt SH, Rojas JJR, Chatterjee M, *et al.*: The genome of woodland strawberry (*Fragaria vesca*). *Nat Genet* 2010, 43:109-116.

14. Xu X, Pan S, Cheng S, Zhang B, Mu D, Ni P, Zhang G, Yang S, Li R, Wang J, Orjeda G, Guzman F, Torres M, Lozano R, Ponce O, Martinez D, la Cruz De G, Chakrabarti SK, Patil VU, Skryabin KG, Kuznetsov BB, Ravin NV, Kolganova TV, Beletsky AV, Mardanov AV, Di Genova A, Bolser DM, Martin DMA, Li G, Yang Y, *et al.*: Genome sequence and analysis of the tuber crop potato. *Nature* 2011, 475:189-195.
15. Wang X, Wang H, Wang J, Sun R, Wu J, Liu S, Bai Y, Mun J-H, Bancroft I, Cheng F, Huang S, Li X, Hua W, Wang J, Wang X, Freeling M, Pires JC, Paterson AH, Chalhoub B, Wang B, Hayward A, Sharpe AG, Park B-S, Weisshaar B, Liu B, Li B, Liu B, Tong C, Song C, Duran C, *et al.*: The genome of the mesopolyploid crop species *Brassica rapa*. *Nat Genet* 2011, 43:1035-1039.
16. Young ND, Debellé F, Oldroyd GED, Geurts R, Cannon SB, Udvardi MK, Benedito VA, Mayer KFX, Gouzy J, Schoof H, Van de Peer Y, Proost S, Cook DR, Meyers BC, Spannagl M, Cheung F, De Mita S, Krishnakumar V, Gundlach H, Zhou S, Mudge J, Bharti AK, Murray JD, Naoumkina MA, Rosen B, Silverstein KAT, Tang H, Rombauts S, Zhao PX, Zhou P, *et al.*: The *Medicago* genome provides insight into the evolution of rhizobial symbioses. *Nature* 2011, 480:520-524.
17. Sato S, Tabata S, Hirakawa H, Asamizu E, Shirasawa K, Isobe S, Kaneko T, Nakamura Y, Shibata D, Aoki K, Egholm M, Knight J, Bogden R, Li C, Shuang Y, Xu X, Pan S, Cheng S, Liu X, Ren Y, Wang J, Albiero A, Dal Pero F, Todesco S, Van Eck J, Buels RM, Bombarely A, Gosselin JR, Huang M, Leto JA, *et al.*: The tomato genome sequence provides insights into fl eshy fruit evolution. *Nature* 2012, 485:635-641.
18. Town CD, Cheung F, Maiti R, Crabtree J, Haas BJ, Wortman JR, Hine EE, Althoff R, Arbogast TS, Tallon LJ, Vigouroux M, Trick M, Bancroft I: Comparative genomics of *Brassica oleracea* and *Arabidopsis thaliana* reveal gene loss, fragmentation, and dispersal after polyploidy. *Plant Cell* 2006, 18:1348-1359.
19. Bancroft I, Morgan C, Fraser F, Higgins J, Wells R, Clissold L, Baker D, Long Y, Meng J, Wang X, Liu S, Trick M: Dissecting the genome of the polyploid crop oilseed rape by transcriptome sequencing. *Nat Biotechnol* 2011, 29:762-766.
20. Prochnik S, Marri PR, Desany B, Rabinowicz PD, Kodira C, Mohiuddin M, Rodriguez F, Fauquet C, Tohme J, Harkins T, Rokhsar DS, Rounsley S: The cassava genome: current progress, future directions. *Tropical Plant Biol* 2012, 5:88-94.
21. Varshney RK, Chen W, Li Y, Bharti AK, Saxena RK, Schlueter JA, Donoghue MTA, Azam S, Fan G, Whaley AM, Farmer AD, Sheridan J, Iwata A, Tuteja R, Penmetsa RV, Wu W, Upadhyaya HD, Yang S-P, Shah T, Saxena KB, Michael T, McCombie WR, Yang B, Zhang G, Yang H, Wang J, Spillane C, Cook DR, May GD, Xu X, *et al.*: Draft genome sequence of pigeonpea (*Cajanus cajan*), an orphan legume crop of resource-poor farmers. *Nat Biotechnol* 2011, 30:83-89.
22. Bennetzen JL, Schmutz J, Wang H, Percifi eld R, Hawkins J, Pontaroli AC, Estep M, Feng L, Vaughn JN, Grimwood J, Jenkins J, Barry K, Lindquist E, Hellsten U, Deshpande S, Wang X, Wu X, Mitros T, Triplett J, Yang X, Ye C-Y, Mauro-Herrera M, Wang L, Li P, Sharma M, Sharma R, Ronald PC, Panaud O, Kellogg EA, Brutnell TP, *et al.*: Reference genome sequence of the model plant *Setaria*. *Nat Biotechnol* 2012, 30:555-561.
23. Garcia-Mas J, Benjak A, Sanseverino W, Bourgeois M, Mir G, González VM, Hénaff E, Camara F, Cozzuto L, Lowy E, Alioto T, Capella-Gutiérrez S, Blanca J, Cañizares J, Ziarsolo P, Gonzalez-Ibeas D, Rodríguez-Moreno L, Droege M, Du L, Alvarez-Tejado M, Lorente-Galdos B, Melé M, Yang L, Weng Y, Navarro A, Marques-Bonet T, Aranda MA, Nuez F, Picó B, Gabaldón T, *et al.*: The genome of melon (*Cucumis melo* L.). *Proc Natl Acad Sci U S A* 2012, 109:11872-11877.
24. Xu Q, Chen L-L, Ruan X, Chen D, Zhu A, Chen C, Bertrand D, Jiao W-B, Hao B-H, Lyon MP, Chen J, Gao S, Xing F, Lan H, Chang J-W, Ge X, Lei Y, Hu Q, Miao Y, Wang L, Xiao S, Biswas MK, Zeng W, Guo F, Cao H, Yang X, Xu X-W, Cheng Y J, Xu J, Liu J-H, *et al.*: The draft genome of sweet orange (*Citrus sinensis*). *Nat Genet* 2012, 45:59-66.
25. Wang K, Wang Z, Li F, Ye W, Wang J, Song G, Yue Z, Cong L, Shang H, Zhu S, Zou C, Li Q, Yuan Y, Lu C, Wei H, Gou C, Zheng Z, Yin Y, Zhang X, Liu K, Wang B, Song C, Shi N, Kohel RJ,

Percy RG, Yu JZ, Zhu Y-X, Wang J, Yu S: The draft genome of a diploid cotton *Gossypium raimondii*. *Nat Genet* 2012, 44:1098-1103.

26. Paterson AH, Wendel JF, Gundlach H, Guo H, Jenkins J, Jin D, Llewellyn D, Showmaker KC, Shu S, Udall J, Yoo M-J, Byers R, Chen W, Doron-Faigenboim A, Duke MV, Gong L, Grimwood J, Grover C, Grupp K, Hu G, Lee T-H, Li J, Lin L, Liu T, Marler BS, Page JT, Roberts AW, Romanel E, Sanders WS, Szadkowski E, *et al.*: Repeated polyploidization of *Gossypium* genomes and the evolution of spinnable cotton fi bres. *Nature* 2012, 492:423-427.
27. Mayer KFX, Waugh R, Langridge P, Close TJ, Wise RP, Graner A, Matsumoto T, Sato K, Schulman A, Muehlbauer GJ, Stein N, Ariyadasa R, Schulte D, Poursarebani N, Zhou R, Steuernagel B, Mascher M, Scholz U, Shi B, Langridge P, Madishetty K, Svensson JT, Bhat P, Moscou M, Resnik J, Close TJ, Muehlbauer GJ, Hedley P, Liu H, Morris J, *et al.*: A physical, genetic and functional sequence assembly of the barley genome. *Nature* 2012, 491:711-716.
28. Brenchley R, Spannagl M, Pfeifer M, Barker GLA, D'Amore R, Allen AM, McKenzie N, Kramer M, Kerhornou A, Bolser D, Kay S, Waite D, Trick M, Bancroft I, Gu Y, Huo N, Luo M-C, Sehgal S, Gill B, Kianian S, Anderson O, Kersey P, Dvorak J, McCombie WR, Hall A, Mayer KFX, Edwards KJ, Bevan MW, Hall N: Analysis of the bread wheat genome using whole-genome shotgun sequencing. *Nature* 2012, 491:705-710.
29. Bentley DR, Balasubramanian S, Swerdlow HP, Smith GP, Milton J, Brown CG, Hall KP, Evers DJ, Barnes CL, Bignell HR, Boutell JM, Bryant J, Carter RJ, Keira Cheetham R, Cox AJ, Ellis DJ, Flatbush MR, Gormley NA, Humphray SJ, Irving LJ, Karbelashvili MS, Kirk SM, Li H, Liu X, Maisinger KS, Murray LJ, Obradovic B, Ost T, Parkinson ML, Pratt MR, *et al.*: Accurate whole human genome sequencing using reversible terminator chemistry. *Nature* 2008, 456:53-59.
30. Margulies M, Egholm M, Altman WE, Attiya S, Bader JS, Bemben LA, Berka J, Braverman MS, Chen Y-J, Chen Z, Dewell SB, Du L, Fierro JM, Gomes XV, Godwin BC, He W, Helgesen S, Ho CH, Irzyk GP, Jando SC, Alenquer MLI, Jarvie TP, Jirage KB, Kim J-B, Knight JR, Lanza JR, Leamon JH, Lefkowitz SM, Lei M, Li J, *et al.*: Genome sequencing in microfabricated high-density picolitre reactors. *Nature* 2006, 441:120.
31. Maccallum I, Przybylski D, Gnerre S, Burton J, Shlyakhter I, Gnirke A, Malek J, McKernan K, Ranade S, Shea TP, Williams L, Young S, Nusbaum C, Jaff e DB: ALLPATHS 2: small genomes assembled accurately and with high continuity from short paired reads. *Genome Biol* 2009, 10:R103.
32. Varshney RK, Song C, Saxena RK, Azam S, Yu S, Sharpe AG, Cannon S, Baek J, Rosen BD, Tar'an B, Millan T, Zhang X, Ramsay LD, Iwata A, Wang Y, Nelson W, Farmer AD, Gaur PM, Soderlund C, Penmetsa RV, Xu C, Bharti AK, He W, Winter P, Zhao S, Hane JK, Carrasquilla-Garcia N, Condie JA, Upadhyaya HD, Luo M-C, *et al.*: Draft genome sequence of chickpea (*Cicer arietinum*) provides a resource for trait improvement. *Nat Biotechnol* 2013, 31:240-246.
33. D'Hont A, Denoeud F, Aury J-M, Baurens F-C, Carreel F, Garsmeur O, Noel B, Bocs S, Droc G, Rouard M, Da Silva C, Jabbari K, Cardi C, Poulain J, Souquet M, Labadie K, Jourda C, Lengellé J, Rodier-Goud M, Alberti A, Bernard M, Correa M, Ayyampalayam S, Mckain MR, Leebens-Mack J, Burgess D, Freeling M, Mbéguié-A-Mbéguié D, Chabannes M, Wicker T, *et al.*: The banana (*Musa acuminata*) genome and the evolution of monocotyledonous plants. *Nature* 2012, 488:213-217.
34. Peng Z, Lu Y, Li L, Zhao Q, Feng Q, Gao Z, Lu H, Hu T, Yao N, Liu K, Li Y, Fan D, Guo Y, Li W, Lu Y, Weng Q, Zhou C, Zhang L, Huang T, Zhao Y, Zhu C, Liu X, Yang X, Wang T, Miao K, Zhuang C, Cao X, Tang W, Liu G, Liu Y, *et al.*: The draft genome of the fast-growing non-timber forest species moso bamboo (*Phyllostachys heterocycla*). *Nat Genet* 2013, 45:456-461.
35. Choulet F, Wicker T, Rustenholz C, Paux E, Salse J, Leroy P, Schlub S, Le Paslier M-C, Magdelenat G, Gonthier C, Couloux A, Budak H, Breen J, Pumphrey M, Liu S, Kong X, Jia J, Gut M, Brunel D, Anderson JA, Gill BS, Appels R, Keller B, Feuillet C: Megabase level sequencing reveals contrasted organization and evolution patterns of the wheat gene and transposable element spaces. *Plant Cell* 2010, 22:1686-1701.

36. Mayer KFX, Martis M, Hedley PE, Simková H, Liu H, Morris JA, Steuernagel B, Taudien S, Roessner S, Gundlach H, Kubaláková M, Suchánková P, Murat F, Felder M, Nussbaumer T, Graner A, Salse J, Endo T, Sakai H, Tanaka T, Itoh T, Sato K, Platzer M, Matsumoto T, Scholz U, Dolezel J, Waugh R, Stein N: Unlocking the barley genome by chromosomal and comparative genomics. *Plant Cell* 2011, 23:1249-1263.
37. Griffi ths S, Sharp R, Foote TN, Bertin I, Wanous M, Reader S, Colas I, Moore G: Molecular characterization of *Ph1* as a major chromosome pairing locus in polyploid wheat. *Nature* 2006, 439:749-752.
38. Allen AM, Barker GLA, Berry ST, Coghill JA, Gwilliam R, Kirby S, Robinson P, Brenchley RC, D'Amore R, McKenzie N, Waite D, Hall A, Bevan M, Hall N, Edwards KJ: Transcript-specifi c, single-nucleotide polymorphism discovery and linkage analysis in hexaploid bread wheat (*Triticum aestivum* L.). *Plant Biotechnol J* 2011, 9:1086-1099.
39. Berkman PJ, Skarshewski A, Lorenc MT, Lai K, Duran C, Ling EYS, Stiller J, Smits L, Imelfort M, Manoli S, McKenzie M, Kubaláková M, Simková H, Batley J, Fleury D, Dolezel J, Edwards D: Sequencing and assembly of low copy and genic regions of isolated *Triticum aestivum* chromosome arm 7DS. *Plant Biotechnol J* 2011, 9:768-775.
40. Paux E, Sourdille P, Salse J, Saintenac C, Choulet F, Leroy P, Korol A, Michalak M, Kianian S, Spielmeyer W, Lagudah E, Somers D, Kilian A, Alaux M, Vautrin S, Bergès H, Eversole K, Appels R, Safár J, Simková H, Dolezel J, Bernard M, Feuillet C: A physical map of the 1-gigabase bread wheat chromosome 3B. *Science* 2008, 322:101-104.
41. Williams LJS, Tabbaa DG, Li N, Berlin AM, Shea TP, MacCallum I, Lawrence MS, Drier Y, Getz G, Young SK, Jaff e DB, Nusbaum C, Gnirke A: Paired-end sequencing of Fosmid libraries by Illumina. *Genome Res* 2012, 22:2241-2249.
42. Nystedt B, Street NR, Wetterbom A, Zuccolo A, Lin Y-C, Scofi eld DG, Vezzi F, Delhomme N, Giacomello S, Alexeyenko A, Vicedomini R, Sahlin K, Sherwood E, Elfstrand M, Gramzow L, Holmberg K, Hällman J, Keech O, Klasson L, Koriabine M, Kucukoglu M, Käller M, Luthman J, Lysholm F, Niittylä T, Olson A, Rilakovic N, Ritland C, Rosselló JA, Sena J, *et al.*: The Norway spruce genome sequence and conifer genome evolution. *Nature* 2013, 497:579-584.
43. Dendrome [http://dendrome.ucdavis.edu/]
44. Jannoo N, Grivet L, Chantret N, Garsmeur O, Glaszmann J-C, Arruda P, D'Hont A: Orthologous comparison in a gene-rich region among grasses reveals stability in the sugarcane polyploid genome. *Plant J* 2007, 50:574-585.
45. Grivet L, Arruda P: Sugarcane genomics: depicting the complex genome of an important tropical crop. *Curr Opin Plant Biol* 2002, 5:122-127.
46. Swaminathan K, Chae WB, Mitros T, Varala K, Xie L, Barling A, Glowacka K, Hall M, Jezowski S, Ming R, Hudson M, Juvik JA, Rokhsar DS, Moose SP: A framework genetic map for *Miscanthus sinensis* from RNAseq-based markers shows recent tetraploidy. *BMC Genomics* 2012, 13:142.
47. Luo MC, Deal KR, Akhunov ED, Akhunova AR, Anderson OD, Anderson JA, Blake N, Clegg MT, Coleman-Derr D, Conley EJ, Crossman CC, Dubcovsky J, Gill BS, Gu YQ, Hadam J, Heo HY, Huo N, Lazo G, Ma Y, Matthews DE, McGuire PE, Morrell PL, Qualset CO, Renfro J, Tabanao D, Talbert LE, Tian C, Toleno DM, Warburton ML, You FM, *et al.*: Genome comparisons reveal a dominant mechanism of chromosome number reduction in grasses and accelerated genome evolution in Triticeae. *Proc Natl Acad Sci U S A* 2009, 106:15780-15785.
48. Ming R, Liu SC, Lin YR, da Silva J, Wilson W, Braga D, van Deynze A, Wenslaff TF, Wu KK, Moore PH, Burnquist W, Sorrells ME, Irvine JE, Paterson AH: Detailed alignment of saccharum and sorghum chromosomes: comparative organization of closely related diploid and polyploid genomes. *Genetics* 1998, 150:1663-1682.
49. Huang X, Kurata N, Wei X, Wang Z-X, Wang A, Zhao Q, Zhao Y, Liu K, Lu H, Li W, Guo Y, Lu Y, Zhou C, Fan D, Weng Q, Zhu C, Huang T, Zhang L, Wang Y, Feng L, Furuumi H, Kubo T, Miyabayashi T, Yuan X, Xu Q, Dong G, Zhan Q, Li C, Fujiyama A, Toyoda A, *et al.*: A map of rice genome variation reveals the origin of cultivated rice. *Nature* 2012, 490:497-501.
50. Huang X, Zhao Y, Wei X, Li C, Wang A, Zhao Q, Li W, Guo Y, Deng L, Zhu C, Fan D, Lu Y, Weng Q, Liu K, Zhou T, Jing Y, Si L, Dong G, Huang T, Lu T, Feng Q, Qian Q, Li J, Han B:

Genome-wide association study of fl owering time and grain yield traits in a worldwide collection of rice germplasm. *Nat Genet* 2011, 44:32-39.

51. Lai J, Li R, Xu X, Jin W, Xu M, Zhao H, Xiang Z, Song W, Ying K, Zhang M, Jiao Y, Ni P, Zhang J, Li D, Guo X, Ye K, Jian M, Wang B, Zheng H, Liang H, Zhang X, Wang S, Chen S, Li J, Fu Y, Springer NM, Yang H, Wang J, Dai J, Schnable PS, *et al.*: Genome-wide patterns of genetic variation among elite maize inbred lines. *Nat Genet* 2010, 42:1027-1030.
52. Winfi eld MO, Wilkinson PA, Allen AM, Barker GLA, Coghill JA, Burridge A, Hall A, Brenchley RC, D'Amore R, Hall N, Bevan MW, Richmond T, Gerhardt DJ, Jeddeloh JA, Edwards KJ: Targeted re-sequencing of the allohexaploid wheat exome. *Plant Biotechnol J* 2012, 10:733-742.
53. Saintenac C, Jiang D, Akhunov ED: Targeted analysis of nucleotide and copy number variation by exon capture in allotetraploid wheat genome. *Genome Biol* 2011, 12:R88.
54. Varshney R, Graner A, Sorrells M: Genomics-assisted breeding for crop improvement. *Trends Plant Sci* 2005, 10:621-630.
55. Kashkush K, Feldman M, Levy AA: Gene loss, silencing and activation in a newly synthesized wheat allotetraploid. *Genetics* 2002, 160:1651-1659.
56. Shivaprasad PV, Dunn RM, Santos BA, Bassett A, Baulcombe DC: Extraordinary transgressive phenotypes of hybrid tomato are infl uenced by epigenetics and small silencing RNAs. *EMBO J* 2012, 31:257-266.
57. Feldman M, Levy AA: Genome evolution in allopolyploid wheat – a revolutionary reprogramming followed by gradual changes. *J Genet Genomics* 2009, 36:511-518.
58. Meissner A, Gnirke A, Bell GW, Ramsahoye B, Lander ES, Jaenisch R: Reduced representation bisulfi te sequencing for comparative high-resolution DNA methylation analysis. *Nucleic Acids Res* 2005, 33:5868-5877.
59. Schmitz RJ, Schultz MD, Urich MA, Nery JR, Pelizzola M, Libiger O, Alix A, McCosh RB, Chen H, Schork NJ, Ecker JR: Patterns of population epigenomic diversity. *Nature* 2013, 495:193-198.
60. Chao S, Zhang W, Akhunov E, Sherman J, Ma Y, Luo M-C, Dubcovsky J: Analysis of gene-derived SNP marker polymorphism in US wheat (*Triticum aestivum* L.) cultivars. *Mol Breeding* 2008, 23:23-33.
61. Chao S, Dubcovsky J, Dvorak J, Luo M-C, Baenziger SP, Matnyazov R, Clark DR, Talbert LE, Anderson JA, Dreisigacker S, Glover K, Chen J, Campbell K, Bruckner PL, Rudd JC, Haley S, Carver BF, Perry S, Sorrells ME, Akhunov ED: Population- and genome-specifi c patterns of linkage disequilibrium and SNP variation in spring and winter wheat (*Triticum aestivum* L.). *BMC Genomics* 2010, 11:727.
62. Kover PX, Valdar W, Trakalo J, Scarcelli N, Ehrenreich IM, Purugganan MD, Durrant C, Mott R: A multiparent advanced generation inter-cross to fi nemap quantitative traits in *Arabidopsis thaliana*. *PLoS Genet* 2009, 5:e1000551.
63. McMullen MD, Kresovich S, Villeda HS, Bradbury P, Li H, Sun Q, Flint-Garcia S, Thornsberry J, Acharya C, Bottoms C, Brown P, Browne C, Eller M, Guill K, Harjes C, Kroon D, Lepak N, Mitchell SE, Peterson B, Pressoir G, Romero S, Oropeza Rosas M, Salvo S, Yates H, Hanson M, Jones E, Smith S, Glaubitz JC, Goodman M, Ware D, *et al.*: Genetic properties of the maize nested association mapping population. *Science* 2009, 325:737-740.
64. Morrell PL, Buckler ES, Ross-Ibarra J: Crop genomics: advances and applications. *Nat Rev Genet* 2011, 13:85-96.
65. Meuwissen TH, Hayes BJ, Goddard ME: Prediction of total genetic value using genome-wide dense marker maps. *Genetics* 2001, 157:1819-1829.
66. Xu SS: Estimating polygenic eff ects using markers of the entire genome. *Genetics* 2003, 163:789-801.
67. Heff ner EL, Sorrells ME, Jannink J-L: Genomic selection for crop improvement. *Crop Sci* 2009, 49:1.
68. Eathington SR, Crosbie TM, Edwards MD, Reiter RS, Bull JK: Molecular markers in a commercial breeding program. *Crop Sci* 2007, 47:S154-S163.
69. The ENCODE Project Consortium: The ENCODE (ENCyclopedia Of DNA Elements) project. *Science* 2004, 306:636-640.

70. Uauy C, Paraiso F, Colasuonno P, Tran RK, Tsai H, Berardi S, Comai L, Dubcovsky J: A modifi ed TILLING approach to detect induced mutations in tetraploid and hexaploid wheat. *BMC Plant Biol* 2009, 9:115.
71. Shukla VK, Doyon Y, Miller JC, DeKelver RC, Moehle EA, Worden SE, Mitchell JC, Arnold NL, Gopalan S, Meng X, Choi VM, Rock JM, Wu Y-Y, Katibah GE, Zhifang G, McCaskill D, Simpson MA, Blakeslee B, Greenwalt SA, Butler HJ, Hinkley SJ, Zhang L, Rebar EJ, Gregory PD, Urnov FD: Precise genome modifi cation in the crop species *Zea mays* using zinc-fi nger nucleases. *Nature* 2009, 459:437-441.
72. Jinek M, Chylinski K, Fonfara I, Hauer M, Doudna JA, Charpentier E: A programmable dual-RNA-guided DNA endonuclease in adaptive bacterial immunity. *Science* 2012, 337:816-821.
73. Mali P, Yang L, Esvelt KM, Aach J, Guell M, DiCarlo JE, Norville JE, Church GM: RNA-guided human genome engineering via Cas9. *Science* 2013, 339:823-826.
74. Higgins JA, Bailey PC, Laurie DA: Comparative genomics of fl owering time pathways using *Brachypodium distachyon* as a model for the temperate grasses. *PLoS ONE* 2010, 5:e10065.
75. Yan L, Fu D, Li C, Blechl A, Tranquilli G, Bonafede M, Sanchez A, Valarik M, Yasuda S, Dubcovsky J: The wheat and barley vernalization gene *VRN3* is an orthologue of FT. *Proc Natl Acad Sci U S A* 2006, 103:19581-19586.

CHAPTER 4

A Decade of Progress in Organic Cover Crop-Based Reduced Tillage Practices in the Upper Midwestern USA

Erin M. Silva [1,*] and Kathleen Delate [2]

[1] *Department of Plant Pathology, University of Wisconsin-Madison, 1630 Linden Dr., Madison, WI 53706, USA*

[2] *Departments of Agronomy and Horticulture, Iowa State University, 106 Horticulture Hall, Ames, IA 50011, USA*

AUTHOR CONTRIBUTIONS

Erin M. Silva and Kathleen Delate equally conceived the concept for the paper and contributed to the writing of all sections.

ABSTRACT

The organic industry continues to expand in the United States (U.S.), with 14,093 organic farms in 2014. The upper Midwestern U.S. has emerged as a hub for organic row crop production; however, the management of these organic row crop hectares heavily relies on tillage and cultivation for weed control. Faced with the soil quality challenges related to these practices, and cognizant of the benefits of conventional no-till practices, organic farmers have shown significant interest in the development of Cover Crop-Based

Reduced Tillage (CCBRT) techniques to lessen soil disturbance while achieving successful weed management. To serve this farmer interest, significant research efforts have been conducted in the upper Midwestern U.S., focused on systems-based practices to ensure adequate suppression of weeds, through a combination of agronomic and cover crop species and variety selection. Within this review article, we discuss the agronomic successes that have been achieved in CCBRT using a combination of cereal rye and soybeans, resulting in consistent suppression of weeds while providing fuel and labor savings for farmers, as well as the continued challenges that have persisted with its implementation. Continued investment in research focused on cover crop breeding and management, optimization of CCBRT equipment and fertility management, and a greater understanding of rotation effects will contribute to the further expansion of this technique across organic farms.

Keywords: organic agriculture; cover cropping; reduced tillage; ecosystem services; USA

1. INTRODUCTION

The organic industry continues to expand in the United States (U.S.), with 14,093 certified organic farms in 2014 [1,2,3]. The 2014 National Agricultural Survey of organic production, conducted by the United States Department of Agriculture (USDA), reported 82,328 hectares of organic corn and 39,996 hectares of organic soybean among the 1.4 million organic cropland and vegetable hectares in the U.S. [1].

Management of these organic row crops heavily relies on tillage and cultivation for weed control. While mechanical practices can be effective to manage weeds, these activities prevent organic farmers from fully optimizing the requirement for soil building, as set forth in 7 CFR §205.203 and 205.205 of the National Organic Program (NOP) [4]. In typical organic row crop production in the upper Midwestern U.S., five to six tractor passes with tine weeders, rotary hoes, and/or row cultivators are often necessary for adequate weed control, which can negatively impact soil aggregation and soil organic matter concentrations, while exposing the land to greater risk of erosion. Additionally, the reliance on cultivation as a primary weed management tool poses risks during wet springs, an increasingly common production challenge with heavy rainfalls occurring more frequently as a result of climate change [5]; consistently wet soil conditions can prevent organic producers from implementing timely weed management through cultivation, increasing weed competition and weed seedbanks while negatively impacting yields. Ongoing data from the Wisconsin Integrated Cropping Systems Trial, a long-term management trial begun in 1989, illustrates that while organic management results in comparable yields to conventional management during more moderate production seasons, years exhibiting wet spring conditions result in yields of corn and soybeans falling to approximately 75% of those produced using conventional practices, due to the inability of farmers to conduct timely weed management [6].

In addition to weed management challenges associated with tillage and cultivation, these practices can also increase the risk for soil erosion. Several Midwestern regions of the U.S. which support high concentrations of organic farms, including the Driftless regions of Wisconsin, Iowa, and Minnesota, are characterized by farm fields exceeding 4% or greater slopes. The soils in these regions vary widely, from heavy, poorly drained clay soils to sandy, shallow, droughty soils. These conditions create a landscape susceptible to erosion, negatively impacting both soil and water quality of several key watersheds, including the Mississippi Valley Watershed.

Conventional no-till farming techniques have been promoted for their role in reducing water runoff and soil erosion, as well as maintaining soil carbon [7,8], although the degree and nature to which the conventional no-till systems build soil C, and the degree to which this C is stably stored, can vary [9,10]. No-till systems can also increase infiltration of water into the soil by 25 to 50% compared with conventional tillage systems [11]. In addition, cover crop surface residues can decrease the effect of wind and temperature on soil water evaporation, increase water storage in the soil profile [12,13], scavenge available nitrogen, and prevent soil erosion [14], thus preventing watershed contamination and nutrient losses [14,15].

Faced with the soil quality challenges associated with tillage and cultivation, and cognizant of the benefits of conventional no-till practices, organic farmers have shown significant interest in the development of no- and reduced-tillage techniques suitable for organic production. Within the majority of the reduced tillage systems currently utilized by organic farmers, no-till phases are incorporated throughout the rotation, with tillage limited to establishing the cover crops [16,17,18]. This technique, often referred to as Cover Crop-Based Reduced Tillage (CCBRT) uses mature cover crop residue as a mulch to smother weeds, replacing the standard organic weed management tactics of tillage and cultivation. Winter annual cover crops (typically cereal rye (Secale cereale L.) and hairy vetch (Vicia villosa L.)) are seeded in the fall and terminated mechanically without herbicides in the spring, often using a roller-crimper (Figure 1a) which creates an in situ surface mulch that physically suppresses weeds. At the time of cover crop termination, the cash crop (typically soybean (Glyine max (L.) Merr.) or field corn (Zea mays L.)) is planted directly into the cover crop mulch (Figure 1b), which provides season-long weed suppression without further soil disturbance throughout cash crop production.

(a) (b)

Figure 1. (**a**) Photograph of the roller-crimper design commonly used by farmers and researchers in the upper Midwestern U.S.; (**b**) Photograph of soybeans emerging through the rolled winter cereal rye mulch.

2. ORGANIC COVER CROP-BASED REDUCED TILLAGE IN THE UPPER MIDWESTERN U.S.

The upper Midwestern U.S. maintains a long history in organic farming, and still remains a primary center of the U.S. organic industry [1]. Prior to the rise of organic agriculture, regional environmental leaders, such as Aldo Leopold, an ardent environmentalist, and Gaylord Nelson, a Wisconsin politician, inspired a strong land ethic in upper Midwestern culture. Counter to the argument that the maturation of the organic industry will, by default, lead to a concurrent conventionalization of the industry, the production practices used by upper Midwestern organic farmers demonstrate adherence to the soil-building ethic that serves as a foundation of the organic regulation outlined by the NOP [19].

CCBRT research began in the early 2000s in the upper Midwest, with several research programs, including Iowa, Wisconsin, and Michigan, establishing experimental plots both at land-grant university research stations and on working certified organic farms. Research approaches have predominantly centered on techniques and equipment made popular by the Rodale Institute in Kutztown, PA, where a fall-planted cover crop is mechanically terminated in the spring (cereal grains at anthesis at Zadok's growth stage 60, or legumes at 100% bloom). Cover crop termination is achieved with a roller-crimper, a hollow steel cylinder with metal slats arranged in a chevron pattern, welded at uniform spacing along the length of the drum that can be filled with water for added weight [20,21,22] (Figure 1a). However, research programs have also included cover crop termination treatments that utilize sickle-bar mowing and flail mowing [20,22].

Much of the CCBRT research in the upper Midwestern U.S. has focused on systems-based practices to ensure effective suppression of weeds, primarily through a combination of agronomic practices and cover crop management. Corn and soybean have been the primary commodity crops that have been evaluated within the CCBRT system in this region. However, studies have also addressed other critical aspects of the system, including cash crop yields, economic assessments regarding fuel and labor savings, and reduction in erosion risk.

As research data is generated and practical experiences are reported by organic farmers, CCBRT has also been not only gaining recognition, but is being implemented across the organic agricultural landscape. According to the 2014 USDA Organic Survey, organic farmers in the upper Midwest are increasingly adopting organic no-till management strategies as part of their farming practices. Of the 3319 certified organic farms in of Iowa, Illinois, Michigan, Minnesota, Missouri, Nebraska, and Wisconsin, 957 reported the use of no-till practices [1]. This survey, however, does not provide adequate resolution to determine whether CCBRT practices in row crops are being used, or whether other no-till practices are being reported, such as reseeding grasses and legumes into existing pastures without tillage. On a state level, the reported use of no-till practices ranged from 20 to 37% of organic farms (Iowa: 27% (166 no-till farms, 612 total organic farms); Illinois: 27% (69, 249); Michigan: 20% (67, 332); Minnesota; 37% (189, 512); Missouri: 28% (61, 216); Nebraska: 32% (54, 170); Wisconsin: 29% (351, 1228)) [1].

The remainder of this paper summarizes experiences with CCBRT in the upper Midwestern U.S., incorporating published results from research studies and on-farm observations. Several platforms exist from which data and information are assembled: (1) land grant University research programs, particularly the efforts in Wisconsin and Iowa; and (2) organic farmer networks and participatory research efforts, such as OGRAIN (the Organic Grain Resource and Information Network), led by the University of Wisconsin-Madison. Further, the authors discuss remaining barriers in the CCBRT system which prevent more wide scale adoption in the upper Midwestern US, and further research needs necessary to address those barriers.

3. SUMMARY OF CCBRT RESEARCH IN THE UPPER MIDWESTERN U.S.

3.1. Organic Soybean Production

Weed management in organic cropping systems remains a significant challenge for organic farmers, particularly in the soybean phase of a typical corn–soybean-winter wheat–alfalfa rotation common to organic grain systems in the upper Midwestern U.S. [21,22]. Unlike corn and winter wheat, which develop above-ground biomass early in the growing season and thus more effectively compete with weeds, soybean crops, often planted on the wider row spacing for cultivation, can be relatively slow to canopy, allowing weeds to establish over a prolonged period during the first half of the production season. As such, the development of CCBRT techniques for this phase of the crop rotation to augment mechanical weed management strategies during soybean production addresses a critical production challenge.

CCBRT research in organic soybean production in the upper Midwestern U.S. have integrated the establishment of a cereal grain cover crop (most often cereal rye) in late summer or early fall, typically seeding at a rate of 180–269 kg·ha−1 [20,22] (Table 1). The cereal grain cover crop is then terminated in late spring, with soybean directly seeded through the cereal grain mulch, at rates of 500,000 seeds·ha−1; or more. Other cereal grain cover crops have also been trialed for use in CCBRT, including Winter triticale (Tritocosecale Wittm. Ex A Camus), winter barley (Hordeum vulgare L.), 'McGregor', and winter wheat (Triticum aestivum) [21,22].

While CCBRT has demonstrated its ability to effectively suppress weeds in organic soybean production [20,22], the capacity of the cover crop residue to prevent weeds from establishing varies widely depending on the weed species present in the field. Small-seeded annual broadleaf weeds tend to be suppressed more easily by the mulch and thus are more dominant in tilled organic systems [23,24,25]; if fields have adequate mulch biomass and complete cover crop termination, up to 80% control of common annual weeds in winter rye stands have been reported [26]. Perennial weeds are less sensitive to suppression by the mulch and readily proliferate in the system over time [25,27].

Weed management using the CCBRT technique is comprised of diversified approaches, with efficacy driven by rye biomass levels, cash crop planting date, seeding rate and placement, and cash crop stand establishment. Under typical cropping conditions, rye can produce biomass levels that range from 4000–11,000 kg·ha^{-1} [22] (Table 1).

Table 1. Summary of results of Cover Crop-Based Reduced Tillage (CCBRT) research conducted in the upper Midwestern U.S.A. as reported in peer-reviewed research journals.

Study Location	Cover Crop Variety	Cover Crop Planting Date [1]	Cover Crop Seeding Rate and Row Spacing	Cover Crop Termination Date	Cover Crop Biomass at Termination	Cash Crop Seeding Rate and Row Spacing	Cash Crop Planting Date	Weed Populations/ Biomass	Cash Crop Stand Populations	Soybean Yields	Citation
Soybean ((Glycine max (L.) Merr.))											
Iowa	Cereal rye (*Secale cereale* L.), Variety Not Specified (VNS)/VNS hairy vetch (*Vicia villosa* L.)	12 September and 31 October	72 kg·ha^{-1} cereal rye, 36 kg·ha^{-1} hairy vetch	Zadoks growth stage 60	N/A [2]	494,210 seeds·ha^{-1}; 76 cm rows	23–25 May	7–8 weeds·m^{-2}	226,719–308,733 plants·ha^{-1}	1067–2724 kg·ha^{-1}	[21]
	Winter wheat (*Triticum aestivum* L.) ('Expedition' and 'Arapahoe'/winter pea (*Pisum sativum* subsp. Arvense) cover crop)	12 September and 31 October	63 kg·ha^{-1} winter wheat, 21 kg·ha^{-1} Winter Pea	Zadoks growth stage 60	N/A	494,210 seeds·ha^{-1}; 76 cm rows	23–25 May	9–15 weeds·m^{-2}	209,422–275,357 plants·ha^{-1}	628–5668 kg·ha^{-1}	[21]
Wisconsin	Cereal rye, 'Rymin'	5 October and 7 October	180 kg·ha^{-1}, 19 cm row spacing	6–11 June	4.3–10.8 Mg DM·ha^{-1}	625,200 seeds·ha^{-1}; 19 cm row spacing	18–21 May	3–229 kg·ha^{-1}	377,100–505,600 plants·ha^{-1}	2751–2885 kg·ha^{-1}	[20]
	Hairy vetch, 'VNS'	8 September and 13 September	33.6 kg·ha^{-1}, 19 cm row spacing	28 May–8 June	3.7–5.0 Mg·DM·ha^{-1}	N/A	N/A	4–47 weeds·m^{-2}	N/A	N/A	[22]
	Winter rye, 'VNS'	8 September and 13 September	269 kg·ha^{-1} 19 cm row spacing	28 May–8 June	10.2–10.3 Mg·DM·ha^{-1}	N/A	N/A	1–25 weeds·m^{-2}	N/A	N/A	[22]
	Winter triticale (*Tritocosecale* Wittm. Ex A Camus), 'Fridge'	8 September and 13 September	269 kg·ha^{-1}, 19 cm row spacing	28 May–8 June	6.4–14.6 Mg·DM·ha^{-1}	N/A	N/A	24–26 weeds·m^{-2}	N/A	N/A	[22]
	Austrian winter pea (*Pisum sativum* subsp. arvense), 'VNS'	8 September and 13 September	44.6 kg·ha^{-1}, 19 cm row spacing	28 May–8 June	0–6.3 Mg·DM·ha^{-1}	N/A	N/A	0–18 weeds·m^{-2}	N/A	N/A	[22]
	Winter barley (*Hordeum vulgare* L.), 'McGregor'	8 September and 13 September	269 kg·ha^{-1}, 19 cm row spacing	28 May–8 June	10.3–11.7 Mg·DM·ha^{-1}	N/A	N/A	19–43 weeds·m^{-2}	N/A	N/A	[22]
Corn (Zea mays L.)											
Iowa	Cereal rye (*Secale cereale* L.), Variety Not Specified (VNS)/VNS hairy vetch (*Vicia villosa* L.)	12 September and 31 October	72 kg·ha^{-1} cereal rye, 36 kg·ha^{-1} hairy vetch	Zadoks growth stage 60	N/A	79,073 seeds·ha^{-1}; 76 m rows	23–25 May	N/A	45,302–59,510 plants·ha^{-1}	628–5668 kg·ha^{-1}	[21]
	Winter wheat (*Triticum aestivum* L.) ('Expedition' and 'Arapahoe'/winter pea (*Pisum sativum* subsp. Arvense) cover crop)	12 September and 31 October	63 kg·ha^{-1} winter wheat, 21 kg·ha^{-1} winter pea	Zadoks growth stage 60	N/A	79,073 seeds·ha^{-1}; 76 cm rows	23–25 May	N/A	49,824–51,479 plants·ha^{-1}	640–5567 kg·ha^{-1}	[21]

[1] Ranges of values in a given column reflect data from separate years in a given study, due to significant year effects [2] N/A: designates data not included in publication.

Research has demonstrated that in order to reliably suppress weeds as a surface mulch, the fall-planted cereal grain cover crop must reach biomass levels at the higher end of this range, ideally exceeding 8000 $kg \cdot ha^{-1}$ [26,28]. This allows the mulch to effectively limit weed populations by not only physically interfering with the emergence process, but also preventing the breaking of seed dormancy and inhibiting weed germination through allelopathy [27].

While CCBRT systems can produce soybean yields that are within state averages for organic soybeans, yields continue to lag behind organic systems that are managed using typical tillage practices. Research in Iowa and Wisconsin has demonstrated that, while during some seasons CCBRT yields are not significantly different than typically managed organic soybean, in many circumstances yield reductions in CCBRT systems as compared to typical organic soybean yields are observed, even with planting dates remaining the same, with up to a 24% or more reduction in yields observed [20,21]. Reasons for these observed yield reductions may be multi-fold: (1) cooler soil temperatures under the rye residue delaying soybean germination; (2) slower or inhibited root growth under the rye mulch; (3) nutrient tie-up under the rye mulch; and (4) allelopathic effects of the cereal rye. Whereas soybean biomass at the end of the production season is similar between CCBRT and typically managed organic soybean, early season growth of CCBRT soybeans has been observed to be slower as compared to typical organic production practices. As the weed pressure experienced within the two systems has been observed to be either equivalent or less in the CCBRT, it is unlikely that weed competition is a primary factor resulting in the observed yield differences.

3.2. Organic Corn Production

Experimentation with CCBRT in the corn phase of the crop rotation has primarily differed with respect to the cover crop used in the system. Whereas the same planting strategies are used—a winter-hardy cover crop is seeded in the late summer and terminated during the following spring—CCBRT techniques for corn have integrated the leguminous cover crops, occasionally in combination with a cereal grain to enhance mulch biomass [21,22] Hairy vetch (Vicia villosa L.) has been the most commonly researched and trialed legume cover crop in the corn CCBRT system, although alternative legumes such as Austrian winter pea and field pea (Pisum sativum subsp. arvense) have also been tested [21,22]. Seeding dates of these legume cover crops have been similar to those used in the establishment of cereal grain cover crops (late summer/early fall), with seeding rates ranging from 33.6–44.6 $kg \cdot ha^{-1}$.

As with the effective mechanical termination of cereal grain cover crops, legume cover crops must be terminated at specific maturities. To obtain effective termination, roller-crimping or mowing must occur at 100% bloom to early pod set [29,30]. In the upper Midwest, this growth stage varies from late May (Austrian winter pea) to mid-June (hairy vetch) [22]. While a late May planting date can allow for corn grain and silage production. If appropriate short-season varieties of corn are used, mid-June planting dates substantially decrease the yield potential of organic corn, for both silage and grain. Within the Austrian winter pea cultivars that are commercially available, winter-hardiness remains marginal in the region, thus reducing the feasibility of this cover crop into the CCBRT system. With hairy vetch remaining the only reliable option for an overwintering legume cover crop which produces adequate biomass for an effective weed-suppressive mulch, CCBRT organic corn systems remain challenging if not prohibitive.

CCBRT in the corn phase has also proven more challenging than soybean due to increased risk of insect pest interactions. In research trials at the University of Wisconsin Arlington Agricultural research station, CCBRT corn stands have been decimated by army worm (Mythimna unipuncta Haworth), particularly when cover crop stands have included a cereal grain. This insect pest, which oviposits on the lower leaves of grasses or the base on grass plants, can be attracted to the cereal grain cover crop during its migration from the southern states of the U.S. Depending on the timing of this migration, newly hatched larvae may emerge from the cover crop residue at the time of corn germination, then feeding upon the corn seedlings. Although it has not been observed, depending on annual conditions and insect life cycles, a similar risk exists with seed corn maggot (Delia platura Meigen), a common corn insect pest. which lays eggs on decaying vegetation in late April or early May. Without the option to use chemical insecticide seed treatments, the development of CCBRT strategies must account for ecological-based management solutions, such as avoiding the use of cover crops that are preferred hosts to insect pests and the altering of cash crop planting dates to avoid specific pest cycles, to mitigate the risk of insect damage to the cash crop.

3.3. Economics and Labor/Fuel Savings

Several CCBRT studies conducted in the upper Midwestern U.S. have integrated economic analyses into their data analysis (Table 2). In an analysis conducted by Bernstein et al. [20], returns to labor, capital, and management of the CCBRT treatments using cereal rye were 27% less than in the organic tilled system, primarily due to the 24% yield reduction shown in this particular study. However, the study also documented that labor and fuel inputs were reduced by nearly 50% in the no-till cereal rye treatments. Although the profitability per hectare was greater in the tilled treatment, the return per labor hour was 25% greater in the no-till cereal rye system. These savings not only translate to economic savings, but could have positive impacts on the farmers' quality of life, potentially allowing them to engage in other enterprises or expand their soybean acres. Significant diesel fuel savings were also documented, with 720 L of diesel fuel saved using no-till techniques. Delate et al. [21] found similar results, with significantly less labor costs in the CCBRT systems. As with the Bernstein study, returns to land and management remained less in the CCBRT systems as compared to the typical organic management systems, due to persistent soybean yield reductions.

The CCBRT studies focused on organic corn are much more limited. Delate et al. [21], over two years of investigating CCBRT strategies using hairy vetch/rye and winter wheat/winter pea, found return to management to range from an economic loss (−660 USD·ha^{-1}) to more profitable scenarios (618 USD·ha^{-1}) (Table 2). In both years over both production systems, however, return to management was significantly less than the tilled organic corn systems (1602–1866 USD·ha^{-1}).

3.4. Impact of CCBRT on Soil Quality Parameters

The adoption of reduced-tillage practices such as CCBRT has been promoted as a tool to mitigate soil C loss, build soil organic matter (SOM), and reduce the risk of erosion [31]. However, in large part due to the lack of long-term organic CCBRT experimental sites, the impact of CCBRT techniques on increasing soil C and SOM remains unclear. Clark et al. (2017) [32], investigating the impact of CCBRT techniques on soil parameters in organic row crop production in Missouri, USA, reported no change in soil organic carbon (SOC) under CCBRT management, but concluded this may be due to the short two-year time frame of the experiment. Using the Revised Universal Soil Loss Equation, Version 2 to predict soil loss, Bernstein et al. [20] estimated that the CCBRT soybean plots integrating cereal rye as a cover crop would result in a soil loss of 1.5 Mg·ha^{-1} on a 1.0% slope and 5.6 Mg·ha^{-1} on a 4.5% slope, significantly less than the predicted soil loss on the organic till soybean fields, which ranged from 10.9 Mg·ha^{-1} on a 1.0% slope to 49.3 Mg·ha^{-1} on a 4.5% slope. While not measured directly, these same models predicted changes in soil organic matter (soil conditioning index, on a scale of −2 to +2), were positive in CCBRT rye treatments (+0.4 on 1.0% slope and +0.3 on 4.5% slope) and negative in the tilled rye treatment for both slope grades, (−0.9 on 1.0% slope and >−2.0 on 4.5% slope).

Table 2. Economic analyses of CCBRT systems in the upper Midwestern U.S.

Study Location	Cash Crop	Cover Crop	Cash Crop Row Spacing	Yields $kg \cdot ha^{-1}$	Gross Revenue $USD \cdot ha^{-1}$	Return to Management $USD \cdot ha^{-1}$	Citation
Iowa	Soybean	Cereal rye/hairy vetch	76 cm	1067–2724 [1]	672–1769	−63–993	[21]
		Winter wheat/winter pea		1042–2862	656–1859	36–1198	
		No cover crop (traditional tillage)		2197–3170	1383–2059	742–1377	
	Corn	Cereal rye/hairy vetch	76 cm	628–5668	217–1394	−660–527	
		Winter wheat/winter pea		640–5567	221–1369	−540–618	
		No cover crop (traditional tillage)		7777–9710	2389–2694	1602–1866	
Wisconsin	Soybean	Cereal rye	19 cm	2751–2885	N/A [2]	1598–1687	[20]
		No cover crop (traditional tillage)	76 cm	3618		2162	

[1] Ranges of values in a given column reflect data from separate years in a given study, due to significant year effects.
[2] N/A: designates data not included in publication.

While benefits are predicted with respect to building SOM and minimizing erosion, nutrient availability dynamics may be negatively impacted during the soybean production season. In the aforementioned Bernstein et al. study [20], plant tissue tests demonstrated that while soil nutrient concentrations were similar among CCBRT and tilled soybean treatments, soybean uptake of N, P, and K as measured by tissue tests was several-fold greater in the tilled treatment as compared to the CCBRT treatments. These same results have been demonstrated in subsequent years on other fields at the University of Wisconsin Arlington Agricultural Research Station (data not published).

4. CHALLENGES TO FURTHER ADOPTION OF CCBRT IN THE UPPER MIDWESTERN U.S.

4.1. Adapting the CCBRT System to Organic Rotations

While the use of CCBRT has demonstrated success in the soybean phase of the rotation using cereal grain cover crops, strategizing crop rotations that are both agronomically and economically sound remains a challenge for organic farmers in the upper Midwestern U.S. A typical representative rotation may include corn–soybean-winter wheat–alfalfa, with corn harvested either as a silage crop throughout September or as a grain crop from late September into October. Research has demonstrated that cereal rye planting date in the fall significantly impacts ground cover in early spring and biomass at termination at anthesis (Zadoks stage 60), two factors critical to ensuring weed suppression throughout the soybean production season [27]. Thus, an organic crop rotation that includes a CCBRT soybean phase would be limited to a rotation that incorporates silage corn as opposed to grain corn in the more northern areas of the upper Midwestern U.S., if the farm were to follow a typical organic rotation scheme. While this is a viable option for organic livestock farmers utilizing CCBRT techniques, it creates a less economically viable rotation for organic cash grain farmers.

Interactions of CCBRT with the winter wheat is also of concern with respect to rotation management. While cereal rye can be planted after a winter wheat and oat harvest, this rotation may not be ideal as it results in sequential cereal grain plantings, which can increase the risk of certain plant diseases. A CCBRT rye-soybean phase into a typical corn–soybean-winter wheat rotation can also have negative repercussions on the subsequent winter wheat crop. With planting dates delayed 2–3 weeks from typical organic yields to synchronize with the appropriate cereal rye maturity stage, CCBRT delays soybean maturity, with harvest occurring in late-October versus late September or mid-October. With much of the high-carbon cereal rye residue remaining at soybean harvest, this further creates delays of winter wheat planting, which typically occurs between mid and late October. This delay in field preparation may result in the need to shift to spring-planted cereal grains, such as oat or spring wheat, which typically result in lower yields and quality as compared to winter cereal grains grown in the regions of the upper Midwestern U.S. with higher precipitation.

Additionally, some farmers utilizing CCBRT in a rye/soybean phase of the rotation have experienced contamination issues in their subsequent cereal grain crop. If the termination of the cereal rye is not completely effective, cereal rye plants producing viable seed may emerge within the soybean crop. While not having a negative impact on the soybean phase, these seeds may germinate in the subsequent cash crop year, resulting in volunteer cereal rye plants. If another cereal grain crop, such as winter wheat, is sown into this field, unacceptable levels of contamination by rye seed at harvest may cause the product to be rejected. While this is less of an issue for livestock farmers feeding the grain to their own herds, this can be a significant concern for farmers selling into the commodity grain market.

4.2. Fertility Management in CCBRT

The impact of cereal rye on soil inorganic nitrogen (N) availability and the subsequent impacts on both weeds and the soybean cash crop growth have just begun to be investigated. Fertility management can include both the cereal rye cover crop and the soybean cash crop. To achieve a desirable level of biomass of 8000 $kg{\cdot}ha^{-1}$ or greater, the cereal rye cover crop requires at least 90 to 110 kg of available $N{\cdot}ha^{-1}$, if a desired shoot N concentration of 9–11 g N kg^{-1} is used [33,34]. To ensure that the cereal rye produces enough biomass to reliably suppress weeds, in some systems, fertilization of the cereal rye may be necessary.

Fertility management in high-residue environments presents specific challenges that need to be addressed in the CCBRT system. Research suggests that cold soil temperature conditions, such as those typically observed in the spring in CCBRT systems, can reduce nutrient mineralization and consequently yields. Work conducted in Wisconsin by Andraski and Bundy [35] concluded that soil temperature in no-till systems is the main factor affecting net N mineralization in corn systems, as opposed to N immobilization by residues, and recommended increasing N fertilization rates up to 30 $kg{\cdot}ha^{-1}$ for high residue no-till systems. In the rye/soybean system, CCBRT has been shown to lead to nutrient deficiency in the system as compared to typical organic management. Bernstein et al.

(2011) [20] documented N and sulfur (S) deficiency in soybean tissue collected at the R1 growth stage grown under CCBRT in Wisconsin. Similarly, in 2013, Silva documented N deficiency in CCBRT soybean plots in Wisconsin through tissue testing conducted at the initial flowering stage, with tissue N averaging 3.5% and tissue S averaging 0.3% (unpublished data). Furthermore, in the studies of both Bernstein and Silva, visual symptoms of N deficiency were observed throughout the study period in the no-till treatments, from the vegetative cotyledon (VC) soybean stage to the reproductive 2 (R2) stage.

As continued efforts are dedicated to developing CCBRT for organic corn production, the issue of nutrient management in the system becomes even more critical. Unlike soybean, which can overcome some degree of initial N deficiency of the system as nitrogen fixation by rhizobium bacteria begins to occur, corn yields are more affected by insufficient N availability at critical stages of plant growth. With the incorporation of nitrogen-rich above and below ground biomass, legume cover crops can provide substantial N credits to the subsequent crop; however, in CCBRT systems where the aboveground biomass is not incorporated prior to cash crop planting, the N credit may be less than what might be anticipated from typical legume cover crop management. Options for supplemental N fertility may include topdressing or side-dressing with manure during the corn phase. However, in the high residue system where disturbance of the cover crop mulch can lead to exposed soil and subsequent weed establishment, the integration of these strategies may require equipment modification or novel application techniques.

4.3. Equipment Modifications

Even with cereal rye biomass reaching levels sufficient for effective weed suppression, potential yields of the system can be reduced due to poor soybean stands. Adequate seed placement through the thick cereal rye mulch is critical for the success of the system and is dependent on both environmental and physical factors. Equipment selection, modifications, and settings, as well as soil moisture at the time of planting, impact the ability of the planter to penetrate the mulch and achieve good seed-to-soil contact. Both conservation tillage planters and no-till drills have been used in the CCBRT system; planters, as compared to drills, are able to cut through the mulch while providing more precise seed metering and placement. Drills, however, can reduce the time to canopy closure due to narrow seed spacing [27,36]. If cereal rye residue is greater than 5000 $kg \cdot ha^{-1}$, the planter is the preferred selection due to more effective mulch cutting [27]. Additionally, as the cover crops are typically planted at a high seeding rate, the cereal grains have a propensity to lodge prior to roller-crimping, making consistent seed placement difficult through the thick, uneven mulch using either equipment option.

Additional planter modifications can better ensure success with the CCBRT system. Extra weight can be added to the equipment to help ensure better cutting of the mulch residue and adequate seed placement. Coulters designed specifically for residue cutting, such as straight-edged coulters, can further facilitate planting and adequate seed-to-soil contact through the thick cereal rye residue. Initial evaluations with closing wheels has shown that a single 38-cm spiked closing wheel along with a smooth cast closing wheel on each row unit can help improve closure of the planting row [27]. In Wisconsin, several modifications have been made to the planter dedicated to CCBRT to contribute to more effective seed placement, including frame-mounted no-till coulters, and down-pressure springs, which increase planter weight, equivalent to 91 kg per row unit [37].

4.4. Planting Date Modifications

Research continues to be conducted to explore options beyond the foundational techniques developed by the Rodale Institute, which uses a one-pass operation at cover crop anthesis to roll-crimp the cover crop first and plant the cash crop through the cover crop residue. To improve the planting dates of the cash crop in the upper Midwestern regions where the production season length is limited, further work is being conducted to explore planting the cash crop prior to cover crop termination, providing a two to three-week advantage in cash crop planting. By planting into the standing versus rolled cover crop, seed placement may also be facilitated, mitigating the issue of poor seed to soil contact and lower cash crop stands. Initial work by Bernstein et al. [20] demonstrated yield improvements of over 130 $kg \cdot ha^{-1}$ with earlier versus later planted soybeans in CCBRT systems on narrow row spacing.

5. FUTURE DEVELOPMENTS IN CCBRT RESEARCH

Research data collected over multiple years and multiple sites in the upper Midwestern U.S. indicate that the CCBRT system can be a viable and productive option for organic row crop farmers. Of all the potential cover crop and cash crop combinations, cereal rye and soybean remains the most consistent with respect to weed suppression and yields. While implementing CCBRT in the corn phase currently remains a less viable option, farmers continue to express interest in developing techniques that would allow for acceptable yields. Currently, the most significant barrier to successful CCBRT adoption in this phase is delayed cash crop planting dates. In order to ensure adequate nitrogen availability to the corn crop, as well as to mitigate pest pressure from armyworm or seedcorn maggot, legumes are the preferred cover crop rather than cereal grains. To ensure successful termination of a legume cover crop using mechanical methods, the legume must be rolled-crimped or mowed at 100% bloom or at early pod set. In the upper Midwest, this does not occur until approximately mid-June for hairy vetch, which is one month after typical organic corn planting dates. This results in significantly reduced yields of corn, or in some cases, failure of the crop to fully mature.

Legume cover crop breeding efforts could provide options to allow this system to be more successfully adapted to corn. While hairy vetch is currently the cover crop that has been most frequently tested under CCBRT management due to its ability to overwinter and establish sufficient biomass, other vining legume cover crops, such as Austrian winter pea, could provide more acceptable alternatives. While Austrian winter pea is more variable in its ability to survive upper Midwestern winters, new cultivars with improved cold-hardiness are becoming available. Combined with agronomic practices that enhance overwintering of Austrian winter pea, such as earlier planting dates, deeper planting depth, and sowing with a winterkilled nurse crop such as oat, it may become a more reliable alternative for organic farmers interested in adapting CCBRT to the corn phase of their rotation. Austrian winter pea can offer significant advantages over hairy vetch as a cover crop, including earlier flowering dates and more effective termination, leading to silage corn yields equivalent to those resulting from typical organic management.

In addition to further research and investments in cover crops for suitability for the CCBRT system, cultivar selection and breeding for cash crops adapted to CCBRT remains a component of the system where a dearth of information exists. The value of breeding crops for specific organic environments is increasingly recognized [38]. Similar to the diversity across organic environments imparted by different fertility inputs, pest management strategies, and crop/cover crop rotation, the integration of CCBRT creates yet another unique management aspect to organic crop farms. Soybean and corn cultivars that can withstand cooler soil temperatures, establish through the thick cover crop residue, and withstand a narrow row spacing could provide specific trait advantages to improve performance in the CCBRT system.

Further optimization of equipment adapted to CCBRT can also contribute to the optimization of the system and more consistent, improved results. Manufacturers such as Dawn Biologic in Illinois, USA [39] and Cross Slot® No-Tillage Systems in New Zealand [40] are experimenting with alternatives that may improve proper seed depth and spacing. Further research to optimize existing corn planters and no-till drills is also needed, to provide better recommendations as to the most effective coulters, closing wheels, and other equipment modifications.

Additionally, to more effectively motivate farmers and policymakers to integrate the technique into management and policy decisions, more research must be conducted to understand the long-term impacts of CCBRT on SOC, SOM, and soil microbial communities. While overall organic farming practices have also been shown to increase soil C concentrations [38], the impacts of organic CCBRT on soil C dynamics is less clear. Several European-based studies have estimated gains in soil C stocks with reduction of tillage in organic systems, but have not conducted these estimations in systems where tillage is reduced to the extent of CCBRT [41,42,43,44]. While it can be hypothesized that the CCBRT would similarly improve soil quality parameters, this has not yet been well-documented on long-term CCBRT fields under upper Midwestern production and soil conditions.

6. CONCLUSIONS

As organic systems continue to expand and mature in the US, organic management strategies to reduce production risk while providing continued ecosystem services will become even more critical to the continued growth and success of the industry. A decade of CCBRT research in the upper Midwestern US has demonstrated that CCBRT can provide a strong management tool for organic farmers aiming to improve their weed management practices while minimizing soil erosion risk, building soil organic matter, and incorporating further crop diversity into their rotations. Particularly in the face of climate change, where extreme weather events will occur with increasing frequency and the need for carbon mitigation tools becomes more imperative, CCBRT provides both management advantages and broader ecosystem services. While research continues to occur at land-grant universities, the continued integration of farmer experiences as adoption increases will provide refinement to the system to further reduce risk and allow for successful implementation of CCBRT over a range of environments and rotation strategies.

ACKNOWLEDGMENTS

The authors acknowledge the contributions of all upper Midwestern CCBRT organic farmers and researchers, as well as CCBRT researchers across the U.S.A., for their ongoing dialogue and discussion to further the refinement and adoption of the technique. This material is partially based on work that was supported by the National Institute of Food and Agriculture, USDA, Integrated Organic Program Award No. 2008-51106-19021. Funding was also received from USDA Sustainable Agriculture Research and Educations (SARE) Award No. 2009-28640-19953 and from the Ceres Trust.

REFERENCES

1. U.S. Department of Agriculture, National Agricultural Statistics Survey. 2012 Census of Agriculture: Organic Production Survey (2014); U.S. Department of Agriculture, National Agricultural Statistics Survey (USDA NASS): Washington, DC, USA, 2017.
2. U.S. Department of Agriculture, Economic Research Service. Organic Market Overview. Available online: http://www.ers.usda.gov/topics/natural-resources-environment/organic- agriculture/ organic-trade.aspx#.U2P7pseAftk (accessed on 4 February 2017).
3. U.S. Department of Agriculture, Economic Research Service. Organic Trade. Available online: http://www.ers.usda.gov/topics/natural-resources-environment/organic-agriculture/ organic-trade.aspx#.U2pePceAftk (accessed on 4 February 2017).
4. U.S. Department of Agriculture, Agricultural Marketing Service (USDA-AMS) National Organic Program (NOP). Organic Regulations. Available online: https://www.ams.usda.gov/rules-regulations/organic (accessed on 4 February 2017).
5. U.S. Department of Agriculture (USDA) Midwest Climate Hub. Climate Changes Impacting Midwest Crops. Available online: https://www.climatehubs.oce.usda.gov/ sites/ default/files/todey_moses_climate_changes_ _ag.pdf (accessed on 4 February 2017).
6. Chavas, J.; Posner, J.L.; Hedtcke, J.L. Organic and Conventional Production Systems in the Wisconsin Integrated Cropping Systems Trial: I. Economic and Risk Analysis 1993–2006. Agron. J. 2009, 101, 288–295.
7. McCool, D.K.; Papendick, R.I.; Hammel, J.E. Surface Residue Management. In Crop Residue Management to Reduce Erosion and Improve Soil Quality; Papendick, R.I., Moldenhauer, W.C., Eds.; Conservation Report Number 40; USDA-Agricultural Research Service: Washington, DC, USA, 1995; pp. 10–16.
8. Fu, G.; Chen, S.; McCool, D.K. Modeling the Impacts of No-till Practice on Soil Erosion and Sediment Yield with RUSLE, SEDD, and ArcView GIS. Soil Tillage Res. 2006, 85, 38–49.
9. Luo, Z.; Wang, E.; Sun, O. Can No-Tillage Stimulate Carbon Sequestration in Agricultural Soils? A Meta-Analysis of Paired Experiments. Agric. Ecosyst. Environ. 2010, 139, 224–231.
10. Powlson, D.S.; Stirling, C.M.; Jat, M.L.; Gerard, B.G.; Palm, C.A.; Sanchez, P.A.; Cassman, K.G. Limited Potential of No-Till Agriculture for Climate Change Mitigation. Nat. Clim. Chang. 2014, 4, 678–683.
11. Naderman, G.C. Effects of Crop Residue and Tillage Practices on Water Infiltration and Crop Production. In Cover Crops for Clean Water; Hargrove, W., Ed.; Soil and Water Conservation Society, West Tennessee Experiment Station: Jackson, TN, USA, 1991; pp. 23–24.
12. Brun, L.J.; Enz, J.W.; Larsen, J.K.; Fanning, C. Springtime Evaporation from Bare and Stubble-Covered Soil. J. Soil Water Conserv. 1986, 41, 120–122.
13. Smart, J.R.; Bradford, J.M. No-tillage and Reduced Tillage Cotton Production in South Texas. In Proceedings of the Beltwide Cotton Conference, Nashville, TN, USA, 9–12 January 1996; Dugger, P., Richter, D.A., Eds.; The National Cotton Council of America: Memphis, TN, USA, 1996.; pp. 1397–1401.
14. Clark, A. Managing Cover Crops Profitably; Sustainable Agriculture Network U.S. Department of Agriculture: Beltsville, MD, USA, 2007.
15. Rudisill, A. 2007–2008 Agronomy Guide; The Pennsylvania State University, University Park: State College, PA, USA, 2008. Mirsky, S.B.; Curran, W.S.; Mortensen, D.A.; Ryan, M.R.;

Shumway, D.L. Control of Cereal Rye with a Roller/Crimper as Influenced by Cover Crop Phenology. Agron. J. 2009, 101, 1589–1596.
16. Mirsky, S.B.; Curran, W.S.; Mortensen, D.A.; Ryan, M.R.; Shumway, D.L. Control of Cereal Rye with a Roller/Crimper as Influenced by Cover Crop Phenology. Agron. J. **2009**, 101, 1589–1596.
17. Mirsky, S.B.; Ryan, M.R.; Curran, W.S.; Teasdale, J.R.; Maul, J.; Spargo, J.T.; Moyer, J.; Grantham, A.M.; Weber, D.;Way, T.R.; et al. Conservation Tillage Issues: Cover Crop-based Organic Rotational No-till Grain Production in the mid-Atlantic Region, USA. Renew. Agric. Food Syst. **2012**, 27, 31–40.
18. Mischler, R.A.; Curran, W.S.; Duiker, S.W.; Hyde, J.A. Use of a Rolled-rye Cover Crop for Weed Suppression in No-till Soybeans. Weed Technol. 2010, 24, 253–261.
19. Silva, E.M.; Moore, V.M. Cover Crops as an Agroecological Practice on Organic Vegetable Farms in Wisconsin, USA. Sustainability 2017, 9, 55.
20. Bernstein, E.R.; Posner, J.L.; Stoltenberg, D.E.; Hedtcke, J.L. Organically Managed No-tillage Rye-Soybean Systems: Agronomic, Economic, and Environmental Assessment. Agron. J. 2011, 103, 1169–1179.
21. Delate, K.; Cwach, D.; Chase, C. Organic No-tillage System Effects on Soybean, Corn and Irrigated Tomato Production and Economic Performance in Iowa, USA. Renew. Agric. Food Syst. **2012**, 27, 49–59.
22. Silva, E.M. Management of Five Fall-sown Cover Crops for Organic No-till Production in the Upper Midwest. Agroecol. Sustain. Food Syst. **2014**, 38, 748–763.
23. Mohler, C.L.; Teasdale, J.R. Response of Weed Emergence to Rate of Vicia villosa Roth and Secale cereale L. residue. Weed Res. **1993**, 33, 487–499.
24. Teasdale, J.R.; Mohler, C.L. The Quantitative Relationship between Weed Emergence and the Physical Properties of Mulches. Weed Sci. 2000, 48, 385–392.
25. Bernstein, E.R.; Stoltenberg, D.E.; Posner, J.L.; Hedtcke, J.L. Weed Community Dynamics and Suppression in Tilled and No-tillage Transitional Organic Winter Rye-Soybean Systems. Weed Sci. 2014, 62, 125–137. Walters, S.A.; Young, B.G.; Krausz, R.F. Influence of Tillage, Cover Crop, and Pre-emergence Herbicides on Weed Control and Pumpkin Yield. Int. J. Veg. Sci. 2008, 14, 148–161.
26. Walters, S.A.; Young, B.G.; Krausz, R.F. Influence of Tillage, Cover Crop, and Pre-emergence Herbicides on Weed Control and Pumpkin Yield. Int. J. Veg. Sci. **2008**, 14, 148–161.
27. Mirsky, S.B.; Ryan, M.R.; Teasdale, J.R.; Curran, W.S.; Reberg-Horton, C.S.; Spargo, J.T.; Wells, M.S.; Keene, C.L.; Moyer, J.W. Overcoming Weed Management Challenges in Cover Crop-based Organic Rotational No-till Soybean Production in the Eastern United States. Weed Technol. 2013, 27, 193–203.
28. Ryan, M.R.; Curran, W.S.; Grantham, A.M.; Hunsberger, L.K.; Mirsky, S.B.; Mortensen, D.A.; Nord, E.A.; Wilson, D.O. Effects of Seeding Rate and Poultry Litter on Weed Suppression from a Rolled Cereal Rye Cover Crop. Weed Sci. 2011, 59, 438–444.
29. Creamer,N.G.; Dabney, S.M. Killing Cover Crops Mechanically: Review of Recent Literature and Assessment of New Research Results. Am. J. Altern. Agric. 2002, 17, 32–40.
30. Mischler, R.; Duiker, S.W.; Curran, W.S.; Wilson, D. Hairy Vetch Management for No-Till Organic Corn Production. Agron. J. 2010, 102, 355–362.
31. Hobbs, P.R.; Sayre, K.; Gupta, R. The Role of Conservation Agriculture in Sustainable Agriculture. Philos. Trans. B 2008, 363, 543–555.
32. Clark, K.M.; Boardman, D.; Staples, J.S.; Easterby, S.; Reinbott, T.M.; Kremer, R.J.; Kitchen, N.R.; Veum, K.S. Crop Yield and Soil Organic Carbon in Conventional and No-Till Organic Systems on a Claypan Soil. Agron. J. 2017, 109, 588–599.
33. Graham, R.; Geytenbeek, P.; Radcliffe, B. Responses of Triticale, Wheat, Rye and Barley to Nitrogen Fertilizer. Aust. J. Exp. Agric. 1983, 23, 73–79.
34. Shipley, P.R.; Messinger, J.J.; Decker, A.M. Conserving Residual Corn Fertilizer Nitrogen with Winter Cover Crops. Agron. J. 1992, 84, 869–876.
35. Andraski, T.W.; Bundy, L.G. Corn Residue and Nitrogen Source Effects on Nitrogen Availability in No-till Corn. Agron. J. 2008, 100, 1274–1279.

36. Hock, S.M.; Lindquist, J.L.; Martin, A.R.; Knezevic, S.Z. Soybean Row Spacing and Weed Emergence Time Influence Weed Competitiveness and Competitive Indices. Weed Sci. 2006, 54, 38–46.
37. University of Wisconsin-Madison Integrated Pest and Crop Management. Advances Using the Roller-Crimper for Organic No-Till in Wisconsin. Available online: https://www.youtube.com/watch?v=UtxH4CJa-jk&t=3s (accessed on 4 May 2017).
38. Lyon, A.; Silva, E.M.; Bell, M.; Zystro, J. Seed and Plant Breeding for Wisconsin's Organic Vegetable Sector: Understanding Farmers' Needs and Practices. Agroecol. Sustain. Food Syst. 2015, 39, 601–624.
39. Dawn Biologic. ZRX Electro-Hydraulic Roller/Crimper/Row Cleaner. Available online: http://www.dawnbiologic.com/zrx/ (accessed on 4 May 2017).
40. Cross Slot® No-Tillage Systems. Drills. Available online: http://www.crossslot.com/drills (accessed on 4 May 2017).
41. Gattinger, A.; Müller, A.; Haeni, M.; Skinner, C.; Fliessbach, A.; Buchmann, N.; Mäder, P.; Stolze, M.; Smith, P.; Scialabba, N.E.-H.; et al. Enhanced Top Soil Carbon Stocks under Organic Farming. Proc. Natl. Acad. Sci. USA 2012, 109, 18226–18231.
42. Cooper, J.; Baranski, M.; Stewart, G.; Nobel-de Lange, M.; Bàrberi, P.; Fließbach, A.; Peigné, J.; Berner, A.; Brock, C.; Casagrande, M.; et al. Shallow Non-Inversion Tillage in Organic Farming Maintains Crop Yields and Increases Soil C Stocks: A Meta-analysis. Agron. Sustain. Dev. 2016, 36, 22.
43. Krauss, M.; Ruser, R.; Muller, T.; Hansen, S.; Mader, P.; Gattinger, A. Impact of Reduced Tillage on Greenhouse Gas Emissions and Soil Carbon Stocks in an Organic Grass-Clover Ley-Winter Wheat Cropping Sequence. Agric. Ecosyst. Environ. 2017, 239, 324–333.
44. Mäder, P.; Berner, A. Development of Reduced Tillage Systems in Organic Farming in Europe. Renew. Agric. Food Syst. 2012, 27, 7–11.

CHAPTER 5

Past, Present and Future Molecular Approaches to Improve Yield in Wheat

Sue Kyung Kim[1], Jin-Ho Kim[1] and Won-Cheoul Jang[1, 2*]

[1] Department of Chemistry, College of Natural Sciences, Dankook University, Cheonan, South Korea

[2] Institute of Tissue Regeneration Engineering (ITREN), Dankook University, Cheonan, South Korea

ABSTRACT

This chapter addresses the development and use of molecular markers for yield enhancement in wheat. Since their key goal for breeding is to maximize yield, extensive efforts have been made toward the improvement of yield. Agronomic traits related to yield, yield-related, disease resistance, and abiotic stresses are considered to be quantitative traits (QTLs), also known as complex traits, because they are controlled by numerous genes and are affected by environmental factors. Researchers have been studying such traits in the past decades for the development of molecular markers which can be used in various wheat breeding studies mainly involving restriction fragment length polymorphism (RFLP), simple sequence repeat (SSR), single nucleotide polymorphism (SNP), random amplified polymorphic DNA (RAPD), and amplified fragment length polymorphism (AFLP). Furthermore, the advent of next-generation sequencing (NGS) has accelerated the discovery of agronomically important genes. All of the technologies have enabled great advances for increasing the productivity of wheat. Here, the past history of first-generation sequencing, present status of second-generation sequencing, and future potential of translational genomics linked to the yield will be discussed.

Keywords: molecular markers, yield, quantitative traits (QTLs), next-generation sequencing (NGS)

1. INTRODUCTION

As the world's population is projected to reach approximately 9 billion by 2050, grain production of major staple crops needs to double to meet global food needs [1]. Together with rice and maize, wheat is one of the major staple crops widely grown in many countries, providing one-fifth of the calories and the protein for the world's population. In addition, bioethanol is made primarily from wheat in Europe. Wheat starch is a major component for the production of bread, porridge, cakes, biscuits, and cereals which is a highly versatile crop for the human diet. In 2013, wheat was the third most produced cereal crop (713 million tons), after maize (1016 million tons) and rice (545 million tons). There are two distinct types of wheat, spring wheat and winter wheat, cultivated in many countries based on growing seasons, of which spring wheat is planted in most countries except in the United States and Northern Europe where the predominant crop is winter wheat. The global consumption of wheat has increased at a much faster rate than all other crops, because of the scale-up cultivation in developing countries, particularly in China [2]. Currently, out of the total cultivation area of more than 217 million hectares, the European Union countries has the largest area, followed by China, India, Russia, United States, and Canada [3]. Therefore, there has been an extensive effort over the past decades to increase wheat production through the application of molecular techniques which are powerful tools for enhancing effectiveness in breeding.

The most common or bread wheat species, *Triticum aestivum*, is an annual grass cultivated in temperate zones worldwide that belongs to the genus *Triticum* of the tribe *Triticeae*, and the family Poaceae. The most economically important cereals in the Poaceae family are maize, wheat, rice, barley, and millet. In the genus *Triticum*, there are approximately 25 species including wild and domesticated consisting of a series of diploid, tetraploid, and hexaploid forms, including the diploid einkorn wheat, *Triticum monococcum* (AA genome), the diploid wild wheat, *Triticum urartu* (AA genome), the allotetraploid emmer wheat, *Triticum turgidum* var. *durum* (AABB genome), the allohexaploid common wheat, *T. aestivum* L. (AABBDD genome), and the autoallohexploid *Triticum zhukovskyi* (AAAAGG genome).

Of these, *T. aestivum* has 42 chromosome pairs that are derived from two rounds of polyploidization events. Its genome size is approximately 17 Gb, composed of A, B, and D genomes created from the hybridization of three different species. The first hybridization between *T. urartu* (2n = 2x = 14) and a B genome species that has not yet been identified occurred 0.20–1.3 million years ago (MYA) to form the tetraploid *Triticum dicoccoides* (2n = 4x = 48) [4, 5]. The B genome is still unclear and may be extinct, but cytological evidence suggests that the S genome of *Aegilops speltoides* is a closely related species or an ancestral progenitor to the B genome of wheat [6, 7]. The second hybridization event resulted in the complete form of the hexaploid genome *T. aestivum* (2n = 6x = 42), which occurred between the domesticated *Triticum dicoccum* or *T. durum*, a wild goatgrass, and *Aegilops Tauschii* about 8000–10,000 years ago [8–10]. Like most allopolyploid plants, wheat also has the diploid-like chromosome pairing behavior during meiosis, preventing multivalent formation created by multiple homologous or homoeologous chromosomes [11].

In wheat breeding, a strong emphasis has been put toward the improvement of grain yield as it the most important goal in wheat breeding. There have been concerns about the stagnation or decline of the staple crops in some parts of the world. It has been detected that 37% of wheat areas have experienced the yield stagnation [12]. If the breeder develops an improved wheat variety, having a superior of trait, but produces low yields, producers unlikely will grow it because the yield is necessary for economic feasibility. The grain yield is a complex character with low heritability which is influenced not only by genes but also by the effects of the environment. In wheat, it has been documented that the higher yield is inversely related to the protein content and can also delay maturity. Furthermore, abiotic stress factors including drought, salinity, extreme temperatures, and acidity contribute the most to yield loss, ranging between 60 and 82% [13]. As a consequence, extensive efforts have been made to identify the QTLs associated with the yield and its related traits which can be deployed by breeders through marker-assisted selection (MAS). The first report of the genome-wide assessment of molecular marker-based map in the nuclear genome of wheat began in the 1989 with the use of restriction fragment length polymorphism (RFLP) [14]. Subsequent analyses have been performed for construction of genetic maps to improve the efficiency of conventional breeding based on amplified-fragment-length polymorphism (AFLP), single-nucleotide polymorphism (SNP), diversity arrays technology (DArT), simple sequence repeat (SSR) or microsatellite, random DNA marker (RDM), gene targeted marker (GTM), and functional marker (FM). Based on these markers, there are 180 genetic maps extrapolated in wheat, most of which are developed by SSR markers. For example, molecular markers such as SSR and DArT have been used to detect QTLs for fusarium head blight (FHB) resistance, which can further be implemented in breeding studies [14]. Since 2007, a number of research studies have been taken to identify QTLs related to yield based on different mapping populations such as kernel length, kernel width, spike length, spike number, the grain number of spike, sterile spikelet number per spike, fertile spikelet number per spike, and thousand kernel weight [15–22]. However, the development of molecular markers and their applications in breeding have been relatively difficult in wheat because of its three closely related subgenomes and a large genome size consisted of high amounts (80%) of repetitive sequences. In 2012, the availability of wheat whole genome sequences has provided a framework for understanding of polyploidization, and domestication by comparing its sequences with ancestral and progenitor genomes, enabling us to understand the genetic diversity of wheat, which may help accelerate breeding programs [11]. Up to now, there are a total of 217,907 loci and 273,739 transcripts identified, of which 104,091 have been assigned as coding genes and 10,156 as long ncRNAs, according to Ensembl Plants (www.plants.ensembl.org). The chapter addresses molecular areas of research for yield improvement in wheat, focusing on finding QTLs for traits that affect yield. There are three objectives of this chapter as follows: (1) to explore the use of major molecular markers that have been used to identify yield and its-related QTLs in the past; (2) the current progress of molecular markers for linkage map construction; (3) to assess genomic studies of wheat; and (4) to discuss the potential of translational genomics in wheat using well-studied grasses such as rice and barley.

2. FIRST-GENERATION SEQUENCING RESOURCES FOR YIELD IMPROVEMENT IN WHEAT

2.1. Restriction fragment length polymorphism (rflp) markers

Earlier molecular studies in wheat have shown that the RFLP markers are the common tools used as the oldest method of molecular markers for the construction of genetic maps. RFLP are typically inherited as simple Mendelian codominant markers, and are not influence by the environment, which could serve as highly heritable genetic markers for the study of inheritance of a trait. Chao et al. created the first linkage maps of the homoeologous group 7 chromosomes on the wheat genome using 18 cDNA clones across six mapping populations from ten varieties [14]. They mapped 31 RFLP loci on chromosomes 7A, 7B, and 7D. In 1991, detailed linkage maps were constructed with a total size of 1800 cm consisting of

197 RFLP loci [23]. Ma et al. [24] and Anderson et al. [25] identified RFLP markers associated with two Hessian fly resistance genes from *Triticum taushii*, and preharvest sprouting genes from two recombinant inbred populations of white wheat (*T. aestivum* L. em. Thell.). Based on comparative mapping among cereal species such as wheat, rye, and barley, the development of detailed RFLP maps of the homoeologous group-2 chromosomes has established and showed that they had collinear relationships, indicating the high degree of conservation in gene orders [26]. The RFLP markers flanking the resistance to cereal cyst nematode were identified to be *Xglk605* and *Xcdo588*, which were mapped at 7.3 and 8.4 cm from the *Cre* locus [27]. Later, RFLP-based genetic maps belonging to different homoeologous groups were generated including homoeologous group 1 [28], group 2 [29], group 3 [30], and group 5 [31], and group 6 [32]. During this period, the powdery mildew genes were detected by RFLP probes. Ma et al. [33] reported that the powdery mildew resistant gene *Pm2* was located at 3.5 cm away from the RFLP marker *BCD1871*, and both *Pm1* and *Pm4a* showed cosegregation of markers. In addition, the *Pm1a* and *Pm2* were positioned on the RFLP maps at 3 cm away from *Xwhs178-7A*, at 2.7 cm away from *Whs296*, respectively [34]. To increase the frequency of polymorphism detection in wheat, a non-intervarietal cross (W7984 X Opata85) developed at the International Maize and Wheat Improvement Center (CIMMYT) was used creating a dense map which consisted of about 1000 RFLP loci on group 1 [28], group 2 [29], group 3 [30], and group 6 [35]. Other disease-resistant genes of *Lr9* responsible for leaf rust and *Sr22* responsible for stem rust were characterized and mapped using RFLP markers [36, 37]. Furthermore, the feasibility of the 37 RFLP probes on group 5 chromosomes from Thatcher near isogenic lines (NILs) for leaf rust resistance gene *Lr1* was conducted by Feuillet et al., of which three were investigated to be linked the gene [38]. Galiba et al. analyzed the cross between a frost-sensitive, vernalization-insensitive substitution line, *Triticum spelta* 5A and a frost-tolerant, vernalization-sensitive line, Cheynne 5A and showed cosegregation with *Vrn1* and *Xpsr426* RFLP marker [39]. There was also a linkage detected between *Vrn1* and *Xwg644* in accordance to Korzun et al. [40]. The mapping of a single locus controlling the aluminum tolerance gene *Alt2* was completed on the Chinese Spring chromosome arm 4DL derived from homoeologous recombination between *T. aestivum* cv. Chinese Spring chromosome 4D and *Triticum turgidum* cv. Cappelli chromosome 4B.

Starting from 1998, QTLs for yield and yield-related traits were investigated. The dwarfing genes, *Rht-B1* and *Rht-D1*, associated with plant height in wheat were firstly mapped on the short arms of chromosomes 4BS (*Xfba1-4B*) and 4DS (*Xfba211-4D*) [41]. A year later, thirteen RFLP probes and one morphological marker locus, *Eps*, were used to develop a genetic linkage map and identified that an RFLP marker *Xcdo549* on the short arm of chromosome 3A was associated with plant height, kernel number spike-1, and 1000-kernel weight [42]. Araki et al [43] reported one QTL for yield, *Qyld.ocs-4A.1*, and other yield-related traits of spikelet number ear, *QSpn.ocs-4A.1*, and grain weight/ear *QGwe.ocs.-4A.1*, on chromosome 4A which were detected by the *Xbcd1738* marker. As wheat lodging can result in yield losses, nine QTLs for lodging resistance were detected with the genetic distance between the flanking RFLP markers, of which seven coincided with QTLs for morphological traits [44]. Since the year 2000, there have been many studies published for loci associated with grain yield, heading date, disease, and spike morphology. Kato et al. conducted mapping of QTLs for grain yield and its components on chromosome 5A and confirmed that the grain-yield QTLs were closely linked to QTLs for yield components [45]. They found RFLP markers associated with grain yield, tiller number/plant, ear grain weight, 50-grain weight, and spikelet number/ear based on a homozygous population of single-chromosome recombinant lines. Genetic mapping of QTLs conferring resistance to stripe (yellow) rust and powdery mildew was performed by Singh et al. [46] and Tao et al. [47]. The resistance gene, *Yr28*, derived from *Ae. tauschii* and the adult-plant resistance (APR) gene *Yr18* were mapped on chromosome arm 4DS and 7DS, respectively. They found that *Yr18* was closely correlated with leaf-rust gene Lr34. In addition, the RFLP *xbcd135* and *xbcd266* loci mapped at a genetic distance of 1.6 and 4.8 cm were identified to be closely linked to *Pm6*, a gene conferring resistance to the powdery mildew. Later, two of the powdery resistant genes, *Pm26* and *Pm29*, were mapped on the RFLP linkage map [48, 49]. Several RFLP markers within six major QTL regions have been identified to be tightly linked traits related to compactness such as spike length and number of spikelets [50]. Based on a set of 114 recombinant lines (RILs) of the 'International Triticeae Mapping Initiative' mapping population, morphological, agronomical, and disease resistance traits have been studied. Börner et al. [51] mapped 211 QTLs distributed over 20 chromosomes, of which they detected 64 major QTLs with a LOD score of >3.0 conferring glume color, leaf erectness, peduncle length, ear emergence time, flowering time, plant height, ear length, winter hardiness, grain-filling, grain number, thousand-grain-weight, fusarium resistance, powdery mildew resistance, and leaf rust resistance. The mapping of an earliness *per se* gene in wheat was carried out by Bullrich et al. [52]. This gene was located between the RFLP *Xcdo393* and *Xwg241* on the log arm of chromosome 1A, which showed a large effect on heading date with a phenotypic variance of 0.47. Despite RFLP markers are highly reproducible and used as the primary way for most genetic work in wheat, it has been difficult to use RFLP markers due to the low levels of polymorphism detected in wheat.

2.2. Random amplified polymorphic dna (rapd) markers

The applications of RAPD markers have been beneficial to improve breeding programs in wheat because they are simple and fast PCR based, require no prior knowledge of target DNA sequence, and are analyzed either by the presence or absence of an amplicon via agarose gel electrophoresis. In wheat, RAPD has been used since 1990 [53]. Devos and Gale confirmed that a degree of polymorphism detected by six RAPD primers was comparable with RFLP markers [54]. They identified four RAPD markers with bread wheat cultivar 'Chinese Spring.' The application of both bulk segregant analysis (BSA) and RAPD has started in 1994 by Eastwood et al. [55]. Using BSA on DNA enriched for low-copy sequences by RAPD markers, the *Cre3* gene resistance to cereal cyst nematode (CCN) in *Triticum tauschii*, was mapped on the long arm of chromosome 2D (Ccn-D2) [55]. More efforts have been made to distinguish the wheat varieties resistant to cadmium stress [56], powdery mildew [57], and common bunt [58]. The RAPD marker *OPC20* was closely correlated with a gene controlling cadmium uptake in western Canadian durum wheat (*T. turgidum* L. var. *durum*) [56]. In addition, a RAPD marker *OPH17-1900* located on the chromosome arm 6VS, was detected that could be used for the detection of a powdery mildew resistance gene, *Pm21*, in breeding [57]. For common bunt, also known as stinking smut and covered smut, the polymorphic RAPD marker, *BW553*, was identified between resistant and susceptible NILs [58]. Furthermore, identification of RAPD markers linked to the *Yr15* gene controlling stripe rust resistance was conducted using 340 RAPD primers, six of which were detected to be polymorphic [59]. The *OPB13* RAPD marker was the only one that produced polymorphism in 123 F_2 individuals and showed that it was linked to *Yr15* through screening a series of NILs each consisted of a different gene for Hessian fly resistance using 1600 random 10-mer primers. Another RAPD marker, *OPE-13*, showed the complete cosegregation with the H21 Hessian fly resistance gene in wheat [60]. Dweikat et al. developed RAPD markers linked to 11 different Hessian fly resistance loci that could be used for determining the presence or absence of specific Hessian fly resistance genes [61]. More RAPD markers including *OPX061050, OPAG04950*, and *OPAI14600*, was observed to be linked to the new powdery mildew resistance *Pm25* gene, where the linkage distance between them was 12.8, 17.2, and 21.6 cm, respectively [62]. The application of the BSA approach on DNA enriched for low-copy sequences was used by Eastwood et al. [55] and William et al. [63], generating an increased level of polymorphism and in repeatability. Hu et al. [64] identified two RAPD markers, *UBC320420* and *UBC638550*, that cosegregated with *Pm1a* and one RAPD marker, *OPF12650*, tightly linked to the resistance gene. In these studies, it has been observed that RAPD markers also have been difficult to use in wheat like RFLP markers due to the very low level of polymorphism. Furthermore, the reproducibility of this RAPD bands have been found to be questionable. Despite this, continuous efforts have been made to develop RAPD genetic markers for the discrimination of species/cultivars from each other [65, 66].

2.3. Amplified fragment length polymorphism (aflp) markers

For assessment of large numbers of polymorphic loci, the AFLP technology has been implemented as a powerful tool because of its advantage of having good levels of reproducibility, insensitivity, fast, and no need of sequence information required for primer design [67]. In 1995, a novel PCR-based assay for plant DNA fingerprinting using AFLP markers has resulted in high levels of DNA polymorphism [68]. In fact, the AFLP technique has observed to be more efficient and less expensive and less labor intensive compared to the RFLP technique in wheat [69]. Earlier AFLP-based marker studies have been found to be informative in assessment of genetic diversity in wheat varieties started in 1998 [70, 71]. Regarding the investigation of traits associated with yield, Goodwin et al. [72] and Hartl et al. [73] initiated the AFLP technique to develop an AFLP marker associated with resistance to Septoria tritici blotch and powdery mildew, respectively. Hartl et al. identified several AFLP markers closely linked to the *Pm1c* and *Pm4a* [73]. Out of 92 AFLP primer combinations, 31 polymorphic fragments detected between the resistance and susceptible lines, of which eight were found to be the most reliable polymorphic markers. One of the AFLP markers, *18M2*, was detected as being highly specific for the *Pm1c* gene, while the 4aM1 AFLP marker was identified at 3.5 cm from *Pm4a*. A couple of years later, an AFLP analysis was conducted using NILs of the strip rust resistance gene *Yr10* and designed AFLP primers that can be useful in detecting the *Yr10* gene [74]. Cao et al. analyzed the 119 individuals of H9020-17-5 x Mingxian169 F_2 population to detect AFLP markers linked to the strip rust resistance gene *YrHua* [75]. They found the two markers, *PM14(301)* and *PM42(249)*, of which *PM14(301)* was converted to PCR marker that could be a useful tool for MAS. In 2006, nine AFLP markers that showed polymorphism between the Argentinian wheat cultivar Sinvalocho MA and the rust leaf susceptible cultivar Gamma 6 were used to construct a linkage map of the *Lr3* gene for leaf rust resistance on chromosome 6BL of wheat [76]. The development of AFLP markers was carried out by Li et al. [77]. They detected seven markers linked to *Lr19* resistance to wheat leaf rust using *Mse I* and *Pst I* based on Thatcher, 23 NILs and F_2 generation of TcLr19 x Thatcher. Dhillon et al. detected putative AFLP markers linked to leaf rust resistance genes *Lr9, Lr19*, and *KLM4-3B* using NILs of wheat [78].

More likely, the AFLP technique has been approached for developing polymorphic markers underlying a trait.

2.4. Simple sequence repeat (ssr) and intersimple sequence repeat (issr) markers

Among all available markers, SSR (or microsatellite) markers have become the best suited tool in plant breeding programs, because they are practical, convenient, easy to use, and inexpensive. Moreover, SSRs with tandem repeats of a motif of <6 bp are the most polymorphic, codominant, easy for scoring banding patterns, and have wide genomic distribution, high reproducibility, and a multiallelic nature [79]. SSR analysis has been conducted in most of the QTL studies for mapping various traits. Up to 2015, there are more than 4000 SSR markers that have been developed and used for the construction of wheat genetic maps [80]. The high level of variability and Mendelian inheritance of SSR DNA markers have been first reported by Devos et al. [81] and Röder et al. [82]. Moreover, Röder et al. [83] and Stephenson et al. [84] placed SSR loci onto the genetic map, providing a starting point for developing a saturated map of the wheat genome. For example, SSR markers have been implemented for tagging and mapping important yield-related genes such as the dwarfing genes *Rh8* [85], *Rht12*, and the vernalization gene *Vrn1* [40], and a gene for preharvest sprouting tolerance [86]. SSR primers also have been used to detect some of the wheat resistance genes. Peng et al. found nine microsatellite loci found to be linked to *YrH52* with recombination frequencies of 0.2–0.35, and LOD scores of 3.56–54.22 [87]. The identification chromosomes with QTLs underlying 1000 grain weight (GW) has conducted by Varshney et al. [88]. They found the SSR marker Xwmc333 as being linked to GW and the major QTL for GW (*QGw1.ccsu-1A*) with a R^2 value of 0.1509. Moreover, the SSR markers of *Xgwm210, Xgwm296*, and *Xgwm455*, were detected to be polymorphic and linked to the leaf rust resistance gene *Lr39*, of which *Xgwm210* was the closest marker mapped 10.7 cm from *Lr39* [89]. Ammiraju et al. [90] found four intersimple sequence repeat (ISSR) markers, *UBC8181000, UBC842600, UBC8431200*, and *UBC814750*, controlling seed size in wheat, which can be measured indirectly by 1000-kernel weight (TKW). These markers showed a signification association with gene effects of 84.662.92, and 2.61%, contributing a total of 31% of the phenotypic variance in seed size. The following year, Zhou et al. [91] identified six SSR markers linked to the major QTL for scab resistance, which were *Xgwm389, Xgwm533, Xgwm493, Xbarc75, Xbarc88*, and *Xbarc147* spanning a region of approximately 20 cm on chromosome 3BS. Additional SSR markers have been reported for strip rust resistance genes, including the *Xgwm501* marker linked to *Yr5* [92], and the *Xpsp3000* marker linked to *Yr10* [93]. A microsatellite linkage map of the powdery mildew resistance gene *Pm5e* on chromosome 7B has constructed with 20 microsatellite loci, consisting of two codominant markers *Xgwm783* and *Wgwm1267* located close to *Pm5e* with a linkage distance of 11.0 cm and 6.6 cm, respectively [94]. The detection of QTL linked to FHB resistance has found that two microsatellite loci, *Xgwm533* and *Xgwm*, were significantly associated with QTL for FHB [95]. In 2004, a SSR-based consensus map has been completed by Somers et al. [96] that has been widely used a reference. Furthermore, five QTLs, *QYld.crc-1A, QYld.crc-2D, QYld.crc-3B, QYld.crc-5A.1*, and *QYld.crc-5A.2*, controlling grain yield have been identified on chromosomes 1A, 2D, 3B, and 5A [97]. Other QTLs associated with 1000 grain weight, spikes meter^{-2}, seed number spike^{-1}, average seed weight spike^{-1}, harvest index, days to heading, days to maturity, and grain filling time have also been detected in the Superb/BW278 mapping population. Liu et al. [98] performed association mapping by genotyping wheat germplasm accession from China using SSR markers and EST-SSR markers, detecting 10 SSR markers on chromosome 4A associated with plant height, spike length, spikelets per spike, spikelet density, grains per spike and thousand-kernel weight. Even though studies involving RFLP, RAPD, and AFLP markers have only been used for identification and mapping of QTLs and genes, they are less likely to be applied into breeding programs because the application of these markers is likely to be less efficient for MAS [99].

3. SECOND-GENERATION SEQUENCING RESOURCES FOR YIELD IMPROVEMENT IN WHEAT

Researchers have focused on discovering SNPs to be used as genetic markers, because they have many advantages. SNPs act as codominant, single-locus, biallelic markers offering a lower error rate, and higher accuracy than SSR markers. However, the nature of polyploidy and having similar sequences among the A, B, and D genomes makes it difficult to identify SNPs [100]. Therefore, progress in SNP detection has been limited, especially related to yield and its related traits. With the advancement in next-generation sequencing (NGS), sequencing has become increasingly popular due to the rapid development of NGS technologies, including SOLiD/Ion Torrent PGM from Life Science, Genome Analyzer/HiSeq 2000/MiSeq from Illumina, and 454 FLX Titanium/GS Junior by Roche [101]. These technologies have led to the implementation of high-density SNP genotyping in wheat [99, 102].

An important milestone in wheat genomic research was accomplished in 2012 with the completion of *de novo* sequencing of bread wheat, the variety Chinese Spring (CS42), facilitating advances in genomic research into the genus *Triticum* and providing insights into the polyploidization and domestication of wheat. Brenchley et al. sequenced the wheat genome using Roche 454 pyro-sequencing technology (GS FLX Titanium and GS FLX+ Platforms [11]. To identify the three component genomes (A, B, and D) of hexaploid wheat, the following technologies have been applied: Illumina sequence assemblies of *Triticum monococcum* (related to the A-genome donor), *Ae. speltoides* complementary DNA assemblies, and 454 sequences from *Ae. tauschii* (the D-genome donor). Sequence analysis revealed that the A, B, and D genomes have estimated to consist of approximately 28,000, 38,000, and 36,000 genes, respectively. As a consequence, they identified the number of genes in wheat to be between 94,000 and 96,000. Like *G. hirsutum* [103] and soybean [104], wheat also has experienced a recent whole genome duplication (WGD) at 0.5 MYA and about 3.5 MYA. Brenchley et al. also examined gene conservation between wheat and its most closely related species *Brachypodium distachyon* and detected a high degree of conservation between the two species [11]. A large set of SNPs (132,000 SNPs) in A, B, and D genes will enhance future studies aimed at identifying QTLs and discovering associations of traits.

Since the completion of the draft sequence of wheat, extensive efforts have put into the identification of various molecular markers influencing yield to increase MAS efficiency. Based on the genotype by sequencing (GBS) approach, the linkage map of wheat comprised of markers including 538 GBS Bin, 258 AFLPs, 175 SSRs, and an EST has been constructed in 2014 [105]. They identified five QTL regions linked to thylakoid membrane damage (TMD), SPAD chlorophyll content (SCC), and plasma membrane damage (PMD), known as indicatives of high temperature tolerance, on chromosomes 6A, 7A, 1B, 2B and 1D and also detected some of the SSR markers associated with these traits such as the SSR marker *Xbarc121* and *Xbarc49* for all three traits and *gwm18* and *Xbarc113* for SCC. More SNP and SSR markers have been investigated using the 9 K Infinium iSelect Beadchips [106]. The SNP distribution between cultivars and landraces has provided impacts on our understanding of crop improvement on the structure of genetic diversity and insight into signatures of selection. Liu et al. reported six SNPs from two genes, *wsnp_CAP11_c209_198467* and *wsnp_JD_c4438_5568170*, showed significant association with soil-borne wheat mosaic virus (SBWMV) resistant which can be used in MAS to improve SBWMV resistance in wheat breeding [107]. Using the high-density Illumina iSelect 90K SNP assay, a linkage map spanning 3609.4 cm was constructed based on 5636 polymorphic SNP markers, with an average length of 171.9 cm per chromosome and marker density of 0.64 cm [108]. Association of agronomic traits with 1,366 SNP markers in durum wheat has been performed by Hu et al. [109]. By genotyping 150 accessions of durum wheat germplasm based on the Illumina Bead Array platform and Golden Gate Assay, a large amount of SNP markers were detected associated with key yield-related traits including plant height, number of effective spikes, length of main spike, number of spikelets per plant, panicle neck length of main spike, grain number per plant, grain weight per plant, and 1000-grain weight. These SNP markers have enhanced the previous QTL analyses and can be utilized for MAS to improve yield in wheat.

4. FUTURE DIRECTIONS: TRANSLATIONAL GENOMICS

The wheat genome is very complex as it is a polyploid species consisted of three diploid progenitor genomes. Recent advances in genome sequencing technologies have accelerated efforts to complete genomes of many crop species, which has opened the door for discovery and knowledge of the genetic basis of a number of important agronomic traits, with the final aim of crop improvement and production. Moreover, the completion of sequences has enabled assessment of translational genomics which is an effective way for researchers and breeders to transfer knowledge of genetic and genomic information among related species, such as rice and wheat [110]. The translational genomics tool is known as 'model to crop' translation that can be contributed to the implementation of genetic and genomics in crop species [111]. There are three well-characterized model grass species rice [112, 113], *Brachypodium* [114], and barley [115], which can be used to accelerate the application of translational genomics among the Poaceae family. The large amount of insufficient knowledge such as biological processes that are controlled by genes can be filled from well-studied species through translational biology approach [110]. In translational genomics, comparative genomics studies can take advantage from available genomes and can provide information on the extrapolation of knowledge of gene functions among species [110]. Genome sequences are currently available for wheat, but information QTL mapping as compared to other model grass species have been less reported. Comparative analysis based on the candidate gene approach (CGA) is known as the most powerful tool for exchange information among species. We conducted a brief comparative analysis of yield-related genes in Arabidopsis, barley, rice, wheat, and maize. The 37 Arabidopsis genes controlling flowering, growth, stems, leaves, flowers, and fruits, have been used to investigate homologous sequences in the barley, rice, wheat, and maize genomes using BLAST [116]. Using the detected homologs among the grass species, we found a number of homologous sequences with the sequence identity values ranging between 36.39 and 60.39 based on MUSCLE v3.51 [117] (Table 1).

Table 1. Comparative analysis of yield-related genes in Arabidopsis, wheat, barley, and maize.

Species		Arabidopsis	Barley		Rice		Wheat		Maize	
Trait		**Locus**	**Locus**	**Identity**	**Locus**	**Identity**	**Locus**	**Identity**	**Locus**	**Identity**
Flowering	Early flowering	At1g28380	KR706151	42.43	XM_015770121	36.86	KX161741.1	44.78	EU241899.1	41.78
		At5g44040	HM133570	43.85	XM_015777033	37.49	KJ711537	43.1	KP202720.1	42.53
	Late flowering	At1g01580	EU331897.1	44.13	XM_015787003	38.04	KJ711539.1	43.16		
		At1g04400	AB476614.1	43.88	XM_015756636	36.72				
Growth	Dwarf/small	At5g19530	AY750996.1	42	AB630963.1	43.05	AY747606.1	43.1		
		At1g02730	EU331690.1	43.14	AY747605.1	43.1	KR816810.1	42.96		
		At4g10180	KT247893.1	42.47			KT750252.1	42.82		
	Growth defective	At4g01690					AY244509.2	42.97		
		At1g02910					JF965395.1	42.79		
		At1g65260								
Stems	Waxy	At1g09560	AK366020.1	58.91	FJ487950.1	58.98	KU376264.1	58.75		
		At1g02205	AK360068.1	58.91	FJ487949.1	58.86	KU376267.1	58.98		
	Fasciation	At1g64670	AJ567377.	60.29	FJ501983.1	59.64	EU981913.	59.82		

			2				1			
			AK354338.1	57.95	EU981914.2	60.39				
			AK356998.1	59.25						
	Short petiole	At4g10180	AK360706.1	36.39	AY585350.1	54.03	M11336.1	55.81		
		At3g03860	AK353983.1	37.12	GQ389628.1	58.62	DQ457416.2	54.2		
	Twisted petiole	At4g27060			EF190873.1	58.4	EU189093.1	54.23		
Leaves	Abnormal shape	At3g16830			JX828333.1	58.57	AY831792.1	56.52		
		At1g61940					JX878122.1	58.41		
	Rough surface	At5g04660								
		At5g11060								
		At1g32640								
Flowers	Carpel	At4g32551	AK368072.1		CT831121.1		AY887064.1		NM_001112060.1	
		At5g11320	AK253078.1		AK100856.1		BT008997.1		AY898650.1	
	Stamen	At5g48390	AB085818.1				AK334664.1		DQ343238.1	
		At1g63990								
		At1g05160								
	Petal	At4g32551								
		At1g5								

		5320								
	Sepal	At2g20860								
Fruits	Short	At2g02000	EU333863.1		AP014958.1		KP749902.1		FJ573211.1	
		At1g69180	AK250398.1				KP749901.1		NM_001151022.1	
		At1g68560							EU968771.1	
	Abnormal shape	At4g05200								
		At5g04660								
		At1g74720								

The release of genomic sequence of wheat [11], barley [115], and maize [118], provides a new opportunity for translational genomics. Since comparative genomics focuses on comparing genomes among plant species looking for similarities and differences of DNA sequence, protein sequence, and gene orders, information from well-studied and analyzed species can be applied for less studied crops to improve a specific target trait, which can be implemented in crop breeding and improvement. For example, genome-wide comparative analysis of flowering-related genes in Arabidopsis, wheat, and barley has revealed that there are 900 and 275 putative orthologs in wheat and barley, respectively [119]. In addition, they showed many orthologous genes having similar expression profiles in different tissues of wheat and barley based on their *in silico* expression analyses. Such a work will help researchers to investigate candidate genes controlling the time of flowering in rice and barley which can be incorporated into molecular breeding for early flowering in wheat and barley in short-season cropping region [119].

REFERENCE

1. Tilman D., Balzer C., Hill J., Befort B.L. Global food demand and the sustainable intensification of agriculture. Proc Natl Acad Sci U S A. 2011;108(50):20260–20264. DOI: 10.1073/pnas.1116437108
2. Kearney J. Food consumption trends and drivers. Phil. Trans. R. Soc. B. 2010;365:2793–2807. DOI: 10.1098/rstb.2010.0149

3. Curtis B.C. 'Wheat in the World' in B.C. [Internet]. 2002 [Updated: 2002]. Available from: http://www.fao.org/docrep/006/y4011e/y4011e04.htm [Accessed: 9.1.16]
4. Devos K.M., Doležel J., Feuillet C. Genome organization and comparative genomics. Wheat Science and Trade. 2009;327–367. DOI: 10.1002/9780813818832
5. Huang S., Sirikhachornkit A., Faris J.D., Su X., Gill B.S., Haselkorn R., et al. Phylogenetic analysis of the acetyl-CoA carboxylase and 3-phosphoglycerate kinase loci in wheat and other grasses. Plant Mol Biol. 2002;48(5–6)DOI: 805–820.
6. Dvorak J., Akhunov E.D. Tempos of gene locus deletions and duplications and their relationship to recombination rate during diploid and polyploid evolution in the Aegilops-Triticum alliance. Genetics. 2005;171(1):323–332.
7. Kihara H. Discovery of the DD-analyser, one of the ancestors of vulgare wheats. Ag. Hort. (Tokyo). 1944;19:889–890.
8. Feldman M., Lupton F.G.H., Miller T.E. Wheats. In: Smartt J, Simmonds NW, editors. Evolution of crop plants. 2nd ed. 1995: Longman; London. 1995. pp. 184–192.
9. Kilian B., Özkan H., Deusch O., Effgen S., Brandolini A., Kohl J., Martin W., Salamini F. Independent wheat B and G genome origins in outcrossing Aegilops progenitor haplotypes. Mol. Biol. Evol. 2007;24(1):217–227.
10. Bordbar F., Rahiminejad M.R., Saeidi H., Blattner F.R. Phylogeny and genetic diversity of d-genome species of Aegilops and Triticum (Triticeae, Poaceae) from Iran based on microsatellites, ITS, and trnL-F. Pl. Syst. Evol. 2011;291:117–131.
11. Brenchley, R., et al. Analysis of the bread wheat genome using whole-genome shotgun sequencing. Nature. 2012;491(7426):705–710. DOI: 10.1038/nature11650
12. Ray D.K., Ramankutty N., Mueller N.D., West P.C., Foley J.A. Recent patterns of crop yield growth and stagnation. Nat Commun. 2012;3:1293. DOI: 10.1038/ncomms2296.
13. Cakmak I. Role of mineral nutrition in tolerance of crop plants to environmental stress factors. In: Fertigation: optimizing the utilization of water and nutrients; September 20–24; Beijing. China: Fertigation proceedings: Selected papers of the IPI-NATESE-CAU-CAAS; 2005. pp. 35–48.
14. Chao S., Sharp P.J., Worland A.J., Warham E.J., Koebner R.M., Gale M.D. RFLP-based genetic maps of wheat homoeologous group 7 chromosomes. Theor Appl Genet. 1989;78(4):495–504. DOI: 10.1007/BF00290833
15. Li S.S., Jia J.Z., Wei X.Y., Zhang X.C., Li L.Z., Chen H.M. et al. An intervarietal genetic map and QTL analysis for yield traits in wheat. Mol Breed. 2007;20:167–178. DOI: 10.1007/s11032-007-9080-3
16. Sun X.C., Marza F., Ma H.X., Carver B.F., Bai G.H. Mapping quantitative trait loci for quality factors in an inter-class cross of US and Chinese wheat. Theor Appl Genet. 2010;120:1041–1051. DOI: 10.1007/s00122-009-1232-x
17. Ramya P., Chaubal A., Kulkarni K., Gupta L., Kadoo N., Dhaliwal H.S. QTL mapping of 1,000-kernel weight, kernel length, and kernel width in bread wheat (*Triticum aestivum* L.). J Appl Genet. 2010;51:421–429. DOI: 10.1007/BF03208872
18. Carter A.H., Garland-Campbell K., Kidwell K.K. Genetic mapping of quantitative trait loci associated with important agronomic traits in the spring wheat (*Triticum aestivum* L.) 'Louise' × 'Penawawa'. Crop Sci. 2011;51:84–95. DOI: 10.2135/cropsci2010.03.0185
19. Deng S.M., Wu X.R., Wu Y.Y., Zhou R.H., Wang H.G., Jia J.Z., Liu S.B. Characterization and precise mapping of a QTL increasing spike number with pleiotropic effects in wheat. Theor Appl Genet. 2011;122:281–289. DOI: 10.1007/s00122-010-1443-1
20. Heidari B., Sayed-Tabatabaei B.E., Saeidi G., Kearsey M., Suenaga K. Mapping QTL for grain yield, yield components, and spike features in a doubled haploid population of bread wheat. Genome. 2011;54(6):517–527.
21. Cui F., Ding A.M., Li J., Zhao C.H., Wang L., Wang X.Q. et al. QTL detection of seven spike-related traits and their genetic correlations in wheat using two related RIL populations. Euphytica. 2012;186:177–192. DOI: 10.1007/s10681-011-0550-7
22. Liu G., Jia L., Lu L., Qin D., Zhang J., Guan P. Mapping QTLs of yield-related traits using RIL population derived from common wheat and Tibetan semi-wild wheat. Theor Appl Genet. 2014;127:2415–2432. DOI: 10.1007/s00122-014-2387-7

23. Liu Y.G., Tsunewaki K. Restriction fragment length polymorphism (RFLP) analysis in wheat. II. Linkage maps of the RFLP sites in common wheat. Jpn J Genet. 1991;66(5):617–633.
24. Ma Z.Q., Gill B.S., Sorrells M.E., Tanksley S.D. RELP markers linked to two Hessian fly-resistance genes in wheat (*Triticum aestivum* L.) from *Triticum tauschii* (coss.) Schmal. Theor Appl Genet. 1993;85(6–7):750–754. DOI: 10.1007/BF00225015
25. Anderson J.A., Sorrells M.E., Tanksley S.D. RFLP analysis of genomic regions associated with resistance to preharvest sprouting in wheat. Crop Sci. 1993;33(3):453–459. DOI: 10.2135/cropsci1993.0011183X003300030008x
26. Devos K.M., Gale M.D. Extended genetic maps of the homoeologous group-3 chromosomes of wheat, rye and barley. Theor Appl Genet. 1993;85:649–652.
27. Williams K.J., Fisher J.M., Langridge P. Identification of RFLP markers linked to the cereal cyst nematode resistance gene (Cre) in wheat. Theor Appl Genet. 1994;89(7–8):927–930. DOI: 10.1007/BF00224519.
28. Van Deyneze A.E., Dubcovsky J., Gill K.S., Nelson J.C., Sorrells M.E., Dvorak J, et al. Molecular-genetic maps for group 1 chromosomes of Triticeae species and their relation to chromosomes in rice and oat. Genome. 1995;38:45–49.
29. Nelson J.C., van Deynze A.E., Autrique E., Sorrells M.E., Lu Y.H., Merlino M., et al. Molecular mapping of wheat. Homoeologous group 2. Genome. 1995a;38:516–524.
30. Nelson J.C., van Deynze A.E., Autrique E., Sorrells M.E., Lu Y.H., Negre S., et al. Molecular mapping of wheat. Homoeologous group 3. Genome. 1995b;38:516–524.
31. Ogihara Y., Shimizu H., Hasegawa K., Tsujimoto H., Sasakuma T. Chromosome assignment of four photosynthesis-related genes and their variability in wheat species. Theor Appl Genet. 1994;88(3–4):383–394. DOI: 10.1007/BF00223649
32. Jia J., Devos K.M., Chao S., Miller T.E., Reader S.M., Gale M.D. RFLP-based maps of the homoeologous group-6 chromosomes of wheat and their application in the tagging of Pm12, a powdery mildew resistance gene transferred from Aegilops speltoides to wheat. Theor Appl Genet. 1996;92(5):559–565. DOI: 10.1007/BF00224558
33. Ma Z.Q., Sorrells M.E., Tanksley S.D. RFLP markers linked to powdery mildew resistance genes Pm1, Pm2, Pm3, and Pm4 in wheat. Genome. 1994;37(5):871–875.
34. Hartl L., Weiss H., Stephan U., Zeller F.J., Jahoor A. Molecular identification of powdery mildew resistance genes in common wheat (Triticum aestivum L.). Theor Appl Genet. 1995;90:601–606.
35. Marino C.L., Tuleen N.A., Hart G.E., Nelson J.C., Sorrells M.E., Lu Y.H., et al. Molecular genetic maps of the group 6 chromosomes of hexaploid wheat (Triticum aestivum L. em. Thell.). Genome. 1996;39(2):359–366.
36. Autrique E., Tanksley S.D., Sorrells M.E., Singh R.P. Molecular markers for four leaf rust resistance genes introgressed into wheat from wild relatives. Genome. 1995;38(1):75–83.
37. Paull J.G., Pallotta M.A., Langridge P., The T.T. RFLP markers associated with Sr22 and recombination between chromosome 7A of bread wheat and the diploid species *Triticum boeoticum*. Theor Appl Genet. 1994;89(7–8):1039–1045. DOI: 10.1007/BF00224536
38. Feuillet C., Messmer M., Schachermayr G., Keller B. Genetic and physical characterization of the LR1 leaf rust resistance locus in wheat (*Triticum aestivum* L.). Mol Gen Genet. 1995;248(5):553–562.
39. Galiba G., Quarrie S.A., Sutka J., Morgounov A., Snape J.W. RFLP mapping of the vernalization (Vrn1) and frost resistance (Fr1) genes on chromosome 5A of wheat. Theor Appl Genet. 1995;90(7-8):1174–1179. DOI: 10.1007/BF00222940
40. Korzun V., Roder M., Worland A.J., Borner A. Intrachromosomal mapping of genes for dwarfing (Rht12) and vernalization response (Vrn1) in wheat by using RFLP and microsatellite markers. Plant Breed. 1997;116:227–232. DOI: 10.1111/j.1439-0523.1997.tb00987.x
41. Cadalen T., Sourdille P., Charmet G., Tixier H., Gay G., Boeuf C., et al. Molecular markers linked to genes affecting plant height in wheat using a doubled-haploid population. Theor Appl Genet. 1998;96(6):933–940.

42. Shah M.M., Gill K.S., Baenziger P.S., Yen Y., Kaeppler S.M., Ariyarathne H.M. Molecular mapping of loci for agronomic traits on chromosome 3A of bread wheat. Crop Sci. 1999;39:1728–1732.
43. Araki E., Miura H., Sawada S. Identification of genetic loci affecting amylose content and agronomic traits on chromosome 4A of wheat. Theor Appl Genet. 1999;98(6):977–984. DOI: 10.1007/s001220051158
44. Keller M., Karutz Ch., Schmid J.E., Stamp P., Winzeler M., Keller B., et al. Quantitative trait loci for lodging resistance in a segregating wheat×spelt population. Theor Appl Genet. 1999;98(6):1171–1182. DOI: 10.1007/s001220051182
45. Kato K., Miura H., Sawada S. Mapping QTLs controlling grain yield and its components on chromosome 5A of wheat. Theor Appl Genet. 2000;101(7):1114–1121. DOI: 10.1007/s001220051587
46. Singh R.P., Nelson J.C., Sorrells M.E. Mapping Yr28 and other genes for resistance to stripe rust in wheat. Crop Sci. 2000;40:1148–1155. DOI: 10.2135/cropsci2000.4041148x
47. Tao W., Liu D., Liu J., Feng Y., Chen P. Genetic mapping of the powdery mildew resistance gene Pm6 in wheat by RFLP analysis. Theor Appl Genet. 2000;100:564–568.
48. Rong J.K., Millet E., Manisterski J., Feldman M. A new powdery mildew resistance gene: Introgression from wild emmer into common wheat and RFLP-based mapping. Euphytica. 2000;115:121. DOI: 10.1023/A:1003950431049
49. Zeller F.J., Kong L., Hartl L., Mohler V., Hsam S.L.K. Chromosomal location of genes for resistance to powdery mildew in common wheat (*Triticum aestivum* L. em Thell.) 7. Gene Pm29 in line Pova. Euphytica. 2002;123(2):187–194. DOI: 10.1023/A:1014944619304
50. Sourdille P., Tixier M.H., Charmet G., Gay G., Cadalen T., Bernard S. Location of genes involved in ear compactness in wheat (*Triticum aestivum*) by means of molecular markers. Mol Breed. 2000;6(3):247–255. DOI: 10.1023/A:1009688011563
51. Börner A., Schumann E., Fürste A., Cöster H., Leithold B., Röder S., et al. Mapping of quantitative trait loci determining agronomic important characters in hexaploid wheat (*Triticum aestivum* L.). Theor Appl Genet. 2002;105(6–7):921–936. DOI: 10. 1007/s00122-002-0994-1
52. Bullrich L., Appendino L., Tranquilli G., Lewis S., Dubcovsky J. Mapping of a thermo-sensitive earliness per se gene on Triticum monococcum chromosome 1A(m). Theor Appl Genet. 2002;105(4):585–593. DOI: 10.1007/s00122-002-0982-5
53. Williams J.G., Kubelik A.R., Livak K.J., Rafalski J.A., Tingey S.V. DNA polymorphisms amplified by arbitrary primers are useful as genetic markers. Nucleic Acids Res. 1990;18(22):6531–6535.
54. Devos K.M., Gale M.D. The use of random amplified polymorphic DNA markers in wheat. Theor Appl Genet. 1992;84(5–6):567–72. DOI: 10.1007/BF00224153
55. Eastwood R.F., Lagudah E.S., Appels R. A directed search for DNA sequences tightly linked to cereal cyst nematode resistance genes in *Triticum tauschii*. Genome. 1994;37(2):311–319.
56. Penner G.A., Bezte L.J., Leisle D., Clarke J. Identification of RAPD markers linked to a gene governing cadmium uptake in durum wheat. Genome. 1995;38(3):543–547.
57. Qi L., Cao M., Chen P., Li W., Liu D. Identification, mapping, and application of polymorphic DNA associated with resistance gene Pm21 of wheat. Genome. 1996;39(1):191–197.
58. Demeke T., Laroche A., Gaudet D.A. A DNA marker for the Bt-10 common bunt resistance gene in wheat. Genome. 1996;39(1):51–55.
59. Sun G.L.,Fahima T., Korol A.B., Turpeinen T., Grama A., Ronin Y.I., et al. Identification of molecular markers linked to the Yr15 stripe rust resistance gene of wheat originated in wild emmer wheat, *Triticum dicoccoides*. Theor Appl Genet. 1997;95(4):622–628. DOI: 10.1007/s001220050604
60. Seo Y.W., Johnson J.W., Jarret R.L. A molecular marker associated with the H21 Hessian fly resistance gene in wheat. Mol Breed. 1997;3(3):177–181. DOI: 10.1023/ A:100 9606304447
61. Dweikat I., Ohm H., Patterson F., Cambron S. Identification of RAPD markers for 11 Hessian fly resistance genes in wheat. Theor Appl Genet. 1997;94(3):419–423. DOI: 10.1007/s001220050431

62. Shi A.N., Leath S., Murphy J.P. Identification of RAPD markers linked to two major genes for powdery mildew resistance in Pm12 wheat line. Phytopathol. 1997;87:S89.
63. William H.M., Hoisington D., Singh R.P., Gonzales-de-Leon D. Detection of quantitative trait loci associated with leaf rust resistance in bread wheat. Genome. 1997;40:253–260.
64. Hu X.Y., Ohm H.W., Dweikat I. Identification of RAPD markers linked to the gene PM 1 for resistance to powdery mildew in wheat. Theor Appl Genet. 1997;94(6):832–840. DOI: 10.1007/s001220050484
65. Myburg A.A., Botha A-M., Wingfield B.D., Wilding W.J.M. Identification and genetic distance analysis of wheat cultivars using RAPD fingerprinting. Cereal Res Commun. 1997;25(4):875–882.
66. Liu Z.Q., Pei Y., Pu Z.J. Relationship between hybrid performance and genetic diversity based on RAPD markers in wheat. *Triticum aestivum* L. Plant Breed. 1999;118(2):119–123.
67. Rafalski J.A., Vogel J.M., Morgante M., Powell W., Andre C., Tingery S.V. Generating and using DNA markers in plants. In: Birren B., Lai E., editors. Non-Mammalian Genomic Analysis: A Practical Guide. London: Academic Press; 1996. pp. 75–134.
68. Vos P., Hogers R., Bleeker M., Reijans M., van de Lee T., Hornes M., et al. AFLP: a new technique for DNA fingerprinting. Nucleic Acids Res. 1995;23(21):4407–4414.
69. Ma Z.-Q., Lapitan N.L.V. A comparison of amplified and restriction fragment length polymorphism in wheat. Cereal Res Commun. 1998;26(1):7–13.
70. Barrett B.A., Kidwell K.K. AFLP-based genetic diversity assessment among wheat cultivars from the Pacific Northwest. Crop Sci. 1998;38(5):1261–1271. DOI: 10.2135/cropsci1998.0011183X003800050025x
71. Barrett B.A., Kidwell K.K., Fox P.N. Comparison of AFLP and pedigree-based genetic diversity assessment methods using wheat cultivars from the Pacific Northwest. Crop Sci. 1998;38(5):1271–1278. DOI: 10.2135/cropsci1998.001183x003800050026x
72. Goodwin S.B., Hu X., Shaner G. An AFLP marker linked to a gene for resistance to Septoria tritici blotch in wheat. In: Slinkard A.E., editor. Proceedings of the 9th International Wheat Genetics Symposium; August 2–7; Saskatoon, Saskatchewan. Univ. of Saskatchewan: Univ. Extension Press; 1998. pp. 108–110.
73. Hartl L., Mohler V., Zeller F.J., Hsam S.L.K., Schweizer S. Identification of AFLP markers closely linked to the powdery mildew resistance genes Pm1c and Pm4a in common wheat (*Triticum aestivum* L.). Genome. 199;42(2):322–329. DOI: 10.1139/g98-129
74. Shao Y., Niu Y., Zhu L., Zhai W., Xu S., Wu L. Identification of an AFLP marker linked to the stripe rust resistance geneYr10 in wheat. Chinese Sci Bull. 2001;46(17):1466–1468. DOI: 10.1007/BF03187033
75. Cao Z.J., Wang X.P., Wang M.N., Cao S.H., Jing J.X., Shang H.S., et al. Genetic analysis and molecular markers of a novel stripe rust resistance gene YrHua in wheat originated from Psathyrostachys huashanica Keng. Yi Chuan Xue Bao. 2005;32(7):738–743.
76. Diéguez M.J., Altieri E., Ingala L.R., Perera E., Sacco F., Naranjo T. Physical and genetic mapping of amplified fragment length polymorphisms and the leaf rust resistance Lr3 gene on chromosome 6BL of wheat. Theor Appl Genet. 2006;112(2):251–257. DOI: 10.1007/s00122-005-0122-0
77. Li X., Yang W., Li Y., Liu D., Yan H., Meng Q., et al. Identification of AFLP Markers Linked to Lr19 Resistance to Wheat Leaf Rust. Agric Sci Chi. 2007;6(3):311–315. DOI: 10.1016/S1671-2927(07)60050-9
78. Dhillon N.K., Dhaliwal H.S. Identification of AFLP markers linked to leaf rust resistance genes using near isogenic lines of wheat. Am J Plant Sci. 2011;2(5):683–687. DOI: 10.4236/ajps.2011.25082
79. Bruford M.W., Wayne R.K. Microsatellites and their application to population genetic studies. Curr Opin Genetics Dev. 1993;3: 939–943.
80. Han B., Wang C., Tang Z., Ren Y., Li Y., Zhang D., et al. Genome-wide analysis of microsatellite markers based on sequenced database in Chinese spring wheat (*Triticum aestivum* L.). PLoS One. 2015;10(11):e0141540. DOI: 10.1371/journal.pone.0141540

81. Devos K.M., Bryan G.J., Collins A.J., Stephenson P., Gale M.D. Application of two microsatellite sequences in wheat storage proteins as molecular markers. Theor Appl Genet. 1995;90(2):247–252. DOI: 10.1007/BF00222209
82. Röder M.S., Plaschke J., König S.U., Börner A., Sorrells M.E., Tanksley S.D., et al. Abundance, variability and chromosomal location of microsatellites in wheat. Mol Gen Genet. 1995;246(3):327–333.
83. Röder M.S., Korzun V., Wendehake K., Plaschke J., Tixier M.-H., Leroy P., et al. A microsatellite map of wheat. Genetics. 1998;149(4):2007–2023.
84. Stephenson P., Bryan G., Kirby J., Collins A., Devos C., Busso C., et al. Fifty new microsatellite loci for the wheat genetic map. Theor Appl Genet. 1998;97(5):946–949. DOI: 10.1007/s001220050975
85. Korzun V., Röder M.S., Ganal M.W., Worland A.J., Law C.N. Genetic analysis of the dwarfing gene (Rht8) in wheat. Part I. Molecular mapping of Rht8 on the short arm of chromosome 2D of bread wheat (*Triticum aestivum* L.). Theor Appl Genet. 1998;96(8):1104–1109. DOI: 10.1007/s001220050845
86. Roy J., Prasad M., Varshney R., Balyan H.S., Blake T.K., Dhaliwal H.S., et al. Identification of a microsatellite on chromosome 6B and a STS on 7D of bread wheat showing association with preharvest sprouting tolerance. Theor Appl Genet. 1999;99(1):336–340. DOI: 10.1007/s001220051241
87. Peng J., Fahima T., Röder M., Li Y.C., Dahan A., Grama A., et al. Microsatellite tagging of the stripe-rust resistance gene YrH52 derived from wild emmer wheat, *Triticum dicoccoides*, and suggestive negative crossover interference on chromosome 1B. Theor Appl Genet. 1999;98(6):862–872. DOI: 10.1007/s001220051145
88. Varshney R., Prasad M., Roy, J., Kumar N., Harjit-Singh, Dhaliwal H.S., et al. Identification of eight chromosomes and a microsatellite marker on 1AS associated with QTL for grain weight in bread wheat. Theor Appl Genet. 2000;100(8):1290--1294. DOI: 10.1007/s001220051437
89. Raupp W.J., Sukhwinder-Singh, Brown-Guedira G.L., Gill B.S. Cytogenetic and molecular mapping of the leaf rust resistance gene Lr39 in wheat. Theor Appl Genet. 2001;102(2):347–352. DOI: 10.1007/s001220051652
90. Ammiraju J.S.S., Dholakia B.B., Santra D.K., Singh H., Lagu M.D., Tamhankar S.A., et al. Identification of inter simple sequence repeat (ISSR) markers associated with seed size in wheat. Theor Appl Genet. 2001;102(5):726–732. DOI: 10.1007/s001220051703
91. Zhou W.-C., Kolb F.L., Bai G.-H., Domier L.L., Boze L.K., Smith N.J. Validation of a major QTL for scab resistance with SSR markers and use of marker-assisted selection in wheat. Plant Breed. 2003;122(1):40–46. DOI: 10.1046/j.1439-0523.2003.00802.x
92. Sun Q., Wei Y., Ni Z., Xie C., Yang T. Microsatellite marker for yellow rust resistance gene Yr5 in wheat introgressed from spelt wheat. Plant Breed. 2002;121(6):539–541. DOI: 10.1046/j.1439-0523.2002.00754.x
93. Wang L., Ma J., Zhou R., Wang X., Jia J. Molecular tagging of the yellow rust resistance gene Yr10 in common wheat, P.I.178383 (*Triticum aestivum* L.). Euphytica. 2002;124(1):71–73. DOI: 10.1023/A:1015689817857
94. Huang X., Wang L., Xu M., Röder M. Microsatellite mapping of the powdery mildew resistance gene Pm5e in common wheat (*Triticum aestivum* L.). Theor Appl Genet. 2003;106(5):858–865. DOI: 10.1007/s00122-002-1146-3
95. Del Blanco I.A., Frohberg R.C., Stack R.W., Berzonsky W.A., Kianian S.F. Detection of QTL linked to Fusarium head blight resistance in Sumai 3-derived North Dakota bread wheat lines. Theor Appl Genet. 2003;106:1027–1031.
96. Somers D.J., Isaac, P., Edwards, K. A high-density microsatellite consensus map for bread wheat (*Triticum aestivum* L.). Theor Appl Genet. 2004;109(6):1105–1114. DOI: 10.1007/s00122-004-1740-7
97. Cuthbert J.L., Somers D.J., Brûlé-Babel A.L., Brown P.D., Crow G.H. Molecular mapping of quantitative trait loci for yield and yield components in spring wheat (*Triticum aestivum* L.). Theor Appl Genet. 2008;117(4):595–608. DOI: 10.1007/s00122-008-0804-5

98. Liu L., Wang L., Yao J., Zheng Y., Zhao C. Association mapping of six agronomic traits on chromosome 4A of wheat (*Triticum aestivum* L.). Mol Plant Breed. 2010;1(5): pp. 1–10. DOI: 10.5376/mpb.2010.01.0005
99. Akhunov E., Nicolet C., Dvorak J. Single nucleotide polymorphism genotyping in polyploid wheat with the Illumina Golden Gate assay. Theor Appl Genet. 2009;119:507–517. DOI: 10.1007/s00122-009-1059-5
100. Chao S., Zhang W., Akhunov E., Sherman J., Ma Y., Luo M.-C., et al. Analysis of gene-derived SNP marker polymorphism in US wheat (*Triticum aestivum* L.) cultivars. Mol Breed. 2009;23(1):23–33. DOI: 10.1007/s11032-008-9210-6
101. Liu L., Li Y., Li S., Hu N., He Y., Pong R., et al. Comparison of Next-Generation Sequencing Systems. J Biomed Biotechnol. 2012;2012:251364. DOI: 10.1155/2012/251364
102. Bérard A., Le Paslier M.C., Dardevet M., Exbrayat-Vinson F., Bonnin I., Cenci A., et al. High-throughput single nucleotide polymorphism genotyping in wheat (Triticum spp.). Plant Biotechnol J. 2009;7(4):364–374. DOI: 10.1111/j.1467-7652.2009.00404.x
103. Li F., Fan G., Lu C., Xiao G., Zou C., Kohel R.J., et al. Genome sequence of cultivated Upland cotton (Gossypium hirsutum TM-1) provides insights into genome evolution. Nature Biotechnol. 2015;33:524–530. DOI: 10.1038/nbt.3208
104. Schmutz J., Cannon S.B., Schlueter J., Ma J., Mitro T., Nelson W., et al. Genome sequence of the palaeopolyploid soybean. Nature. 2010;463:178–183. DOI: 10.1038/nature08670
105. Talukder S.K., Babar M.A., Vijayalakshmi K., Poland J., Prasad P.V.V., Bowden R., et al. Mapping QTL for the traits associated with heat tolerance in wheat (Triticum aestivum L.). BMC Genetics. 2014;15:97. DOI: 10.1186/s12863-014-0097-4
106. Cavanagh C.R., Chao S., Wang S., Huang B.E., Stephen S., Kiani S., et al. Genome-wide comparative diversity uncovers multiple targets of selection for improvement in hexaploid wheat landraces and cultivars. Proc Nat Acad Sci USA. 2013;110:8057–8062. DOI: 10.1073/pnas.1217133110
107. Liu S., Yang X., Zhang D., Bai G., Chao S., Bockus W. Genome-wide association analysis identified SNPs closely linked to a gene resistant to Soil-borne wheat mosaic virus. Theor Appl Genet. 2014;127(5):1039–1047. DOI: 10.1007/s00122-014-2277-z
108. Gao F., Wen W., Liu J., Rasheed A., Yin G., Xia X., et al. Genome-wide linkage mapping of QTL for yield components, plant height and yield-related physiological traits in the Chinese wheat cross Zhou 8425B/Chinese spring. Front Plant Sci. 2015;6:1099. DOI: 10.3389/fpls.2015.01099
109. Hu X., Ren J., Ren X., Huang S., Sabiel S.A., Luo M., et al. Association of agronomic traits with SNP markers in Durum Wheat (*Triticum turgidum* L. durum (Desf.)). PLoS One. 2015;10(6):e0130854. DOI: 10.1371/journal.pone.0130854
110. Valluru R., Reynolds M.P., Salse J. Genetic and molecular bases of yield-associated traits: a translational biology approach between rice and wheat. Theor Appl Genet. 2014;127(7):1463–1489. DOI: 10.1007/s00122-014-2332-9
111. Salentijn E.M.J., Pereira A., Argenent G.C., Linden G., Krens F., Smulders J.M., et al. Plant translational genomics: from model species to crops. Mol Breed. 2007;20(1):1–13. DOI: 10.1007/s11032-006-9069-3
112. Yu J., Hu S., Wang J., Wong G., Li S., Liu B., et al. A draft sequence of the rice genome (*Oryza sativa* L. ssp. *indica*). Science. 2002;296(5565):79–92. DOI: 10.1126/science.1068037
113. Goff S., Ricke D., Lan T.-H., Presting G., Wang R., Dunn M., et al. A draft sequence of the rice genome (*Oryza sativa* L. ssp. *japonica*). Science. 2002;296(5565):92–100. DOI: 10.1126/science.1068275
114. The International brachypodium initiative. Genome sequencing and analysis of the model grass Brachypodium distachyon. Nature. 2010;463:763–768. 10.1038/nature08747
115. The international barley genome sequencing consortium. A physical, genetic and functional sequence assembly of the barley genome. Nature. 2012;491:7426. DOI: 10.1038/nature11543
116. Altschul S.F., Gish W., Miller W., Myers E.W., Lipman D.J. Basic local alignment search tool. J. Mol. Biol. 1990;215(3):403–410. DOI: 10.1016/S0022-2836(05)80360-2
117. Edgar R.C. MUSCLE: multiple sequence alignment with high accuracy and high throughput. Nucleic Acids Res. 2004;32(5):1792–1797. DOI: 10.1093/nar/gkh340

118. Schnable P.S., Ware D., Fulton R.S., Stein J.C., Wei F., Pasternak S., et al. The B73 maize genome: complexity, diversity, and dynamics. Science. 2009;326(5956):1112–1115. DOI: 10.1126/science.1178534
119. Peng F.Y., Hu Z., Yang R.-C. Genome-wide comparative analysis of flowering-related genes in arabidopsis, wheat, and barley. Int J Plant Genomics. 2015;2015(874361). DOI: 10.1155/2015/874361

CHAPTER 6

Approaches to Enhance Salt Stress Tolerance in Wheat

Mirza Hasanuzzaman[1,2], Kamrun Nahar[3,4], Anisur Rahman[1,3], Taufika Islam Anee[1,3], Mazhar Ul Alam[3,5], Tasnim Farha Bhuiyan[3,4], Hirosuke Oku[2] and Masayuki Fujita[3]*

[1] *Department of Agronomy, Faculty of Agriculture, Sher-e-Bangla Agricultural University, Dhaka, Bangladesh*

[2] *Molecular Biotechnology Group, Center of Molecular Biosciences (COMB), Tropical Biosphere Research Center, University of the Ryukyus, Okinawa, Japan*

[3] *Laboratory of Plant Stress Responses, Faculty of Agriculture, Kagawa University, Kagawa, Japan*

[4] *Department of Agricultural Botany, Faculty of Agriculture, Sher-e-Bangla Agricultural University, Dhaka, Bangladesh*

[5] *Institute of Seed Technology, Sher-e-Bangla Agricultural University, Dhaka, Bangladesh*

ABSTRACT

Wheat is consumed as a staple food by more than 36% of world population. Wheat provides nearly 55% of the carbohydrates and 20% of the food calories consumed globally. The productivity of wheat is often adversely affected by salt stress which is associated with decreased germination percentage, reduced growth, altered reproductive behavior, altered enzymatic activity, disrupted photosynthesis, damage of ultrastructure of cellular components, hormonal imbalance, and oxidative stress. Different approaches have been adopted to improve plant performance under salt stress: introduction of genes, screening of better performing genotypes, and crop improvement through conventional breeding methods which are often not so successful and suitable due to time-consuming or reduction of plant vigor with the succession of time. Uses of exogenous phytoprotectants, seed priming, nutrient management, and application of plant hormone are convenient for improving plant performances. This chapter reviews the mechanism of damage of wheat plants under salt stress and also the recent approaches to improve growth and productivity of salt-affected wheat plants emphasizing the use of exogenous phytoprotectants from the available literature.

Keywords: abiotic stress, antioxidant defense, cereal crop, stress tolerance, phytohormones

1. INTRODUCTION

Worldwide, more than 20% of the cultivable land is affected by salinity. Due to climate change and anthropogenic activities, the salt affected area is tended to increase day by day [1]. Abiotic stresses (including salinity) are accountable for more than 50% yield reduction [2]. In contrary, due to rapid increase of global population, food production should be increased by more than 70% by 2050 [3]. Wheat (*Triticum* spp.) ranks first in the world's grain production. Wheat is consumed as staple food by more than 36% of world population. Wheat provides nearly 55% of the carbohydrates and 20% of the food calories consumed globally [4, 5]. The productivity of wheat is often adversely affected by salt stress. The yield of wheat starts to decline at 6–8 dS m^{-1} [6]. Under salt stress, hyperosmotic and hyperionic (ion toxicity) stresses occur due to low water potential of soil and excess sodium ion accumulation within the plant. Ionic stress is also associated with nutritional imbalance [7, 8]. Decreased germination percentage, reduced growth, altered reproductive behavior, and reduced yield are general effects on plants under salt stress [9]. Altered enzymatic activity, disrupted photosynthesis, oxidative stress, disrupted biomembrane structure and function, damage of ultrastructural cellular components, and hormonal imbalance are some reasons for decreasing overall growth and development of plants under salt stress [10–12].

Salt stress tolerance is a polygenic trait regulated by multiple factors/genes. Exclusion of Na^+, cytosolic K^+ retention and maintenance of K^+/Na^+ homeostasis, osmotic adjustment, transpiration efficiency, and enhanced antioxidant defense system are vital for better plant performance under salt stress [13–15]. Different approaches have been adopted to improve plant performance under salt stress; introduction of genes, screening of better performing genotypes, and crop improvement through conventional breeding methods which are often not so successful and not suitable due to time consuming or reduction of plant vigor with the succession of time. Uses of exogenous phytoprotectants, seed priming, nutrient management, and application of plant hormones are convenient for improving plant performances. These approaches are being also popular for stress management practices including the salt stress [16–25]. In this chapter, we will review the recent research works on different approaches of salt stress tolerance in wheat plants emphasizing the use of exogenous phytoprotectants.

2. WHEAT RESPONSES TO SALT STRESS

Salinity is one of the most devastating abiotic stresses having enormous negative effects on morphological, physiological, and biochemical attributes of plant including germination, growth, water uptake, photosynthesis, nutrient uptake, enzymatic activities, and yield. A number of studies revealing the effects of salt stress on different wheat cultivars, among which some are tolerant and some susceptible to salt stress. Higher salinity causes lower germination rate, photosynthesis, transpiration, and higher accumulation of Na^+ and Cl^- ions which disturb the normal metabolic processes of wheat plants (Table 1 ; Figure 1).

Table 1. Responses of *T. aestivum* plants to salt stress.

Cultivars	Salinity level	Effects	Reference
MH-97 and AUQAB-2000	15 dS m^{-1}	Decreased DW and FW of root and shoot Increased activities of catalase (CAT) and superoxide dismutase (SOD) Increased protein and AsA contents	Afzal et al. [43]
KRL-19 and WH-542	100 mM NaCl, 6 d	Decreased leaf relative water content (RWC) and leaf osmotic potential Increased H_2O_2, malondialdehyde (MDA) contents and electrolyte leakage	Mandhania et al. [37]

		Increased activities of CAT, glutathione reductase (GR), SOD, ascorbate peroxidase (APX), and peroxidase (POD) Increased Na^+accumulation, decreased K^+ accumulation, and increased K^+/Na^+ ratio	
S-24 and MH-97	150 mM NaCl, 7 d	Decreased shoot and root FW Decreased grain yield, 1000 grain weight and transpiration rate Increased Na^+ and Cl^-contents and decreased $K^+$and Ca^{2+} contents in both leaf and root	Arfan et al. [19]
MH-97	150 mM NaCl	Increased mean germination time Decreased FW and DW of shoot and leaf area Increased Na^+ and Cl^-contents, and decreased $K^+$and Ca^{2+} contents and K^+/Na^+ ratio Decreased NO_3^- and PO_4^{3-} contents	Wahid et al. [35]
Inqlab-91 and SARC-1	125 mM NaCl, 7 d	Increased Na^+ and Cl^-contents, and decreased $K^+$and Ca^{2+} contents Decreased FW and DW	Afzal et al. [28]
Banysoif 1	320 mM NaCl, 155 d	Decreased contents of photosynthetic pigments and rate of transpiration Increased Pro content and decreased amino acid content Higher accumulation of Na^+ in root, shoot, and spike	Tammam et al. [36]
Hirmand, Chamran, Hamoon, Bolani, Sorkhtokhm, and Kavir	12.5 dS m^{-1}	Delayed and decreased seed germination Increased Na^+accumulation and decreased K^+ content in both shoot and root	Akbarimoghaddam et al. [30]

Tatara-96, Ghaznavi-98, Fakhri Sarhad, Bakhtawar-92, Pirsabaq-2004, and AUQAB-2000	120 mM NaCl	Decreased shoot FW and DW Increased Na^+ and K^+contents, and decreased K^+/Na^+ ratio	Jamal et al. [32]
Tajan, Rasoul, Atrak, and Kouhdasht	16 dS m^{-1}	Decreased grain yield and 1000-grain weight Increased Na^+ and Cl^-contents, and decreased $K^+$and Ca^{2+} contents	Asgari et al. [40]
Caxton	200 mM NaCl, 8 d	Decreased germination percentage	Fuller et al. [31]
MH-97 and Inqlab-91	15 dS m^{-1}	Decreased net CO_2assimilation rate, stomatal conductance, and transpiration rate Decreased shoot FW	Iqbal and Ashraf [44]
Dan-4589	80 mM (NaCl and Na_2SO_4 at a molar ratio of 9:1)	Increased Na^+ content and decreased K^+ content Decreased rate of photosynthesis and transpiration Decreased chl content and intercellular CO_2concentration	Guo et al. [33]
HD 2329 and Kharchia-65	200 mM NaCl, 9 d	Decreased activity of SOD and increased activities of POD, APX, CAT, and GR	Singh et al. [45]
Transgenic lines: T1, T4, and T6	200 mM NaCl, 4 d	Decreased net photosynthetic rate, stomatal conductance, and increased intercellular CO_2concentration in leaves Decreased chl and carotenoid contents	Tian et al. [46]
Yangmai 16	0.75% NaCl	Higher accumulation of Na^+ and decreased K^+/Na^+ratio	Zhang et al. [47]
Jimai 22	100 mM NaCl, 10d	Increased MDA content Increased activities of SOD, POD, CAT, GR, APX, and dehydroascorbate reductase (DHAR)	Zou et al. [34]

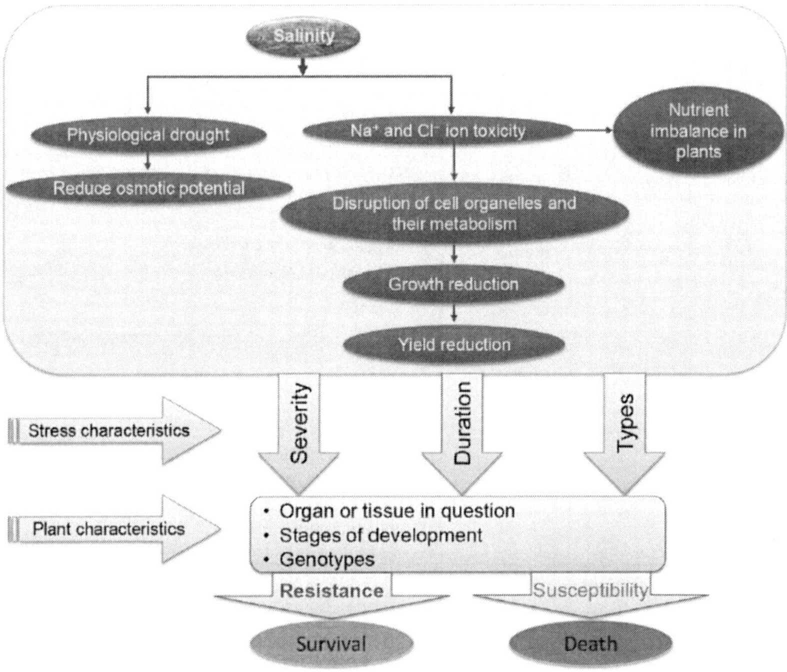

Figure 1. General scheme of salt stress responses and adaptation in plants.

2.1. Germination

Germination is one of the most important and vital processes of plant life cycle. It is the determinant of the subsequent growth and yield characteristics of plants. Available literature showed that wheat seeds tended to germinate at a lower rate and consumed longer time when exposed to salt stress. The reasons underlying this fact are higher concentrations of salt create lower osmotic potential of germination media which hampers the imbibition of water by seed, creates an imbalance in the normal activities of enzymes responsible for nucleic acid and protein metabolism, causes hormonal imbalance, and deteriorates the food reserves of seed [26]. But, there are many other factors related to the plant and environment which also have effects on germination process. These include age of seed, seed dormancy, seed coat hardness, seed polymorphism, vigority of seedling, moisture, temperature, gasses, and light, etc. [27]. Germination also varies with different cultivars considering whether tolerant or susceptible type. Afzal et al. [28] reported that under saline condition (125 mM NaCl), wheat seeds required longer time for germination than seeds under nonsaline condition. Similar results were presented by Ghiyasi et al. [29] with upto 16 dS m^{-1} salinity levels. Mean germination time increased and germination rate and germination index decreased with increasing levels of salinity. Akbarimoghaddam et al. [30] reported delayed and decreased wheat seed germination at 12.5 dS m^{-1} salinity. Fuller et al. [31] also showed a distinct relationship of the decreasing germination percentage with increasing salinity levels (upto 200 mM NaCl).

2.2. Growth

Salt stress affects the growth of wheat seedlings remarkably. Root and shoot length, plant height, leaf area, number of effective tillers, and number of spike, etc. are considered to be growth parameters. There are many reports that show the evidence of hampering these characters under saline condition. Moreover, in the seedling stage, plants are more sensitive to adverse environmental conditions. So, in this stage, high salinity may even cause death of seedlings. Fresh and dry mass of shoot, leaf area, etc. of both sensitive and tolerant cultivars declined under salt stress in wheat seedlings [19]. Length, fresh weight (FW), and dry weight (DW) of both root and shoot of wheat seedlings were negatively affected by different levels of salinity as 150 mM NaCl [20]; 125 mM NaCl [28], 16 dS m^{-1} salinity [29], and 120 mM NaCl [32]. Guo et al. [33] showed decreased growth of leaves of wheat seedlings and roots under salt stress, compared to the nonstressed control. Similarly, reduced shoot length, root length, wet weight, and DW after 10 d with 100 mM NaCl treatment were observed by Zou et al. [34].

2.3. Photosynthesis

Photosynthesis is the major physiological process for plant survival and greatly influenced by environmental factors. As salinity reduces water potential and increases accumulation of Na^+ and Cl^-ions in the chloroplast, the rate of photosynthesis gets inhibited [26]. According to the experiment conducted by Arfan et al. [19], exposure to salt stress reduced the transpiration rate, net CO_2assimilation rate, stomatal conductance, and substomatal CO_2 concentration of both cultivars. Similarly, net photosynthetic rate, transpiration rate, stomatal conductance, and substomatal CO_2 concentration were decreased significantly at 150 mM NaCl stress [35]. Tammam et al. [36] reported that amount of photosynthetic pigments were significantly deceased in seedlings under 320 mM NaCl stress. Reduction of stomatal conductance and transpiration rate were also reported by Guo et al. [33]. Significant decrease of chlorophyll (chl) content was recorded in wheat seedlings at 100 mM NaCl, for 10 d [34].

2.4. Water relation

Availability of moisture in plants is a crucial factor for all physiological and metabolic processes of plants. Higher salt concentrations induce osmotic stress to plants, which ultimately causes low water potential. Relative water content (RWC) declined by 3.5 and 6.7%, compared to their controls in the salt-tolerant and salt-sensitive cultivars, respectively, after 6 d of 100 mM NaCl exposure [37]. They also reported lowering of osmotic potential with increasing salt concentrations. Arfan et al. [19] showed reduced water use efficiency (WUE) of both sensitive and tolerant cultivars under saline condition. Leaf water potential also decreased under salt stress of 150 mM NaCl [35] and 16 dS m^{-1} [38]. Percentage of water content decreased in root, but increased in shoot and spike of Banysoif 1 cultivar of wheat [36]. Lv et al. [39] recorded lower RWC in leaves of *T. monococcum* seedlings exposed to salt stress of 320 mM NaCl.

2.5. Cellular damage

Inconsistent growth and improper uptake of water and nutrients ultimately result in deterioration of cell membrane properties of plants. Lipid peroxidation, accumulation of hydrogen peroxide (H_2O_2), and increased membrane permeability are some common phenomenon of wheat seedlings under salt stress. Mandhania et al. [37] reported higher damage of cellular membranes of salt-sensitive cultivar due to higher H_2O_2 accumulation and lipid peroxidation which enhanced the electrolyte leakage compared to the tolerant one. Higher accumulation of H_2O_2 in salt-stressed wheat seedlings was also proved by Wahid et al. [35] which was responsible for the increased relative membrane permeability. Lipid peroxidation increased by 68% under NaCl treatment of 100 mM for 10 d compared to control [34].

2.6. Ion uptake

Higher accumulation of Na^+ and Cl^- ions interferes with the uptake of other necessary ions which disturbs plant processes. Salt-sensitive cultivars tend to uptake more Na^+ compared to the tolerant one and this uptake rate increases with increasing concentration of salt [37]. Lower accumulation of NO_3^- and PO_4^{3-} ions were recorded by Wahid et al. [35]. They also reported higher uptake of Na^+ and Cl^-, and reduced uptake of K^+ and Ca^{2+} by salt stressed wheat seedlings. Similar results were published by Afzal et al. [28] in wheat seedlings exposed to 125 mM of NaCl stress for 7 d. But, Jamal et al. [32] reported increased uptake of Na^+ and K^+ both ions, and decreased K^+/Na^+ ratio in wheat shoots when exposed to 120 mM of NaCl. On the other hand, both Asgari et al. [40] and Afzal et al. [41] recorded significant decrease of K^+ uptake under saline condition (15–16 dS m^{-1}). Under medium salinity, higher accumulation of both Na^+ and Cl^-, and lower uptake of K^+, Ca^{2+}, and Zn^{2+} ions were reported by Guo et al. [33].

2.7. Yield

All the above mentioned factors are responsible directly or indirectly for the subsequent yield reduction of wheat plants. Yield of almost all crops, except some halophytes, is reduced under salt stress. The amount of yield reduction may vary upon the sensitivity and tolerance of the wheat cultivars. Chinnusamy et al. [42] indicated that above the threshold level of salinity of 6 dS m^{-1}, wheat yield can reduce at a rate of 7.1% per dS m^{-1} increase of salinity. Asgari et al. [40] reported that the spikes number per plant, spike length, number of spikelets per spike, straw weight, grain yield, 1000-grain weight, and harvest index declined with the increasing level of salinity, which ultimately caused yield loss. A significant decrease in number of grains per spike, 1000-grain weight, and grain yield were reported in both tolerant and sensitive cultivars of wheat seedlings under 15 dS m^{-1} salinity [41].

3. SALT-INDUCED OXIDATIVE STRESS IN WHEAT

Salt stress can lead to stomatal closure, which reduces CO_2 availability in the leaves and inhibits carbon fixation, exposing chloroplasts to excessive excitation energy which in turn increase the generation of reactive oxygen species (ROS) such as superoxide ($O_2^{\bullet-}$), H_2O_2, hydroxyl radical ($OH^\bullet$), and singlet oxygen ($^1O_2$) [26, 48, 49] (Figure 2). Since, salt stress is complex and imposes a water deficit because of osmotic effects on a wide variety of metabolic activities [50]; this water deficit leads to the formation of ROS that are highly reactive and may cause cellular damage through oxidation of lipids, proteins, and nucleic acids [51]. If there is a serious imbalance in any cellular compartment between the production of ROS and antioxidant defense, oxidative stress and damage occur [52] (Figure 2). Enhanced production of ROS under salinity stress induces phytotoxic reactions such as lipid peroxidation, protein degradation, and DNA mutation [53]. When a plant faces harsh conditions, ROS production overcomes scavenging systems and oxidative stress will burst. In many plant studies, it was observed that production of ROS increased under saline conditions [54] and ROS-mediated membrane damage has been demonstrated to be a major cause of the cellular toxicity by salinity in different crop plants ([49]; Table 2). Long-term salinity treatments (5.4 and 10.6 dS m^{-1}, 60 d) caused significant increase in H_2O_2 and lipid peroxidation in wheat seedlings, which were higher in salt-sensitive cultivar than salt-tolerant cultivar [55]. Increased lipid peroxidation and H_2O_2 levels with increased salinity stress in *T. aestivum* were observed in our study [24]. Wheat seedlings exposed to 300 mM NaCl resulted in 60 and 73% increase in H_2O_2 and MDA contents. Salt stress also decreased ascorbic acid (AsA) content by 52%. According to Zou et al. [34], *T. aestivum* leaves showed 35% increase in MDA content upon 100 mM NaCl treatment for 5 d which further increased by 68% after 10 d of treatments. Rao et al. [56] observed dose dependent increase in lipid peroxidation in wheat exposed to salt (2, 4, 8, and 16 EC) and these effects were variable among the cultivars. They found increased MDA content in cultivars, ZARDANA (55.9%), ROHTAS-90 (42.26%), SAUGHAT-90 (51%), and SHAHEEN-94 (52%), and hence they were designated as salt sensitive, whereas PUNJAB-85 (33%), BHAKAR 2002 (35%), PIRSBAK-05 (31%), and AUQAB (28%) showed decreased levels of lipid peroxidation and were categorized as salt tolerant [57].

Table 2. Salt-induced oxidative stress in *T. aestivum* compared to control.

Name of cultivars	Dose and duration of salt	Level of oxidative stress	References
Pradip	300 mM NaCl, 4 d	Increased H_2O_2 and MDA content by 60 and 73%, respectively	Hasanuzzaman et al. [24]
Kharchia-65	6.85 dS m^{-1}NaCl	Enhanced lipid peroxidation (TBARS) and H_2O_2 content by 21 and 38%, respectively	Sairam and Srivastava [58]
Jimai 22	100 mM NaCl, 10 d	MDA content increased by 68.3% in leaves	Zou et al. [34]
ZARDANA BAKHAR-2002, SAUGHAT-90, and AUQAB-2000	16 dS m^{-1}NaCl	Lipid peroxidation enhanced by 56, 35, and 51% in ZARDANA BAKHAR-2002, and SAUGHAT-90 cultivars, respectively	Rao et al. [56]

		DPPH radical scavenging activity decreased by 47% in AUQAB-2000	
Kızıltan-91	100 mM NaCl, 5 d	Increased lipid peroxidation level by 53%	Seckin et al. [59]
Yongliang 4	150 mM NaCl, 16 h	Increased MDA content in leaves by 50%	Zhang et al. [60]
Altındane	100 mM NaCl, 3 d	Elevated MDA, $O_2^{\bullet-}$, and H_2O_2contents by 26, 43, and 53%, respectively	Gorcek and Erdal [61]
Waha	150 mM NaCl, 14 d	MDA content increased by 10% and fourfold increase of H_2O_2content	Fercha [62]
WenmaiNo.6	150 mM NaCl, 4 d	Increased MDA, $O_2^{\bullet-}$, and H_2O_2contents by 47, 38, and 33%, respectively	Qiu et al. [63]
	150 mM NaCl, 72 h	Increased MDA and H_2O_2contents by 52 and 47%, respectively	Genişel and Erdal [64]

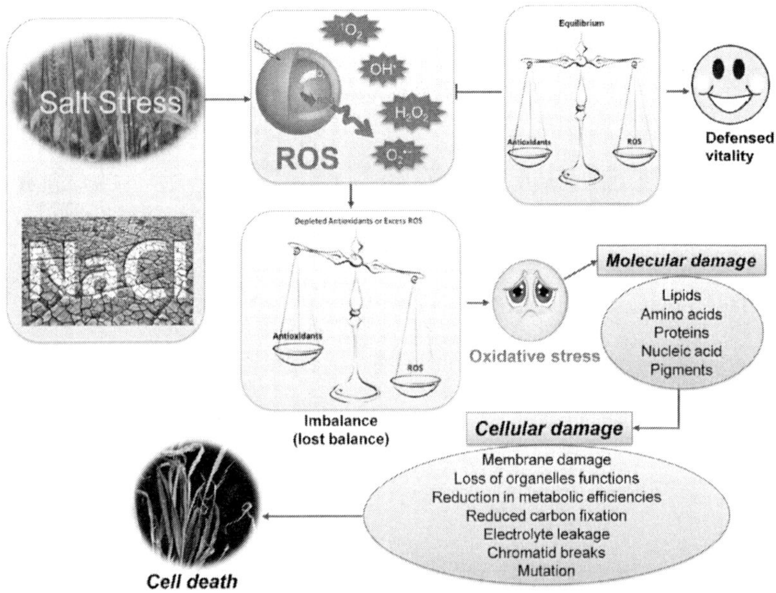

Figure 2. Generalized scheme of salt-induced oxidative stress in plants.

Plants have antioxidative mechanism to fight against stress under adverse conditions. So, they naturally produce higher amount of antioxidant enzymes, e.g., CAT, GR, SOD, APX, POD, and DHAR, etc. to minimize the damage due to stress. Mandhania et al. [37] reported that the activities of CAT, GR, SOD, APX, and POD enzymes increased with the increasing concentration of salt irrespective to tolerance or sensitivity of the cultivar. In another experiment with sensitive and tolerant type of cultivars, ascorbic acid (AsA) content and activities of SOD, CAT, and POD also increased in both under salt stress [20]. But, in another experiment by Singh et al. [45], SOD activity was recorded to decrease with the increasing concentration of salt in a salt-sensitive cultivar named HD2329; while activities of POD, APX, CAT, and GR increased with the same treatments. Significantly, higher activities of SOD and POD were presented by Zou et al. [34] with NaCl treatment of 100 mM for 10d, but they showed insignificant increase of CAT and APX activities, and significant decrease of GR and DHAR activities under same treatment. The activities of SOD and POD were increased with increasing the levels of salt concentrations in *T. monococcum* seedlings [39].

4. SALT TOLERANCE APPROACHES

Considering the adverse effects of salt stress in wheat plant, biologists are trying to find out the salt-tolerant strategies in plants by different approaches. Many researchers found positive effect in using exogenous phytoprotectants in alleviating salt-induced damages in wheat. In this section, some of the evidences are discussed.

4.1. Use of osmoprotectants

To prevent the adverse effects of various environmental stresses including salt stress, plants demonstrate a variety of adaptive mechanisms both at the cellular and organismal levels. Under salt stress conditions, to cope with the salt-induced osmotic, ionic as well as oxidative stresses, plant synthesizes and accumulates organic compatible solutes or osmolytes [48, 65, 66]. Accumulation of these compatible solutes is one of the most important physiological strategies employed by plants under salt stress conditions. Osmoprotectants or osmolytes are small, highly soluble, uncharged, and nontoxic organic molecules which help to survive organisms in extreme osmotic stresses. Osmoprotectants comprise of (i) α-amino acids such as proline (Pro) and ectoine; (ii) ammonium compounds such as glycine betaine (GB), β-alanine betaine, dimethylsulfoniopropionate (DMSP), and choline; and (iii) polyols, sugars, and sugar alcohols such as trehalose (Tre), sorbitol, and mannitol, etc. These osmoprotectants perform vital functions in osmotic adjustment, stabilizing proteins and membranes. Thus, enhanced salt stress tolerance is observed in plants overexpressing the osmoprotectants biosynthetic and metabolic genes. Enhanced salt exposure causes increased biosynthesis of osmoprotectants (Pro, GB, Tre, ecotine, and sorbitol, etc.) which provides enhanced osmotic stress tolerance generated from salt stresses [67, 68] (Table 3). For mitigating salt-induced damages, in recent decades, the use of exogenous osmoprotectants has been found effective [12, 69]. Several research findings demonstrated that the use of osmoprotectants provided significant protection against adverse effects of salt stresses in *T. aestivum* seedlings (Table 3). At the same time, several research studies proved Pro as a potent protectant against the adverse effects of salt. Proline acts not only in osmotic adjustment as a compatible solute, but also in scavenging ROS, chelating metal, activating detoxification pathways, balancing cells redox status, buffering cytosolic pH, storing energy (carbon and nitrogen), stabilizing subcellular membranes and structures including photosystem II (PS II), and as signaling molecule [70–74]. Raza et al. [75] demonstrated the effect of exogenous GB (50 mM and 100 mM) in moderately salt-sensitive (MH-97) and salt-tolerant (S-24) wheat cultivars grown under salt stress (15 dS m^{-1}

NaCl). Glycine betaine treatment ameliorated the salt-induced photosynthetic reduction as well as increased the photosynthetic capacity, water use efficiency, and osmotic adjustment where salt-tolerant (S-24) cultivar showed better performance against salt stress compared to moderately salt-sensitive (MH-97) cultivar. Later, with the same experimental procedure, they again suggested that the exogenous GB modulated the activities of antioxidant enzymes such as SOD, CAT, and POD which contributed significantly to salt stress tolerance in *T. aestivum* [76]. It has been reported that accumulation of Pro protects *T. aestivum* from the salt-induced damages by maintaining a higher K^+/Na^+ ratio and reducing ionic toxicity [38], increasing the major antioxidant enzymes (CAT, APX, SOD, and POD) activities [77]. In *T. aestivum*, GB (10 mM and 30 mM) supplementation with salt stress (150 mM NaCl) increased the germination percentage, shoot Ca content, total chl content, and thus confer salt stress tolerance [78]. Khan et al. [79] reported that increased grain yield in *T. aestivum* associated with the increased Pro, chl content and K^+/Na^+ ratio. Overexpression of GB in transgenic *T. aestivum* lines T1, T4, T6, and Shi 4185 (wild type line) caused enhanced salt stress (200 mM NaCl) tolerance by enhancing ROS scavenging, osmotic adjustment and regulating ion homeostasis [80]. Salt stresses (10 dS m^{-1} NaCl) were imposed in two wheat cultivars (cv. Seher and Lasani). In both wheat cultivars, salt stresses caused significant reduction in the germination percentage, chl contents and growth. Exogenous Pro (50 and 100 mM) application alleviated the adverse effects of salt stress by improving the germination percentage, seedling growth and chl contents of wheat plants but 100 mM Pro was found more effective compared to 50 mM Pro [81]. Mahboob et al. [82] reported that the supplementation of Pro (50 and 100 mM) ameliorated the salt (60 and 120 mM NaCl) induced reduction of plant growth, photosynthetic pigments and ionic balance by increasing shoot and root length, chl *a*, *b* contents, FW and DW of seedlings and endogenous Pro, GB, and K^+/Na^+ ratio in *T. aestivum* seedlings. Exogenous Pro (60 ppm) upregulated the endogenous hormones (gibberelic acid (GA_3) and indole acetic acid (IAA)), ammonium compounds (GB and choline) and downregulated the MDA content and growth inhibitor abscisic acid (ABA) in salt stressed *T. aestivum* [83]. Salt (50, 150, and 300 mM) induced disruption of photosynthetic pigments and protein polypeptide synthesis in *T. aestivum* were prevented by the exogenously applied Pro (50 ppm) and at the same time by protecting the turnover machinery of proteins [84]. Besides osmotic adjustment, GB is also involved in ROS scavenging, stabilizing macromolecules (nucleic acids, proteins, and lipids) and various components of photosynthetic machinery such as PS II complexes and RuBisCO and acts as reservoir of carbon and nitrogen sources [85–87]. Upon salt exposure (150 mM NaCl), reduced lipid peroxidation, increased glutathione (GSH) and GB concentrations, enhanced plasma membrane protection, increased cell solute potential and improved ion homeostasis were observed when caryopsis of *T. aestivum* were primed with GB (25, 50, 100 mM) [88]. Increasing the K^+/Na^+ and Ca^+/Na^+ ratios, reducing MDA content, protecting photosynthetic apparatus, improving plasma membrane integrity and stabilizing macromolecules (proteins, PS II and transporters) GB (20 mM) imparted in salt stress tolerance in *T. aestivum* [83]. Exogenous GB (5 mM) application improved chl *a*, total chl and K^+ content of roots, increased root length, plant height, FW and DW of *T. aestivum* under salt stresses (100 and 200 mM NaCl) [89]. Rao et al. [57] suggested that the enhanced production of Pro and GB in six salt-tolerant cultivars (*T. aestivum* cv. AUQAB-2000, PUNJAB-85, PIRSABAK-05, BAKHAR-2002, FARKHARE-SARHAD and KAGHAN-94) alleviated the damaging effects of salt stress by activating their antioxidant enzymes. Endogenous Pro and GB mediated salt stress (8 EC, 16 EC) mitigation in fifteen *T. aestivum* cultivars were further reported by Rao et al. [57]. They suggested that the five cultivars of wheat (SEHAR-2006, LU26-CTR, NARC-2009, BARS-2009, PIRSABAK-09) showed obvious salt stress tolerance by increasing the production of Pro and GB. Yan and Zheng [90] demonstrated that pretreatment with Tre (2, 20, and 40 mM) alleviated the adverse effects of salt stress (3 g L^{-1} NaCl) in *T. aestivum* cv. Yangmai-19. Various beneficial effects were observed in different physiological parameters. Increased relative growth rate, relative chl content, N content, DW and biomass plant^{-1}were observed with Trehalose supplementation. Trehalose application also improved Pro accumulation, K^+ accumulation and K^+/Na^+ ratio. In addition, Tre has functions in stabilizing the biomolecules and structures like membrane lipids, proteins under salt stress [91–93]. Salt-sensitive wheat cultivar (*T. aestivum* cv. Kızıltan-91) under salt stress (100 mM NaCl) showed physiological alteration. However, pretreatment with exogenous mannitol (100 mM) reversed the deleterious salt effects by increasing antioxidant enzymes (such as SOD, POD, CAT, APX, and GR) activities, appearance of SOD and POD isozyme activity bands and reducing lipid peroxidation [59].

Table 3. Protective effects of various exogenously applied osmoprotectants under salt stress in *T. aestivum*

Cultivars	Salinity doses and duration	Doses of osmolytes	Protective effects	References
ESW-9525 and kherman	60 and 120 mM NaCl, 7 d	50 and 100 mM Pro, foliar spray	Increased shoot and root length Increased FW and DW of seedlings Increased chl *a*, *b* contents	Mahboob et al. [82]

			Improved Pro, GB, K^+contents, and K^+/Na^+ratio	
Seher and Lasani	10 dS m^{-1}NaCl, 6 d	50 mM and 100 mM Pro, foliar spray	Improved gaseous exchange parameters (net CO_2 assimilation rate, stomatal conductance, substomatal CO_2concentration, and transpiration rate) Increased chl *a*, *b*, and total chl contents	Talat et al. [81]
MH-97 and S-24	15 dS m^{-1}NaCl	50 mM and 100 mM GB, foliar spray	Improved WUE Increased photosynthetic capacity Increased stomatal conductance	Raza et al. [75]
Kızıltan-91	100 mM NaCl, 5 d	100 mM mannitol, pretreatment, 24 h	Increase activities of SOD, POD, CAT, APX, and GR Reduced lipid peroxidation and membrane damage	Seckin et al. [59]
Gomeza 7	150 mM NaCl, 38 d	25, 50, and 100 mM GB, caryopsis priming, 24 h	Reduced lipid peroxidation Increased the GSH and GB concentrations Enhanced plasma membrane (PM) protection Increased the cell solute potential Improved ion homeostasis	Salama et al. [88]
Sakha 93 and Gimmez a7	10.04 dS m^{-1}(soil), 35–65 d 7.33 dS m^{-1}(irrigation water), 35–65 d	60 ppm Pro, foliar spray	Increased chl *a* and *b* Increased endogenous hormones (GA and IAA) Increased GB and choline Decreased MDA content and ABA	Hendawey et al. [83]
Sakha 93 and Gimmez a7	10.04 dS m^{-1}(soil) 7.33 dS m^{-1}(irrigation water), 35 to 65	20 mM GB, foliar spray	Increased K^+/Na^+ and Ca^+/Na^+ ratios Improved K, Ca, and Zn	Hendawey et al. [83]

	d		contents Reduced MDA content Protected photosynthetic apparatus Improved PM integrity and stabilization of macromolecules (proteins, PS II, and transporters)	
Yangmai -19	3 g L^{-1} NaCl	2, 20, and 40 mM Tre, seed soaking	Increased relative growth rate, relative chl content, N content, DW, and biomass per plant Increased Pro accumulation Increased K^+accumulation and K^+/Na^+ ratio, and decreased Na^+ content	Yan and Zheng [90]
S-24 and MH-97	150 mmol L^{-1}NaCl, 15 d	10 and 30 mM of GB, presoaking	Increased germination percentage Increased shoot Ca content Increased total chl content	Akhter et al. [78]
Gomeza	150 mM NaCl, 21 d	5 and 10 mM choline, caryopsis priming	Increased K^+, Ca^{2+}, and decreased Na^+ and Cl^- in both shoot and root Improved PM permeability Remarkably reduced lipid peroxidation Increased GB accumulation and improved ion homeostasis	Salama et al. [94]
S-24 and MH-97	15 dS m^{-1}NaCl, 47 d	50 and 100 mM GB, foliar spray	Modulated activities SOD, CAT, and POD Significantly increased K^+/Na^+ ratio in leaves and roots, and Ca^{2+}/Na^+ ratio in leaves Enhanced endogenous GB and K^+	Raza et al. [76]

Priming of *T. aestivum* seeds with choline (5 and 10 mM) reduced the damaging effects of NaCl (150 mM) by increasing the K^+, Ca^{2+}, GB accumulation, improved ion homeostasis and decreased $Na^+$and Cl^- in both shoot and root, mitigated PM permeability and reduced lipid peroxidation of leaf [94]. Expression of *mtlD* gene encoding mannitol-1-phosphate dehydrogenase resulted in enhanced salt stress tolerance in *T. aestivum* due to defensive roles of mannitol against salt stress [95]. The *mtlD* gene encoding mannitol-1-phosphate dehydrogenase transformation in *T. aestivum* cv. Giza 163 conferred salt stress tolerance by inducing mannitol and reducing sugars in tissues of plant [96]. Kerepesi et al. [97] demonstrated that increased fructan contents in salt resistant (Sa) and moderately salt-tolerant (Ch) varieties of *T. aestivum* showed increased tolerance against salt stress (200 mM NaCl). Sharbatkhari et al. [98] investigated the role of fructan in salt-tolerant (Bam) and salt-sensitive (Ghods) cultivars of *T. aestivum*. They found higher fructan accumulation and remobilization in salt-tolerant Bam cultivar, which contributed to the higher salt stress tolerance by increasing the photosynthetic capacity and decreasing the salt induced severe yield loss.

4.2. Plant hormone

Plant hormones are chemicals produced within the plants at low concentration involved in regulation of plant development and tolerance towards various stresses including salinity [99]. Now-a-days, various kinds of plant hormones such as ABA, auxin, cytokinins (CK), brassinosteroids and GA_3 are externally used for alleviating various kinds of abiotic stresses including salinity (Table 4). The plant growth hormone auxin increased the germination percentage, shoots DW and maintained ion homeostasis under salt stress condition [100]. Iqbal and Ashraf [101] reported that seed priming with different auxins alleviated salt stress (15 dS m^{-1}) by maintaining hormonal balance and assimilation rate and improved growth and yield of both tolerant and sensitive cultivars under salt stress condition. Seed priming with GA_3 alleviates the drastic effect of salinity and increases grain weight and grain quality by improving photosynthetic pigments, leaf area and plant growth [102]. Foliar application of GA_3 also confers salt stress tolerance by increasing germination percentage, plant growth and upregulating antioxidant enzyme [103]. Seed priming with cytokinin such as kinetin and benzylaminopurine (BAP) increase germination percentage and grain yield by increasing plant growth, productive tiller and 1000-grain weight under salt stress condition [104, 105]. Gurmani et al. [106] noted that, seed priming with ABA improved salt stress tolerance by increasing net assimilation rate, chl content and decreasing Na uptake. It is also evident that phytohormone brassinosteroid plays role in alleviating salt stress. Ali et al. [107] reported that brassinosteroid increased grain yield by improving photosynthetic attribute, assimilation rate and transpiration rate under salt stress condition (150 mM NaCl). Eleiwa et al. [22] also showed brassinosteroid-induced positive response in wheat seedlings under salt stress conditions (Table 4).

Table 4. Protective effects of various exogenously applied phytohormones under salt stress in *T. aestivum*

Cultivars	Salinity dose and duration	Dose of phytohormones	Protective effects	References
MH-97 (salt intolerant), Inqlab-91 (salt tolerant)	(15 ds m^{-1}) 150 mM NaCl, entire growth period	Auxin (Tryptopan) 4.89×10^{-4} mM, 12 h seed priming	Increased CO_2assimilation rate Increased net assimilation rate Increased growth Increased productive tiller and grain yield	Iqbal and Ashraf [44]
MH-97 (salt intolerant), Inqlab-91 (salt tolerant)	150 mM NaCl, entire growth period	4.89×10^{-1} mM auxin (tryptophan), 12 h seed priming	Increased germination percentage Improved ion homeostasis Increased shoot DW	Iqbal and Ashraf [100]
Sohag 3 (sensitive), Giza 168 (tolerant)	50, 100, 150, and 200 mM NaCl),entire life cycle	150 mg L^{-1} GA_3, foliar spray	Improved leaf area, photosynthetic pigment, carobohydrate, protein, amino acid and Pro content, grain weight	Shaddad et al. [102]
MH-97, Inqlab-91	15 dS m^{-1}, 8 d	100, 150 and 200 mg L^{-1}cytokinins (kinetin and BAP), 12 h seed priming	Increased germination rate Increased early seedlings growth such as shoot DW and root DW	Iqbal et al. [105]
MH-97, Inqlab-91	15 dS m^{-1}, entire life cycle	100, 150 and 200 mg L^{-1}cytokinins (kinetin and BAP), 12 h seed priming	Increased plant height, shoot dry biomass Increased fertile tiller, 1000-grain weight,	Iqbal and Ashraf [104]

			grain yield	
Mehran-89	0.13 M NaCl, 8 d	10^{-6} M ABA, 8 d	Increased germination percentage, and shoot and root biomass	Naqvi et al. [108]
Kharchia-65, PUNJAB-85	100 mM NaCl 16 d	10 mM ABA, seed priming 24 h	Increased plant height, root length Improved root and shoot dry weight Increased chl content Increased net assimilation rate Decreased Na uptake	Gurmani et al. [106]
Giza 164	2000–6000 ppm NaCl, irrigation water entire life cycle	0, 50, 100 and 200 mg L^{-1} 28-homobrassinoloide, foliar application,	Increased chl, carotenoids and total pigments Increased plant height, leaf area Improved tiller number, weight of 1000 grain, grain yield and biological yield	Eleiwa et al. [22]
S-24, MH-97	150 mM NaCl, 45 d	0, 0.052, 0.104, 0.156 μM 24-epibrassinolide	Increased photosynthetic attribute and chl content Increased net CO_2assimilation rate, stomatal conductance and transpiration rate Increased root and shoot weight and length Increased number of grain $plant^{-1}$ and grain yield	Ali et al. [107]

4.3. Plant nutrient

Along with other physiological and biochemical functions, plant nutrients play positive roles in alleviating damage effects of abiotic stresses including salinity (Table 5). Exogenous application of K enhanced salt stress tolerance in wheat seedlings by improving photosynthetic pigments, antioxidant enzyme activity, K uptake and decreasing Na uptake [109, 110]. Foliar application of phosphorus (P) also alleviated salt-induced damage by increasing plant biomass, leaf area and decreasing Na uptake [111]. Application of $CaSO_4$ increased plant growth, water status and K and Ca uptake under salt stress condition [112]. Later on, Tian et al. [113] noted that application of $Ca(NO_3)_2$ reduced salt-induced oxidative damage by decreasing lipid peroxidation and electrolyte leakage in wheat seedlings.

Table 5. Protective effects of plant nutrients under salt stress in *T. aestivum*.

Cultivars	Salinity dose and duration	Plant nutrients	Protective effects	References
NAYAB-11 and MILLAT-11	150 mM NaCl, 113 d	50, 100, 150 and 200 mM K_2SO_4, 106 d	Increased root length and biomass Increased plant height and biomass Increased K^+ uptake and decreased Na^+uptake	Kausar and Gull [110]
Gemiza 9, Sakha 93	10, 80, and 120 mM NaCl, 90 d	25 and 150 mg K_2O kg^{-1} soil, 110 d	Increased plant height and biomass Increased chl *a*, chl *b* and carotenoid content Increased SOD and POD activity	El-Lethy et al. [109]
	150 mM NaCl	400 and 800 mg P L^{-1}, foliar application	Increased plant height, root length, root and shoot biomass Increased leaf number, leaf area and chl content Decreased Na uptake and increase K uptake	Khan et al. [111]

PUNJAB-85	50 mM NaCl, 34 d	3 and 6 mM $CaSO_4$	Increased root and shoot biomass Increased root and leaf RWC Increase K and Ca uptake	Zaman et al. [112]
Jimai 22	100 mM NaCl, 15 d	17.5 mM $Ca(NO_3)_2$, 15 d	Decreased $O_2^{\bullet -}$ and H_2O_2 contents Decreased lipid peroxidation, electrolyte leakage Increased SOD, POD, and CAT activities	Tian et al. [113]

4.4. Antioxidant

Antioxidants are important for plants to maintain the ROS level lower. Plant possesses various non-enzymatic antioxidants in their cellular components to protect themselves from oxidative stress. The major antioxidant includes AsA, GSH, tocopherol and some phenolic compounds. Some of these antioxidants showed advanced protection against salt-induced oxidative stress when they were applied exogenously (Table 6). However, these are mostly dose dependent. A number of studies have been reported the positive effects of AsA in mitigating salt stress in wheat. Athar et al. [20] studied the effect of AsA on wheat plants subjected to salt stress. Salt stress (150 mM NaCl) caused reduction in growth and photosynthesis which were associated with decrease in tissue K^+/Na^+ ratio in both sensitive and moderately tolerant varieties. However, root applied AsA (100 mg L^{-1}) counteracted the adverse effects of salt stress on the growth of tolerant variety which was due to the enhanced endogenous AsA level and CAT activity, and higher photosynthetic capacity, and accumulation of K^+ and Ca^{2+} in the leaves. Their study supports the notion that exogenous AsA counteracts the adverse effects of salt stress on growth of wheat by improving photosynthetic capacity of wheat plants against salt-induced oxidative stress and maintaining ion homeostasis, however, these effects were cultivar specific [20]. Ascobin (compound composed of ascorbic acid and citric acid) was found to be effective in mitigating salt-induced damages in wheat as reported by Elhamid et al. [114]. Salt stress markedly increased the lipid peroxidation while the activities of antioxidant enzymes (SOD, CAT, POD, APX and GR) dramatically increased. However, foliar treatment of wheat cultivars with ascobin could partially alleviate the harmful effect of salinity especially at the lower levels of salinity imposed in the two cultivars of wheat at most of the studied parameters [114]. Apart from the dose, mode of application is also a factor to initiate the protective effect by exogenous AsA. In their study Athar et al. [115] found differential effects when AsA was applied through the rooting medium, or as seed soaking or as foliar spray to salt stressed (120 mM NaCl) wheat plants. Exogenous AsA mitigated the adverse effect, e.g. improved leaf ascorbic acid, activities of CAT, POD, and SOD. Root applied AsA caused more enhancements in photosynthetic capacity and more reduction in leaf sodium (Na^+) compared with AsA applied as seed soaking or foliar spray. However, the effects were also cultivar specific [115]. In a hydroponic experiment Khan et al. [116] showed that foliar applied AsA (50 and 100 mg L^{-1}) could not alleviate the adverse effects of salt stress on plants, but it improved the growth of nonstressed plants. Since AsA failed to enhance the antioxidant defense, it enhanced the Na^+ accumulation in the leaves but did not change the K^+ accumulation in the salt-stressed plants. Azzedine et al. [21] observed that the exogenous AsA improved the plant growth under salt stress condition which was partly due to the increased leaf area, improved chl and carotenoid contents, enhanced Pro accumulation, and decreased H_2O_2 content. Melatonin (*N*-acetyl-5-methoxytryptamine) is also considered a potential antioxidant in plants which is distributed in many parts of the plant. Due to its universal hydrophilic and hydrophobic nature and solubility in both water and lipid, it can cross cell membranes easily and enter subcellular compartments and hence, considered as an antioxidant and a modulator in multiple plant developmental processes and various stress responses [117]. In their pot experiment, Sadak et al. [117] observed that wheat seeds presoaked with melatonin (100 and 500 μM) provided better growth, photosynthetic pigments, yield, and quality in wheat under salinity (3.85 and 7.69 dS m^{-1}). Melatonin treatments at different levels caused significant increase in yield and yield attributes, carbohydrate, protein, N, P, K, flavonoids, phenolic contents, and antioxidant activity either in nonstressed and salinity-stressed plants relative to their corresponding controls. Importantly, 500 μM melatonin was more effective than 100 μM. Farouk [118] reported that both AsA and α-tocopherol minimized salt-induced senescence of flag leaves of wheat. This was due to enhanced activities of antioxidant enzymes which led to the lower lipid peroxidation and H_2O_2 accumulation. Exogenous antioxidants also decreased membrane permeability, Na and Cl content. These higher levels of antioxidants and lower level of H_2O_2 in flag leaf might be the prerequisite for delayed leaf senescence in antioxidants-sprayed plants [118].

Table 6. Protective effects of various exogenously applied antioxidants under salt stress in *T. aestivum*

Cultivars	Dose and duration of stress	Antioxidants	Major effects	References
S-24 and MH-97	150 mM NaCl, 58 d	50, or 150 mg L^{-1}AsA	Decreased Na^+content, and increased K^+ and Ca^{2+} content Improved photosynthesis Increased AsA content and CAT activities Improved growth	Athar et al. [20]
Sids 1 and Giza 168	3000 and 6000 mg L^{-1}NaCl, 75 d	200-600 mg L^{-1}ascorbin (ascorbic acid and citric acid 2:1)	Decreased MDA content Decreased activities of antioxidant enzymes	Elhamid et al. [114]
S-24 and MH-97	120 mM NaCl, throughout the growth duration	100 mg L^{-1}AsA	Increased activities of CAT, POD, and SOD Improved photosynthesis Decreased Na^+content	Athar et al. [115]
S-24 and MH-97	150 mM NaCl, 4 weeks	50 and 100 mg L^{-1}AsA	Lower Na^+accumulation Protection of photosynthesis machineries	Khan et al. [116]
Waha	150 mM NaCl, 2 weeks	0.7 mM AsA	Increased leaf area Improved chl and carotenoid contents Enhanced Pro accumulation Decreased H_2O_2content	Azzedine et al. [21]
Giza 168	0.23, 3.85, and 7.69 dS m^{-1} salinity, 75 d	500 µM melatonin	Improved shoot height, number of leaves per plant, FW and DW of shoot	Sadak et al. [117]

			Increased photosynthetic pigments Increased carbohydrate, protein, N, P, K, flavonoids, phenolic contents, and antioxidant activity	
Giza 168	0.8, 7.5, and 11.5 dS m^{-1}salinity, 65 d	100 mg L^{-1}AsA or α-tocopherol	Enhanced antioxidant enzymes activities Reduced H_2O_2accumulation, lipid peroxidation, and membrane permeability Decreased Na^+ and Cl^- contents	Farouk [118]
Huaimai 17	300 mM NaCl, 7 d	100 μM SNP (sodium nitroprusside, a nitric oxide/NO donor)	Improved germination Deceased Na content and increased K content Enhanced CAT and SOD activities	Zheng et al. [119]
S-24	150 mM NaCl, 2 weeks	0–150 μM SNP	Increased FW Increased leaf area Increased photosynthetic parameters	Kausar and Shahbaz [120]
Pradip	100–200 mM NaCl, 48 h	1 mM SNP	Decreased MDA and H_2O_2 content Increased AsA and GSH content Enhanced activities of antioxidant enzymes Increased activities of glyoxalase enzymes	Hasanuzzaman et al. [24]
Sakha	2000–8000 ppm NaCl, 75 d	1.25–5.0 mM Arg (arginine)	Decreased growth Decreased yield components	Qados et al. [121]

			Decreased grain and straw yield Lower amount of Pro, secondary metabolites and mineral contents	
Sepahan and Neyshabour	100 and 200 mM NaCl, 41 d	0.5 and 1.0 mM Spm	Increased chl content Enhanced antioxidant enzymes' activities	Saeidnejad et al. [122]
Zhengmai No. 004	150 mM NaCl, 48 h	0.05 μM H_2O_2	Increased activities of SOD, CAT, APX, and POD Increased AsA and GSH level Decreased MDA and $O_2^{\bullet -}$ level Improved plant height and biomass	Li et al. [123]

4.5. Signaling molecules

Although there are specific signaling roles of phytohormones and antioxidants present in plants, which have been discussed in previous sections, this part will discuss the role of exogenously applied signaling molecules. Among the signaling molecules, nitric oxide (NO) has been widely studied in recent decades, due to its diverse role in tolerance to several abiotic stresses including salinity. Nitric oxide exerts its signaling role through various pathways and through interaction with other molecules (Figure 3) [26]. In the last decade, exogenous application of NO through different donors was found to enhance crop growth and productivity under stressful conditions [26]. Zheng et al. [119] observed great improvement in seed germination of wheat under high salinity (300 mM NaCl). Wheat seeds soaked in SNP solution provided better germination under salinity which was associated with decreased Na^+concentration and increased K^+ concentration in the seeds. Exogenous SNP also helped in increasing starch and amylase content in seeds which increased the weights of coleoptile and radical. Moreover, exogenous NO enhanced the activities of SOD and CAT which decreased the oxidative damages evident with lower level of lipid peroxidation, $O_2^{\bullet -}$, and H_2O_2 [119]. Kausar and Shahbaz [120] found the positive effect of foliar applied NO in mitigating salt stress in wheat. Wheat seedlings grown under 100 mM NaCl exhibited reduced growth and photosynthetic rate. However, NO spray ameliorated the effect by enhancing FW of plants, leaf area, stomatal conductance, and internal CO_2 concentration. However, NO could not take part role in enhancing PS II activity [120]. In our laboratory, we examined the effect of exogenous NO in conferring salt stress tolerance in wheat [24]. Wheat plant exposed to any level of salt (150 and 300 mM NaCl) caused significant increase in oxidative stress (as indicated by MDA and H_2O_2 content). Salt stress-induced oxidative stress was due to the disruption of antioxidant defense. However, the seedlings which were pretreated with NO donor (1 mM SNP) showed enhanced tolerance which was due to increased nonenzymatic antioxidants (AsA and GSH pool) and the activities of monodehydroascorbate reductase (MDHAR), DHAR, GR, glutathione *S*-transferase (GST), GPX, glyoxalase (Gly) I, and Gly II. Therefore, we concluded that both antioxidant defense and glyoxalase systems worked together in enhancing salt stress tolerance as induced by NO [24]. As shown in Figure 3 Arg is one of the precursors of NO production. Few studies have indicated the role of exogenous Arg in salt stress tolerance in wheat. Qados et al. [121] observed that Arg could alleviate the salt-induced adverse effects in wheat. When wheat plants were exposed to different levels of salinity (2000–8000 ppm NaCl), plant mass, relative water content, yield components (spike length, spike weight, and spikelets per spike), grain yield, straw yield, biological yield, and harvest index decreased in dose dependent manners. Salt stress also deteriorated the chemical constituents of the grains. However, when the grains were presoaked with Arg, they provided better growth, yield components, yield as well as the quality aspects (nutrient content) at harvest [121]. Polyamines are often considered as signaling molecules which interact with NO and also exert direct beneficial effects [124–126]. Saeidnejad et al. [122] found the positive effect of spermine (Spm) in mitigating salt stress (100 and 200 mM NaCl) effect in wheat. In general, although seed priming with Spm showed a slight effect on germination process on both susceptible and tolerant cultivars, Spm application was an effective approach in salinity tolerance induction of wheat cultivars mostly through the activation of enzymatic antioxidants and increasing osmolytes production [122]. H_2O_2, which was previously thought to be a toxic substance and a major ROS recently been considered as signaling molecules. The double role of H_2O_2 is now an interesting topic of research of many plant scientists. However, as exogenous application, most of the experiments were conducted using H_2O_2 as priming agents or pretreatments rather than using as cotreatment. Signaling cross talk of H_2O_2 with NO is also well established since last two decades [127]. Exogenous H_2O_2 protected wheat plants from salt-induced damages by enhancing antioxidant defense as reported by Li et al. [123]. The seedlings supplemented with H_2O_2(0.05 μM) decreased the levels of MDA and $O_2^{\bullet -}$, which was associated with the increased activities of SOD, POD, CAT and APX and the concentration of GSH and carotenoid under salt stress (150 mM NaCl). Exogenous H_2O_2 also increased plant height, shoot length, root length, and biomass under saline condition. The results were reversed when H_2O_2 scavenger was used that indicated a clear role of H_2O_2in initiating its signaling role when applied at lower concentration [123].

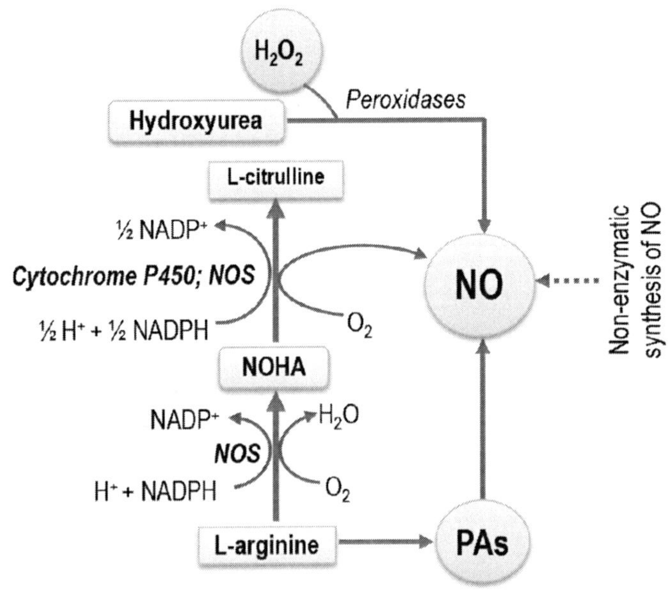

Figure 3. Interaction with PA, H_2O_2, and Arg during NO biosynthesis.

4.6. Seed priming

Seed priming is one of the easiest and cheapest techniques for successful crop production under various abiotic stress conditions including salinity [128, 129]. Seed priming is a presowing, controlled hydration technique that regulates and increases pregermination metabolic activity during early germination stage, but before radical projection [130, 131]. Seed priming has been effectively affirmed to improve germination percentage and seedling establishment in many crops such as wheat, rice, maize, soybean, canola, sunflower, sugarbeet, etc. [29, 132, 133]. Positive effects of seed priming might originate from de novo synthesis of certain germination-promoting substances, enhancing pregermination metabolites [131], early DNA replication, greater ATP availability, enzyme activation, osmotic adjustments [134], and membrane reorganization through restoring their original structures and reducing leakage of metabolites. Along with synchronous and fast emergence, primed seeds show reduced photo and thermodormancy, a wider range of germination temperatures and better capacity to compete with weeds and pathogens [135, 136]. Seed priming can be an easy solution for crops to overcome adverse environmental situations; it is reliable, simple, low cost, and also low risk technique [128, 137]. Various priming techniques such as hydropriming (soaking seed in water), osmopriming (soaking seed in nutrient, hormone, or chemicals), and halopriming (soaking seed in salt solution) have been developed to increase speed of germination, uniform seedling establishment, and crop production [138].

Seed priming has been effectively shown to increase germination and emergence of seeds of many crops in the tropical and subtropical areas, especially under salt stress conditions [139]. Increased germination rates and better seedling establishment resulted in higher levels of salt stress tolerance and crop yields when seeds were primed. Seed priming has recently been applied to overcome the salt stress problem on agricultural land [137]. Several research findings evidenced the role of seed priming to improve salt stress tolerance in wheat (Table 7). Hydropriming for 12 h on six Indian wheat cultivars showed 50% reduction of mean germination time under saline condition [140]. Effect of hydropriming was studied in salt-sensitive (MH-97) and salt-tolerant (AUQAB-2000) cultivars of wheat under salt stress (15 dS m^{-1}) condition [16]. It is well documented that seed osmopriming helps to improve salt stress tolerance in wheat seedlings. Seed osmopriming with PEG-8000 solution showed increased germination percentage, germination index, root and shoot length, and seedling FW and DW than salt-affected wheat seedlings at different salinity levels (4, 8, 12, and 16 dS m^{-1}). It has been reported that seed osmopriming with AsA helped to increase the endogenous AsA content and CAT activity which increased the salt stress tolerances in wheat [141]. Increased germination percentage, early seedling establishment, accumulation of ABA and Pro, and plant growth were featured due to seed osmopriming with 0.05 mM SA in wheat under salt stress condition [142]. Seed halopriming improves plant salt-tolerance by maintaining ion homeostasis mechanism. Salt stress increases the accumulation of Na^+concentrations in the roots and shoots of wheat plants and decreases the uptake of beneficiary nutrients. However, seed halopriming helps to maintain the ion homeostasis by decreasing Na^+ concentration and increasing K^+, Ca^{2+} concentration, and K^+/Na^+ ratio in roots and shoots. Increasing K^+ and Ca^{2+}absorption, K^+/Na^+ ratio due to seed halopriming under salt stress was connected with vigorous seedling growth and crop production, increased photosynthetic activity, and reduced electrolyte leakage. Seed halopriming with $CaCl_2$ helps in the maintenance of ionic balance by reducing the Na^+ and increasing the K^+ absorption consequently improves salt stress tolerances [143]. Salt stress also induced oxidative damage by producing ROS. Seed halopriming detoxifies the ROS by increasing the activity of enzymatic antioxidant such as SOD and CAT [43]. Iqbal and Ashraf [100] demonstrated that halopriming with 100 mM KCl, NaCl, and $CaCl_2$ reduced the salt stress affect on growth and grain production of two wheat cultivars. Priming with phytohormone increased germination with better seedling establishment and tolerance to various stresses including salinity. Seed priming of wheat with IAA increased germination percentage by improving amylase activity [144] and mitigated the growth inhibitory effect of salinity [16]. Seed priming of three wheat cultivars with auxin (0, 1, and 2 mg L^{-1}) increased germination percentage, root and shoot length, seedling FW and DW, and yield under salt stress condition [18]. Priming with SA (100 mg L^{-1}) solution for 24 h enhanced growth, photosynthetic pigments such as chl *a*, chl *b* and also increased total soluble and reducing sugar for maintaining osmotic adjustment during salt stress [145]. Iqbal and Ashraf [101] reported that seed priming with GA (150 mg L^{-1}) played a potential role in alleviating salt stress damages by reducing Na^+ and Cl^-concentrations, Na^+/K^+ ratio, and increasing K^+ and Ca^{2+} contents. Moreover, seed priming with GA increased germination percentage, seedling growth and yield contributing components under salt stress condition.

Table 7. Beneficial effects of seed priming in improving salt stress tolerance in *T. aestivum*.

Cultivar	Priming agent	Duration of priming	Salinity doses and duration	Major responses	References
AUQAB-2000, MH-97	Hydropriming, 50.0 mM $CaCl_2 \cdot 2H_2O$, 50 mg L^{-1}AsA	Seeds soaked for 12 h	15 dS m^{-1}, 12 d	Increased germination percentage Reduced mean germination time Increased root and shoot FW and DW Enhanced activities of CAT, SOD, and POD	Afzal et al. [43]
Gomeza 7	25, 50, and 100 mM GB	Seed soaked for 24 h	150 mM NaCl, 38 d	Decreased lipid peroxidation Increased PM stability and eventually ion homeostasis	Salama et al. [88]
MH-97	1, 40, 80, and 120 μM H_2O_2	Seed soaked for 8 h	150 mM NaCl	Increased photosynthetic capacity Enhanced the leaf gas	Wahid et al. [35]

				exchange Increased K^+/Na^+ ratio	
	0.6 mM AsA and sodium salicylate, 0.3 mM thiamine	Seed soaked for 6 h	40, 80, 120, and 160 mM NaCl, 30 d	Stimulated starch accumulation Inhibited production of soluble protein Reduced water soluble Pro accumulation	Al-hakimi and Hamada [141]
Gomeza 7	5 and 10 mM choline chloride	Seed soaked for 24 h	150 mM NaCl, 21 d	Increased stigmasterol Decreased cholesterol and campesterol Increased the plasma membrane stability	Salama et al. [146]
AUQAB-2000	10 ppm ABA, 50 ppm SA, 50 and 100 ppm AsA	Seed soaked for 12 h	15 dS cm^{-1}	Increased seed germination time Decreased electrolyte leakage by modulating antioxidant enzymes	Afzal et al. [147]
AUQAB-2000	25 ppm IAA, 50 ppm GA_3,100 ppm kinetin, and 1% prostart	Seed soaked in IAA, GA_3, and kinetin for 12 h; and in prostart for 2 h	15 dS cm^{-1}, 21d	Decreased electrolyte leakage Increased invertase, α-amylase and starch synthetase activities which helped in better seedling growth	Afzal et al. [16]
PUNJAB-11	10, 20, 30, 40, and 50 mM Na_2SiO_3	Seed soaked for 12 h	15 dS cm^{-1}	Reduced accumulation of Na^+ Increased Ca^{2+} content Increased germination percentage, and root and shoot length Vigorous seedling establishment	Azeem et al. [148]
DK961	0.06 mM SNP	Seed soaked for 24 h	100 mM NaCl	Increased germination percentage by increasing α-amylase, β-amylase isoenzymes	Duan et al. [149]

				activities Decreased MDA content, Na^+ content Increased SOD, CAT, APX activities	
Kakaba and Paven-76	1 and 2% $CaCl_2$ and KNO_3	Seed soaked for 12 h	5.97, 9.62, 13.28, and 16.9 dS m^{-1}	Increased germination with uniform seedlings Increased tillers per plant Shortened the physiological maturity period	Dugasa et al. [150]
Tatara-96, Ghaznavi-98, Fakhri Sarhad, Bakhtawar-92, Pirsabaq-2004 and AUQAB-2000	30 mM NaCl	Seed soaked	0, 40, 80, and 120 mM NaCl, 55 d	Enhanced the activities of enzymatic antioxidants Maintained ionic balance by increasing K^+ and Ca^{2+}accumulation Increased tillers per plant and grain yield	Jamal et al. [32]
MH-97, Inqlab-91	100, 150, and 200 mg L^{-1}kinetin and BAP	Seed soaked for 12 h	15 dS m^{-1}	Increased germination and early seedling establishment Increased shoot dry weight and grain yield Enhanced the endogenous growth hormones Maintained hormonal homeostasis	Iqbal et al. [105]
MH-97, Inqlab-91	100, 150, and 200 mg $L^{-1}GA_3$	Seeds primed for 12 h	15 dS m^{-1}	Maintained ionic balance by decreasing Na^+ and Cl^-ions in roots and shoots Increased Ca^{2+} and K^+in roots and shoot Increased leaf salicylic acid concentration Increased fertile tiller per plant and grain yield	Iqbal and Ashraf [101]

MH-97, Inqlab-91	2.5 mM Spd and 5 mM Spm	Seeds soaked for 12 h	15 dS m^{-1}	Increased shoot growth and grain yield Enhanced beneficial mineral nutrient uptake by maintaining ion homeostasis Increased biomass production and photosynthesis rate	Iqbal [151]
Inqlab-91 and SARC-1	50 mM NaCl, $CaCl_2$, and $CaSO_4$	Seeds soaked for 12 h	125 mM NaCl	Increased germination percentage by increasing total soluble and reducing sugar Increased shoot and root length under $CaCl_2$ and $CaSO_4$priming Increased biomass production Improved $K^+$and Ca^{2+}accumulation, and reduced Na^+concentration	Afzal et al. [28]
SARC-1 and MH-97	50 mg L^{-1}AsA, $CaCl_2$, kinetin, and SA	Immersed seed in solutions for 12 h	20 dS m^{-1}	Decreased emergence time by inducing biochemical changes and antioxidant enzymes activity Reduced Na^+absorption, and increased K^+ and Ca^{2+}absorption Improved protease and α-amylase activities Enhanced all agronomic and yield characteristics such as plant height, number of tillers, number of spikelets, grain yield, biological yield, and harvest index	Jafar et al. [143]
Caxton	22% PEG-6000	Seed soaked for	50, 100, 150, and 200 mM	Improved germination related metabolic activity	Fuller et al. [31]

		6 h	of NaCl	such as synthesis of nucleic acids, proteins, and enzymes, and enhanced respiratory activity upto 150 mM level of salt stress but at 200 mM salt stress priming effect becomes reduced	
Sakha-93, Gemmiza-9	0.2 mM SNP, 9% diluted sea water, diluted sea water + SNP	Seeds soaked for 10 h	9 dS m^{-1}	Increased leaf pigment concentration Enhanced membrane stability by decreasing lipid peroxidation Increased total soluble sugar, K^+ and Ca^{2+}concentration which decreased Na^+uptake	Maswad a and Abd El-Kader [152]
Inqlab and S-24	100 mg L^{-1}SA	Seeds soaked for 24 h	50 or 100 mM NaCl; 14 d	Increased root and shoot length, root and shoot dry weight, total soluble sugar, and carbohydrate metabolism Increased chl *a* and *b* content	Hamid et al. [145]
Azar 2	3% NaCl, 5% mannitol, 25% sugar beet extract and hydropriming	Seeds soaked for 4, 8, and 10 h	3.6 dS m^{-1}	Increased hypocotyle length, root number and leaf length, shoot and root fresh weight Increased photosynthesis rate Enhanced biomass production	Amoghe in et al. [153]

5. CONCLUSIONS AND FUTURE PERSPECTIVES

Wheat is the most popular and widely consumed cereal crops in the world due to its diverse uses. Most of the cultivated wheat is hexaploid which has some acquired tolerance to salt stress. However, increasing levels of salinity in irrigated lands make wheat production difficult because plant growth and productivity of wheat are severely affected by high salinity. Salt stress adversely affects seed germination, plant growth, photosynthesis, water relations, nutrient uptake, and yield. Oxidative stress is one of the most common effects of salt stress in wheat. However, salt stress effects depend on the dose and duration of stress, and mostly on genotypes. Considering the importance of wheat and the adverse effects of salt stress, plant biologists are trying to develop strategies to improve salt tolerance in wheat. Some of the strategies are related to the genetic manipulation of salt-tolerant traits. Physiologists are also trying to find the adaptive mechanisms to cope with the salt stress. However, the actual physiological mechanism of salt stress tolerance is yet to be revealed. Therefore, coordinated attempts by plant physiologists, breeders, and agronomists are essential to find out a sustainable strategy to enhance salt tolerance in wheat.

ACKNOWLEDGEMENTS

The first author acknowledges Japan Society for the Promotion of Science (JSPS) for funding in his research. We are also highly thankful to Mr. Md. Rumainul Islam, Environment Engineering Program, Chongqing University and Md. Mosfeq-Ul-Hasan, Zhejiang University, Hangzhou, China for providing us several supporting articles. As page limitation precluded us from citing a large number of studies, we apologize to those whose original publications are therefore not directly referenced in this chapter.

REFERENCES

1. Munns R, Tester M. Mechanism of salinity tolerance. Annual Reviews and Plant Biology2008; 59:651–681.
2. Acquaah G. Principles of Plant Genetics and Breeding. Blackwell, Oxford; 2007. p. 385.
3. FAO. High Level Expert Forum—How to Feed the World in 2050. Economic and SocialDevelopment, Food and Agricultural Organization of the United Nations, Rome, Italy; 2009.
4. Wiese MV. Compendium of Wheat Diseases. American Phytopathological Society, USA;1977.
5. Breiman A, Graur D. Wheat evolution. Israel Journal of Plant Sciences 1995; 43:85–98.
6. Royo A, Abió D. Salt tolerance in durum wheat cultivars. Spanish Journal of AgriculturalResearch 2003; 1:27–35.
7. Hajihashemi S, Kiarostami K, Enteshari S, Saboora A. Effect of paclobutrazol on wheatsalt tolerance at pollination stage. Russian Journal of Plant Physiology 2009; 56:251–257.
8. Huang Y, Bie Z, He S, Hua B, Zhen A, Liu Z. Improving cucumber tolerance to major nutrientsinduced salinity by grafting onto *Cucurbita ficifolia*. Environmental and ExperimentalBotany 2010; 69:32–38.
9. Turan MA, Elkarim AHA, Taban N, Taban S. Effect of salt stress on growth, stomatalresistance, proline and chlorophyll concentrations on maize plant. African Journal ofAgricultural Research 2009; 4:893–897.

10. Çelik Ö, Atak Ç. The effect of salt stress on antioxidative enzymes and proline content oftwo Turkish tobacco varieties. Turkish Journal of Biology 2012; 36:339–356.
11. Hasanuzzaman M, Alam MM, Nahar K, Al-Mahmud A, Ahamed KU, Fujita M.Exogenous salicylic acid alleviates salt stress-induced oxidative damage in *Brassica napus* by enhancing the antioxidant defense and glyoxalase systems. Australian Journal of Crop Science 2014; 8:631–639.
12. Hasanuzzaman M, Alam MM, Rahman A, Hasanuzzaman M, Nahar K, Fujita M.Exogenous proline and glycine betaine mediated upregulation of antioxidant defenseand glyoxalase systems provides better protection against salt-induced oxidative stress in two rice (*Oryza sativa* L.) varieties. Biomed Research International 2014;doi:10.1155/ 2014/ 757219
13. Munns R. Genes and salt tolerance: bringing them together. The New Phytologist 2005;167:645–663.
14. Shabala S, Munns R. Salinity stress: physiological constraints and adaptive mechanisms.In: Shabala S (ed) Plant Stress Physiology. Oxford: CAB International; 2012. pp. 59–93.
15. Rahman A, Nahar K, Hasanuzzaman M, Fujita M. Calcium supplementation improves Na+/K+ ratio, antioxidant defense and glyoxalase systems in salt-stressed rice seedlings. Frontiers in Plant Science 2016; doi:10.3389/fpls.2016.00609
16. Afzal I, Basra MAS, Iqbal A. The effects of seed soaking with plant growth regulators on seedling vigor of wheat under salinity stress. Journal of Stress Physiology and Biochemistry 2005; 1:6–14.
17. Akhiyarova GR, Sabirzhanova IB, Veselov DS, Frike V. Participation of plant hormones in growth resumption of wheat shoots following short-term NaCl treatment. Russian Journal of Plant Physiology 2005; 52:788–792.
18. Akbari G, Sanavy SA, Yousefzadeh S. Effect of auxin and salt stress (NaCl) on seed germinationof wheat cultivars (*Triticum aestivum* L.). Pakistan Journal of Biological Scinces2007; 10:2557–2561.
19. Arfan M, Athar HR, Ashraf M. Does exogenous application of salicylic acid through the rooting medium modulate growth and photosynthetic capacity in differently adapted spring wheat cultivated under salt stress? Journal of Plant Physiology 2007;6:685–694.
20. Athar HUR, Khan A, Ashraf M. Exogenously applied ascorbic acid alleviates salt-induced oxidative stress in wheat. Environmental and Experimental Botany 2007; 63:224–231.
21. Azzedine F, Gherroucha H, Baka M. Improvement of salt tolerance in durum wheat byascorbic acid application. Journal of Stress Physiology & Biochemistry 2011; 7:27–37.
22. Eleiwa ME, Bafeel SO, Ibrahim SO. Influence of Brassinosteroids on wheat plant (*Triticumaestivum* L.) production under salinity stress conditions I—growth parameters and photosynthetic pigments. Australian Journal of Basic and Applied Sciences 2011; 5:58–65.
23. Erdal S, Aydın M, Genisel M, Taspınar MS, Dumlupinar R, Kaya O, Gorcek Z. Effects of salicylic acid on wheat salt sensitivity. African Journal of Biotechnology 2011; 10:5713–5718.
24. Hasanuzzaman M, Hossain MA, Fujita M. Nitric oxide modulates antioxidant defenseand the methylglyoxal detoxification system and reduces salinity-induced damage owheat seedlings. Plant Biotechnology Reports 2011; 5:353–365.
25. Hasanuzzaman M, Hossain MA, Fujita M. Selenium-induced up-regulation of the antioxidant defense and methylglyoxal detoxification system reduces salinity induced damage in rapeseed seedlings. Biological Trace Element Research 2011;143:1704–1721.
26. Hasan uzzaman M, Nahar K, Fujita M. Plant responses to salt stress and role of exogenousprotectants to mitigate salt-induced damages. In: Ahmad P, Azooz MM, Prasad MNV(eds) Ecophysiology and Responses of Plants under Salt Stress. New York: Springer;2013. pp. 25–87.
27. Wahid A, Farooq M, Basra SMA, Rasul E, Siddique KHM. Germination of seeds andpropagules under salt stress. In: Pessarakli M (ed) Handbook of Plant and Crop Stress,3rd edn. Boca Raton: CRC Press; 2011. pp. 321–337.

28. Afzal I, Rauf S, Basra SMA, Murtaza G. Halopriming improves vigor, metabolismof reserves and ionic contents in wheat seedlings under salt stress. Plant Soil andEnvironment 2008; 54:382–388.
29. Ghiyasi M, Seyahjani AA, Tajbakhsh M, Amirnia R, Salehzadeh H. Effect of osmoprimingwith polyethylene glycol (8000) on germination and seedling growth of wheat(*Triticum aestivum* L.) seeds under salt stress. Research Journal of Biological Sciences2008; 3:1249–1251.
30. Akbarimoghaddam H, Galavi M, Ghanbari A, Panjehkeh N. Salinity effects on seed germinationand seedling growth of bread wheat cultivars. Trakia Journal of Sciences 2011;9:43–50.
31. Fuller MP, Hamza JH, Rihan HZ, Al-Issawi M. Germination of primed seed under NaClstress in wheat. ISRN Botany 2012; 167801. doi:10.5402/2012/167804
32. Jamal Y, Shafi M, Bakht J, Arif M. Seed priming improves salinity tolerance of wheat
33. varieties. Pakistan Journal of Botany 2011; 43:2683–2686.
34. Guo R, Yang Z, Li F, Yan C, Zhong X, Liu Q, Xia X, Li H, Zhao L. Comparative metabolicresponses and adaptive strategies of wheat (*Triticum aestivum*) to salt and alkali stress.BMC Plant Biology 2015; 15:170. doi:10.1186/s12870-015-0546-x
35. Zou P, Li K, Liu S, He X, Zhang X, Xing R, Li P. Effect of sulfated chitooligosaccharides on wheat seedlings (*Triticum aestivum* L.) under salt stress. Journal of Agricultural andFood Chemistry 2016; 64:2815–2821.
36. Wahid A, Perveen M, Gelani S, Basra SMA. Pretreatment of seed with H2O2 improvessalt tolerance of wheat seedlings by alleviation of oxidative damage and expression ofstress proteins. Journal of Plant Physiology 2007; 164:283–294.Tammam AA, Alhamd MFA, Hemeda MM. Study of salt tolerance in wheat (*Triticum aestium* L.) cultivar Banysoif 1. Australian Journal of Crop Science 2008; 1:115–125.
37. Mandhania S, Madan S, Sawhney V. Antioxidant defense mechanism under salt stress inwheat seedlings. Biologia Plantarum 2006; 50:227–231.
38. Poustini K, Siosemardeh A, Ranjbar M. Proline accumulation as a response to salt stressin 30 wheat (*Triticum aestivum* L.) cultivars differing in salt tolerance. Genetic Resourcesand Crop Evolution 2007; 54:925–934.
39. Lv DW, Zhu GR, Zhu D, Bian YW, Liang XN, Cheng ZW, Deng X, Yan YM. Proteomicand phosphoproteomic analysis reveals the response and defense mechanism in leavesof diploid wheat *T. monococcum* under salt stress and recovery. Journal of Proteomics2016; 143:93–105.
40. Asgari HR, Cornelis W, Damme PV. Salt stress effect on wheat (*Triticum aestivum* L.) growthand leaf ion concentrations. International Journal of Plant Production 2012; 6:195–208.
41. Afzal I, Basra SMA, Cheema MA, Farooq M, Jafar MZ, Shahid M, Yasmeen A. Seed priming:a shotgun approach for alleviation of salt stress in wheat. International Journal ofAgriculture and Biology 2013; 15:1199–1203.
42. Chinnusamy V, Jagendorf A, Zhu JK. Understanding and improving salt tolerance inplants. Crop Science 2005; 45:437–448.
43. Afzal I, Basra SMA, Hameed A, Farooq M. Physiological enhancements for alleviation ofsalt stress in wheat. Pakistan Journal of Botany 2006; 38:1649–1659.
44. Iqbal M, Ashraf M. Salt tolerance and regulation of gas exchange and hormonalhomeostasis by auxin-priming in wheat. Pesquisa Agropecuária Brasileira 2013;48:1210–1219.
45. Singh A, Bhushan B, Gaikwad K, Yadav OP, Kumar S, Rai RD. Induced defence responsesof contrasting bread wheat genotypes under differential salt stress imposition. IndianJournal of Biochemistry Biophysics 2015; 52:75–85.
46. Tian F, Wang W, Liang C, Wang X, Wang G, Wang W. Overaccumulation of glycinebetaine makes the function of the thylakoid membrane better in wheat under salt stress.The Crop Journal 2016; 00174:1–10.
47. Zhang S, Gan Y, Xu B. Application of plant-growth-promoting fungi *Trichoderma longibrachiatum*T6 enhances tolerance of wheat to salt stress through improvement

ofantioxidative defense system and gene expression. Frontiers in Plant Sciences 2016; 7.doi:10.3389/fpls.2016.01405

48. Parida AK, Das AB. Salt tolerance and salinity effect on plants: a review. Ecotoxicologyand Environmental Safety 2005; 60:324–349.
49. Hasanuzzaman M, Nahar K, Fujita M, Ahmad P, Chandna R, Prasad MNV, Ozturk M.Enhancing plant productivity under salt stress—relevance of poly-omics. In: Ahmad P,Azooz MM, Prasad MNV (eds) Salt Stress in Plants: Omics, Signaling and Responses. Berlin: Springer; 2013. pp. 113–156.
50. Greenway H, Munns R. Mechanisms of salt tolerance in nonhalophytes. Annual Reviewof Plant Physiology 1980; 31:149–190.
51. Hasanuzzaman M, Hossain MA, da Silva JAT, Fujita M. Plant responses and toleranceto abiotic oxidative stress: antioxidant defense is a key factor. In: Bandi V, Shanker AK,Shanker C, Mandapaka M (eds) Crop Stress and Its Management: Perspectives andStrategies. Berlin: Springer; 2012. pp 261–316
52. Mittler R. Oxidative stress, antioxidants and stress tolerance. Trends in Plant Science 2002;7:405–410.
53. Tanou G, Molassiotis A, Diamantidis G. Induction of reactive oxygen species andnecrotic death-like destruction in strawberry leaves by salinity. Environmental andExperimental Botany 2009; 65:270–281.
54. Hasegawa P, Bressan RA, Zhu JK, Bohnert HJ. Plant cellular and molecular responses to highsalinity. Annual Review of Plant Physiology and Plant Molecular Biology 2000; 51:463–499.
55. Sairam RK, Roa KV, Srivastava GC. Differential response of wheat genotypes to longterm salinity stress in relation to oxidative stress, antioxidant activity and osmolyte concentration.Plant Science 2002; 163:1037–1046.
56. Rao A, Ahmad SD, Sabir SM, Awan S, Shah AH, Khan MF, Shafique S, Arif S, AbbasSR. Potential biochemical indicators improve salt tolerance in fifteen cultivars of wheat(*Triticum aestivum L.*) from Pakistan. International Journal of Scientific and EngineeringResearch 2013; 4:389–406.
57. Rao A, Ahmad SD, Sabir SM, Awan SI, Hameed A, Abbas SR, Shehzad M, Khan MF,Shafique F, Ahmad Z. Detection of saline tolerant wheat cultivars (*Triticum aestivum* L.using lipid peroxidation, antioxidant defense system, glycinebetaine and proline contents.of Animal and Plant Sciences 2013; 23:1742–1748.
58. Sairam RK, Srivastava GC. Changes in antioxidant activity in sub-cellular fractions oftolerant and susceptible wheat genotypes in response to long term salt stress. PlantScience 2002; 162:897–904.
59. Seckin B, Sekmen AH, Turkan I. An enhancing effect of exogenous mannitol on the antioxidantenzyme activities in roots of wheat under salt stress. Journal of Plant GrowthRegulation 2009; 28:12–20.
60. Zhang X, Shi Z, Tian Y, Zhou Q, Cai J, Dai T, Cao W, Pu H, Jiang D. Salt stress increasescontent and size of glutenin macropolymers in wheat grain. Food Chemistry 2016;197:516–521.
61. Gorcek Z, Erdal S. Lipoic acid mitigates oxidative stress and recovers metabolic distortionsin salt-stressed wheat seedlings by modulating ion homeostasis, the osmo-regulatorlevel and antioxidant system. Journal of Science of Food and Agriculture 2015;95:2811–2817.
62. Fercha A. Some physiological and biochemical effects of NaCl salinity on durum wheat(*Triticum durum* Desf.). Advances in Biological Research 2011; 5:315–322.
63. Qiu Z, Guo J, Zhu A, Zhang L, Zhang M. Exogenous jasmonic acid can enhance toleranceof wheat seedlings to salt stress. Ecotoxicology and Environmental Safety 2014;104:202–208.
64. Genişel M, Erdal S. Alleviation of salt-induced oxidative damage by 5-aminolevulinicacid in wheat seedlings. AIP Conference Proceedings 2016; 1726:020025.doi:10. 1063/ 1.4945851
65. Ashraf M, Foolad MR. Roles of glycine betaine and proline in improving plant abioticstress resistance. Environmental and Experimental Botany 2007; 59:206–216.

66. Ahmad P, Jeleel CA, Azooz MM, Nabi G. Generation of ROS and non-enzymatic antioxidantsduring abiotic stress in plants. Botanical Research International 2009; 2:11–20.
67. Chen H, Jiang JG. Osmotic adjustment and plant adaptation to environmental changesrelated to drought and salinity. Environmental Reviews 2010; 18:309–319.
68. Gill SS, Tuteja N. Polyamines and abiotic stress tolerance in plants. Plant Signaling &Behavior 2010; 51:26–33.
69. Alam MM, Nahar K, Hasanuzzaman M, Fujita M. Trehalose-induced drought stress tolerance:a comparative study among different *Brassica* species. Plant Omics Journal 2014;7:271–283.
70. Verbruggen N, Hermans C. Proline accumulation in plants: a review. Amino Acids 2008;35:753–759.
71. Sharma SS, Dietz KJ. The relationship between metal toxicity and cellular redox imbalance.Trends in Plant Science 2009; 14:43–50.
72. Mattioli R, Costantino P, Trovato M. Proline accumulation in plants: not only stress.Plant Signaling & Behavior 2009; 4:1016–1018.Approaches to Enhance Salt Stress Tolerance in Wheat http://dx.doi.org/10.5772/67247
73. Szabados L, Savouré A. Proline: a multifunctional amino acid. Trends in Plant Science2010; 15:89–97.
74. Hayat S, Hayat Q, Alyemeni MN, Wani AS, Pichtel J, Ahmad A. Role of proline underchanging environments: a review. Plant Signaling & Behavior 2012; 7:1456–1466.
75. Raza SH, Athar H, Ashraf M. Influence of exogenously applied glycinebetaine on the photosynthetic capacity of two differently adapted wheat cultivars under salt stress.Pakistan Journal of Botany 2006; 38:241–251.
76. Raza SH, Athar HR, Ashraf M, Hameed A. Glycinebetaine-induced modulation of antioxidantenzymes activities and ion accumulation in two wheat cultivars differing in salttolerance. Environmental and Experimental Botany 2007; 60:368–376.
77. Ashraf MA, Ashraf M, Shahbaz M. Growth stage-based modulation in antioxidantdefense system and proline accumulation in two hexaploid wheat (*Triticum aestivum* L.)cultivars differing in salinity tolerance. Flora—Morphology, Distribution, FunctionalEcology of Plants 2012; 207:388–397.
78. Akhter N, Akram NA, Shahbaz M. Presowing seed treatments with glycinebetaineand mineral nutrients of wheat (*Triticum aestivum* L.) under saline conditions. PakistanJournal of Agricultural Sciences 2007; 44:236–241.
79. Khan MA, Shirazi MU, Alikhan M, Mujtraba SM, Islam E, Mumtaz S, Shereen A,Anasari RU, Ashraf MY. Role of proline, K/Na ratio and chlorophyll content in salttolerance of wheat (*Triticum aestivum* L.). Pakistan Journal of Botany 2009; 41:633–638.
80. Liang C, Zhan XY, Wang GP, Zou Q, Wang W. Overaccumulation of glycine betainealleviates the negative effects of salt stress in wheat. Russian Journal of Plant Physiology2009; 56:370–376.
81. Talat A, Nawaz K, Hussian K, Bhatti KH, Siddiqi EH, Khalid A, Anwer S, Sharif MU.Foliar application of proline for salt tolerance of two wheat (*Triticum aestivum* L.) cultivars.World Applied Sciences Journal 2013; 22:547–554.
82. Mahboob W, Khan MA, Shirazi MU. Induction of salt tolerance in wheat (*Triticum aestivum*L.) seedlings through exogenous application of proline. Pakistan Journal of Botany2016; 48:861–867.
83. Hendawey MH. Biochemical changes associated with induction of salt tolerance inwheat. Global Science and Research Journal 2015; 10:84–99.
84. Ismail MA. Exogenous proline induced changes in SDS-PAGE protein profile for salttolerance in wheat (*Triticum aestivum* L.) seedlings. Research Journal of Pharmaceutical,Biological and Chemical Science 2014; 5:749–755.
85. Chen TH, Murata N. Glycinebetaine protects plants against abiotic stress: mechanismsand biotechnological applications. Plant, Cell and Environment 2011; 34:1–20.
86. Giri J. Glycinebetaine and abiotic stress tolerance in plants. Plant Signaling & Behavior2011; 6:1746–1751.

87. Ahmad R, Lim CJ, Kwon SK. Glycine betaine: a versatile compound with great potentialfor gene pyramiding to improve crop plant performance against environmentalstresses. Plant Biotechnology Reports 2013; 7:49–57.
88. Salama KHA, Mansour MM, Al-Malawi HA. Glycinebetaine priming improves salt toleranceof wheat. Biologia. 2015; 70:1334–1339.
89. Silini A, Cherif-Silini H, Yahiaoui B. Growing varieties durum wheat (*Triticum durum*)in response to the effect of osmolytes and inoculation by *Azotobacter chroococcum* undersalt stress. African Journal of Microbiology Research 2016; 10:387–399.
90. Yan D, Zheng B. Effects of soaking seeds in trehalose on physiological characteristicsof wheat Yangmai-19 under salt stress. Acta Agriculturae Zhejiangensis 2016;28:1271–1276.
91. Aghdasi M, Smeekens S, Schluepman H. Microarray analysis of gene expressionpatterns in *Arabidopsis* seedlings under trehalose, sucrose and sorbitol treatment.International Journal of Plant Production 2008; 2:309–320.
92. Duman F, Aksoy A, Aydin Z, Temizgul R. Effects of exogenous glycinebetaine and trehaloseon cadmium accumulation and biological responses of an aquatic plant (*Lemnagibba* L.). Water, Air & Soil Pollution 2010; 217:545–556.
93. Luo Y, Li F, Wang GP, Yang XH, Wang W. Exogenously-supplied trehalose protects thylakoid membranes of winter wheat from heat-induced damage. Biologia Plantarum 2010; 54:495–501.
94. Salama KHA, Mansour MMF, Hassan NS. Choline priming improves salt tolerancein wheat *(Triticum aestivum* L.) Australian Journal of Basic and Applied Sciences 2011;5:126–132.
95. Abebe T, Guenzi AC, Martin B, Cushman JC. Tolerance of mannitol-accumulatingtransgenic wheat to water stress and salinity. Plant Physiology 2003; 131: 1748–1755.
96. Ramadan AM, Eissa HF, Hassanein SE, Azeiz AZA, Saleh OM, Mahfouz HT,El-Domyati FM, Madkour MA, Bahieldin A. Increased salt stress tolerance and modifiedsugar content of bread wheat stably expressing the *mtlD* gene. Life Science Journal2013; 10:2348–2362.
97. Kerepesi I, Bányai-Stefanovits E, Galiba G. Fructans in wheat under stress conditions.Acta Biologica Szegediensis 2002; 46:101–102.
98. Sharbatkhari M, Shobbar Z, Galeshi S. Wheat stem reserves and salinity tolerance:molecular dissection of fructan biosynthesis and remobilization to grains. Planta 2016;244:191–202.
99. Ryu H, Cho Y. Plant hormones in salt stress tolerance. Journal of Plant Biology 2015;58:147–155.
100. Iqbal M and Ashraf M. Seed treatment with auxins modulates growth and ion partitioningin salt-stressed wheat plants. Journal of Integrative Plant Biology 2007; 49:1003–1015.
101. Iqbal M, Ashraf M. Gibberellic acid mediated induction of salt tolerance in wheatplants: growth, ionic partitioning, photosynthesis, yield and hormonal homeostasis. Environmental and Experimental Botany 2010; 86:76–85.
102. Shaddad MAK, Abd El-Samad HM, Mostafa D. Role of gibberellic acid (GA3) inimproving salt stress tolerance of two wheat cultivars. International Journal of PlantPhysiology and Biochemistry 2013; 5:50–57.
103. Tabatabei SA. The effect of salicylic acid and gibberellin on enzyme activity and germinationcharacteristics of wheat seeds under salinity stress conditions. InternationalJournal of Agriculture and Crop Sciences 2013; 6:236–240.
104. Iqbal M, Ashraf M. Presowing seed treatment with cytokinins and its effect on growth,photosynthetic rate, ionic levels and yield of two wheat cultivars differing in salt tolerance.Journal of Integrative Plant Biology 2005; 47:1315–1325.
105. Iqbal M, Ashraf M, Jamil A. Seed enhancement with cytokinins: changes in growth andgrain yield in salt stressed wheat plants. Plant Growth Regulation 2006; 50:29–39.
106. Gurmani AR, Bano A, Najeeb U, Zhang J, Khan SU, Flowers TJ. Exogenously applied silicateand abscisic acid ameliorates the growth of salinity stressed wheat (*Triticum aestivum*L.) seedlings through Na+ exclusion. Australian Journal of Crop Science 2013; 7:1123–1130.

107. Ali Q, Athar H, Ashraf M. Modulation of growth, photosynthetic capacity and waterrelations in salt stressed wheat plants by exogenously applied 24-epibrassinolide. PlantGrowth Regulation 2008; 56:107–116.
108. Naqvi SSM, Mumtaz S, Shereen A, Khan MA, Khan AH. Role of abscisic acid inregulation of wheat seedling growth under salinity stress. Biologia Plantarum 1997;39:453–456.
109. El-Lethy SR, Abdel hamid MT, Reda F. Effect of potassium application on wheat (*triticumaestivum* l.) cultivars grown under salinity stress. World Applied Sciences Journal2013; 26:840–850.
110. Kausar A, Gull M. Effect of potassium sulphate on the growth and uptake of nutrientsin wheat (*Triticum aestivum* L.) under salt stressed conditions. Journal of AgriculturalScience 2014; 6:1–12.
111. Khan A, Ahmad I, Shah A, Ahmad F, Ghani A, Nawaz M. Amelioration of salinitystress in wheat (*Triticum aestivum* L.) by foliar application of phosphorus. Phyton,International Journal of Experimental Botany 2013; 82:281–287.
112. Zaman B, Niazi BH, Athar M, Ahmad M. Response of wheat plants to sodium and calciumion interaction under saline environment. International Journal of EnvironmentalScience and Technology 2005; 2:7–12.
113. Tian X, He M, Wang Z, Zhang J, Song Y, He Z, Dong Y. Application of nitric oxideand calcium nitrate enhances tolerance of wheat seedlings to salt stress. Plant GrowthRegulation 2015; 77: 343–356.
114. Elhamid EMA, Sadak MS, Tawfik MM. Alleviation of adverse effects of salt stress inwheat cultivars by foliar treatment with antioxidant 2—changes in some biochemicalaspects, lipid peroxidation, antioxidant enzymes and amino acid contents. AgriculturalSciences 2014; 5:1269–1280.
115. Athar HUR, Khan A, Ashraf M. Inducing salt tolerance in wheat by exogenouslyapplied ascorbic acid through different modes. Journal of Plant Nutrition 2009; 32:1–19.
116. Khan A, Ahmad MSA, Athar HUR, Ashraf M. Interactive effect of foliarly applied ascorbicacid and salt stress on wheat (*Triticum aestivum* L.) at the seedling stage. PakistanJournal of Botany 2006; 38:1407–1414.
117. Sadak MS. Mitigation of salinity adverse effects of on wheat by grain priming withmelatonin. International Journal of ChemTech Research 2016; 9:85–97.
118. Farouk S. Ascorbic acid and α-tocopherol minimize salt-induced wheat leaf senescence.Journal of Stress Physiology & Biochemistry 2011; 7:58–79.
119. Zheng C, Jiang D, Liu F, Dai T, Liu W, Jing Q, Cao W. Exogenous nitric oxide improvesseed germination in wheat against mitochondrial oxidative damage induced by highsalinity. Environmental and Experimental Botany 2009; 67:222–227.
120. Kausar F, Shahbaz M. Interactive effect of foliar application of nitric oxide (NO) andsalinity on wheat (*Triticum aestivum* L.). Pakistan Journal of Botany 2013; 45:67–73.
121. Qados AMSA. Effect of arginine on growth, yield and chemical constituents of wheatgrown under salinity condition. Academic Journal of Plant Science 2009; 2:267–278.
122. Saeidnejad AH, Kafi M, Dashti, M. Ameliorative effects of spermine application onphysiological performance and salinity tolerance induction of susceptible and tolerantcultivars of wheat (*Triticum aestivum*). Archives of Agronomy and Soil Science 2016;62:1337–1346.
123. Li J, Qiu Z, Zhang X, Wang L. Exogenous hydrogen peroxide can enhance tolerance ofwheat seedlings to salt stress. Acta Physiologiae Plantarum 2011; 33:835–842.
124. Nahar K, Hasanuzzaman M, Alam MM, Rahman A, Suzuki T, Fujita M. Polyamine andnitric oxide crosstalk: antagonistic effects on cadmium toxicity in mung bean plantsthrough upregulating the metal detoxification, antioxidant defense, and methylglyoxal detoxification systems. Ecotoxicology and Environmental Safety 2015; 126:245–255.
125. Nahar K, Rohman MM, Hasanuzzaman M, Alam MM, Rahman A, Suzuki T, Fujita M.Physiological and biochemical mechanism of spermine-induced cadmium stress tolerancein mung bean (*Vigna radiata* L.). Environmental Science and Pollution Research2016; 23:21206–18.

126. Nahar K, Hasanuzzaman M, Alam MM, Rahman A, Suzuki T, Fujita M. Polyaminesconfer salt tolerance in mung bean by reducing sodium uptake, improving nutrienthomeostasis, antioxidant defense and methylglyoxal detoxification systems. Frontiersin Plant Science 2016; doi:10.3389/fpls.2016.01104
127. Neill SJ, Desikan R, Clarke A, Hurst RD, Hancock JT. Hydrogen peroxide and nitricoxide as signalling molecules in plants. Journal of Experimental Botany 2002; 53:1237–47.
128. Iqbal M, Ashraf M. Changes in growth, photosynthetic capacity and ionic relations inspring wheat (*Triticum aestivum* L.) due to pre-sowing seed treatment with polyamines.Plant Growth Regulation 2005; 46:19–30.
129. Bakare SO, Ukwungwu MN. On-farm evaluation of seed priming technology inNigeria. African Journal of General Agriculture 2009; 5:93–97.
130. Bewley JD, Bradford KJ, Hilhorst HWM, Nonogaki H. Seeds: Physiology ofDevelopment, Germination and Dormancy, 3rd ed. New York: Springer; 2013; p. 392.
131. Hussian I, Ahmad R, Farooq M, Rehman A, Amin M. Seed priming improves the performanceof poor quality wheat seed under drought stress. Applied Science Reports2015; 7:12–18.
132. Kaya MD, Okcu G, Atak M, Cikili Y, Kolsarici O. Seed treatments to overcome saltdrought stress during germination in sunflower (*Helianthus annuus* L.). EuropeanJournal of Agronomy 2006; 24:291–295.
133. Salehzade H, Shishvan MI, Ghiyasi M, Forouzin F, Siyahjani AA. Effect of seed primingon germination and seedling growth of wheat (*Triticum aestivum* L.). Research Journalof Biological Science 2009; 4:629–631.
134. Bradford KJ. Water relations in seed germination. In: Kigel J, Galili G (eds) SeedDevelopment and Germination. New York: Marcel Dekker, Inc.; 1995; pp. 351–396.
135. Ellis RH, Butcher PD. The effects of priming and 'natural' differences in quality amongstonion seed lots on the response of the rate of germination to temperature and the identificationof the characteristics under genotypic control. Journal of Experimental Botany1988; 39:935–950.
136. Hill H, Bradford KJ, Cunningham J, Taylor AG. Primed lettuce seeds exhibit increasedsensitivity to moisture during aging. Acta Horticulturae 2008; 782:135–141.
137. Tavili A, Zare S, Moosavi SA, Enayati A. Effects of seed priming on germination characteristicsof *Bromus* species under salt and drought conditions. American-EurasianJournal of Agriculture and Environmental Science 2011; 10:163–168.
138. Jisha KC, Vjayakumari K, Puthur JT. Seed priming for abiotic stress tolerance: an overview.Acta Physiologiae Plantarum 2013; 35:1381–1396.
139. Yari L, Abbasian A, Oskouei B, Sadeghi H. Effect of seed priming on dry matter, seedsize and morphological characters in wheat cultivar. Agricultural and Biology Journalof North America 2011; 2:232–238.
140. Saiki TP, Barman B, Ferrara OG. Participatory evaluation by farmers of on-farm seedpriming in wheat in Assam, India. 2006; 13th Australian Agronomy Conference. Perth,WA.
141. Al-hakimi AMA, Hamada AM. Counteraction of salinity stress on wheat plants bygrain soaking in ascorbic acid, thiamin or sodium salicylate. Biologia Plantarum 2001;44:253–261.
142. Shakirova FM, Sakhabutdinova AR, Bezrukova MV, Fatkhutdinova RA, FatkhutdinovaDR. Changes in the hormonal status of wheat seedlings induced by salicylic acid andsalinity. Plant Science 2003; 164:317–322.
143. Jafar MZ, Farooq M, Cheema MA, Afzal I, Basra SMA, Wahid MA, Aziz T, Shahid M.Improving the performance of wheat by seed priming under saline conditions. Journalof Agronomy and Crop Science 2012; 198:38–45.
144. Gulnaz AJ, Iqbal J, Azam F. Seed treatment with growth regulators and crop productivity.II. Response of critical growth stages of wheat (*Triticum aestivum* L.) under salinitystress. Cereal Research Communications 1999; 27:419–426.
145. Hamid M, Ashraf MY, Rehman KU, Arashad M. Influence of salicylic acid seed primingon growth and some biochemical attributes in wheat grown under saline conditions.Pakistan Journal of Botany 2008; 40:361–367.

146. Salama KHA, Mansour MMF, Al-Malawi HA. Choline priming-induced plasma membranelipid alterations contributed to improved wheat salt tolerance. Acta PhysiologiaePlantarum 2015b; 37:1–7.
147. Afzal I, Basra SMA, Ahmad N, Farooq M. Optimization of hormonal priming techniquesfor alleviation of salinity stress in enhanced wheat (*Triticum aestivum* L.). Caderno dePesquisa Sér Bio Santa Cruz do Sul 2005; 17:95–109.
148. Azeem M, Iqbal N, Kausar S, Javed MT, Akram MS, Sajid MA. Efficacy of silicon primingand fertigation to modulate seedling's vigor and ion homeostasis of wheat (*Triticumaestivum* L.) under saline environment. Environmental Science and Pollution Research2015; 22:14367–71.
149. Duan P, Ding F, Wang F, Wang BS. Priming of seeds with nitric oxide donor sodiumnitroprusside (SNP) alleviates the inhibition on wheat seed germination by salt stress.Journal of Plant Physiology and Molecular Biology 2007; 33:244–50.
150. Dugasa T, Abebie B, Tomer RPS, Barnabas J. Tolerance of *Triticum aestivum* L. (Breadwheat) varieties for growth yield in high salinity soils of Ethiopia. International Journalof Scientific Research 2016; 5:139–153.
151. Iqbal RM. Effect of salinity on ion partitioning in spring wheat. Pakistan Journal ofBiological Sciences 2005; 8:302–306.
152. Maswada HF, Abd El-Kader NIK. Redox halopriming: a promising strategy for inducingsalt tolerance in bread wheat. Journal of Agronomy and Crop Science 2014; 202:37–50.
153. Amoghein MB, Amoghein RS, Tobeh A, Jamaati-e-Somarin S. The effect of osmoprimingand hydropriming on the different index of germination & early growth of wheat undersalty stress. International Research Journal of Applied Basic Science 2013; 4:1924–1931.

CHAPTER 7

Wheat functional genomics and engineering crop improvement

Michael Francki and Rudi Appels

Department of Agriculture, Crop Improvement Institute, Locked Bag No 4, Bentley WA 6983, Australia. †CRC for Molecular

Plant Breeding - Western node, School of Biological Science, Murdoch University, Murdoch WA 6150, Australia.

ABSTRACT

Genetic mapping and determination of the organization of the wheat genome are changing the wheat-breeding process. New initiatives to analyze the expressed portion of the wheat genome and structural analysis of the genomes of *Arabidopsis* and rice are increasing our knowledge of the genes that are linked to key agronomically important traits.

The growing world-wide demand for wheat is placing pressure on breeding programs to produce elite cultivars that can adapt to a range of environments without compromising agronomic performance, grain quality or disease resistance. Wheat-breeding efforts focus on developing new varieties with improved disease resistance (to nematode, fungal and/or viral infection), tolerance to abiotic stresses (such as heavy-metal toxicity, drought and cold tolerance) and numerous grain quality attributes that affect baking and other uses of the final product. The combination of existing knowledge and resources with modern biotechnology and functional genomics is providing the opportunity to study the genetic, biochemical and physiological basis of these complex traits. Current efforts aim to address the major challenge of capturing the information from both wheat and model organisms, such as rice and *Arabidopsis,* in order to define genes that underpin the unique attributes of wheat. The resources being developed using biotechnology include comprehensive mapping initiatives and genome-wide expression studies; these need to be implemented together with wheat-breeding programs, in conjunction with high-throughput screening, to efficiently develop new, improved wheat varieties.

1. WHEAT CHROMOSOME EVOLUTION

Bread wheat (*Triticum aestivum* L.) is an allohexaploid species containing three different ancestral genomes (designated A, B, and D), each of which has seven pairs of chromosomes (2n = 6x = 42). The total wheat genome is in excess of 16,000 megabases in size [1], and approximately 80% of it consists of repetitive (non-coding) DNA sequences, interspersed amongst low-copy-number or singleton genes. It is undoubtedly one of the largest and most complex genomes of all crop species.

The wheat genome can tolerate major chromosomal rearrangements and deletions [2,3] without major detrimental effects on the plant. Polyploidy provided a basis for chromosome engineering experiments [4] even before the cloning of DNA and the successful generation of transgenic plants. Significant improvements in wheat have been made by introgressing (transfer by hybridization and repeated back-crossing) chromosome segments from related grass species and the subsequent transfer of desirable genes. The most common introgression involves the short arm of rye chromosome 1 ($_1$RS) translocated onto the long arm of wheat chromosome $_1$B ($_1$BL), so that several disease-resistance genes from rye are transferred to wheat [5,6]. This whole-arm translocation is caused by a rare centromeric mis-division between wheat and rye univalent homeologous (closely related but not homologous) 1B chromosomes. The authors of recent studies have argued that combining genomes from different species can lead to chromosomal instability and rapid genome evolution [7], but the results of introducing chromosomes from related grasses into wheat have shown that many features of the introduced DNA remain stable. For example, we have analyzed a number of wheat lines harboring 1BL.1RS translocations that show no changes in 1BL.1RS centric-fusion breakpoints [8] or other chromosomal regions (M.F. and W. Berzonsky, unpublished observations).

Genome sequence analyses have shed further light on the structural modifications that have occurred during the evolution of the wheat genome, particularly in intergenic regions. Alignment of a single four-gene region of *Triticum monococcum* (closely related to the wheat A genome), barley, and rice [9,10,11] indicated that gene order was maintained. Small inversions or duplications could occur, as indicated by gene 2 in this comparison, which was inverted in barley relative to its orientation in *T. monococcum* and rice, and by gene 4, which was duplicated in barley. The intron-exon structure of the predicted genes was conserved and could even be recognized in the distantly related *Arabidopsis* genome; the intergenic regions were quite divergent, however, as a result of the presence of different retrotransposable elements. In many instances the retrotransposable elements were nested, as described by San Miguel *et al.* [10], and estimates for the time of insertion, based on sequence divergence in the long terminal repeats of these elements, were less than 5 million years. The activity of some retrotransposable elements in rice as a result of cell culture [12,13] is consistent with the possibility that these elements provide an important means of genome modification.

2. LINKING PHENOTYPIC TRAITS TO GENES

Genetic mapping of wheat has provided the basis for identifying quantitative trait loci (QTL) that control complex traits [14]. The International Triticeae Mapping Initiative (ITMI) [15] has overseen the coordination of resources from different organizations world-wide to provide genetic mapping information. Langridge *et al.* [16] recently reviewed the development of a variety of genetic maps of wheat that serve a purpose for a number of applications: restriction fragment length polymorphisms (RFLPs), amplified fragment length polymorphisms (AFLPs) and simple sequence repeats (SSRs) are the predominant marker systems used, and together these provide numerous markers linking phenotypes to genes distributed across the wheat genome. As part of a major initiative to evaluate the expressed portion of the wheat genome, the US National Science Foundation (NSF) has funded an ongoing project to develop and sequence in excess of 100,000 expressed sequences tags (ESTs) [17]. ESTs are partial sequences derived from a range of cDNA libraries, providing a broad representation of expressed sequences from different tissue types during normal development or after treatment with abiotic and biotic stresses. To identify expressed sequences controlling specific traits, genetic mapping using ESTs as RFLP markers is hampered by the low-level polymorphism within wheat cultivars - polymorphism being a prerequisite for genetic-mapping studies. Well-characterized deleted regions of wheat chromosomes have therefore been used to determine the physical position of all ESTs and to assign them to specific chromosome regions. This leads to a correlation of genetic and physical maps that will create a reference by which ESTs can be linked to QTLs. The NSF initiative is targeting 10,000 selected ESTs to specific chromosomal regions using deletion stocks of wheat [18].

Analyses of wheat ESTs from the mid-endosperm stage of seed development by our group and others [19,20,21] have revealed new predicted classes of seed storage protein that are most closely related to genes coding for avenin seed storage proteins from oat and foam-stabilizing proteins from barley. It is most likely that these new classes of gene encode proteins in the low-molecular-weight class of prolamin seed storage proteins described in earlier literature (discussed in [20,21]). Although this family of genes is extensive, we have localized some representatives to the long arm of chromosome 4A [20] using deletion stocks of wheat, and this chromosomal location is consistent with that of some low-molecular-weight prolamins determined by earlier researchers [22,23]. The wheat genes encoding foam-stabilizing-like proteins have been expressed in bacteria and the properties of the protein recovered in this way indicate they have significant influences on wheat-flour processing properties, an important functional attribute of seed storage proteins (B.C. Clarke and F. Bekes, personal communication).

3. MODEL SYSTEMS FOR IDENTIFYING CANDIDATE GENES CONTROLLING COMPLEX TRAITS

The completion of the *Arabidopsis* genome sequence [24] provided valuable information for the analysis of genome evolution across higher plants. For example, wheat starch properties are important in industrial and commercial uses, and specific starch-granule properties affect the quality of particular final products. The starch synthase III gene is involved in the production of the amylopectin class of starch, and the ratio of amylose to amylopectin has an impact on uses of the end product. A comparative analysis of the starch synthase III genes in wheat and *Arabidopsis* show that the exons are largely conserved, with some insertion events in wheat having introduced two additional exons and an enlarged exon 3 [25]. The large degree of conservation indicates that isoforms of starch synthase III have specific roles to play in starch synthesis across higher plants and that, in some instances, *Arabidopsis* genes can be used to identify corresponding genes in wheat.

Comparative genomics is contributing substantially to the analysis of genetically complex crops such as wheat. Although the much-discussed colinearity of grass genomes [26] is an over-simplification [27], because numerous inversion and insertion/deletion events have occurred during evolution, the concept of colinearity has provided a useful framework for analysis of the wheat genome. Many aspects of the genome, such as intergenic regions and introns, have changed in sequence since wheat and other cereals diverged from their progenitor species, while the genic regions have been more conserved. Useful synteny between rice and wheat has been uncovered for several genes and for small regions of the genome [27], an important example being in the cloning of the *Rht-1* genes of wheat. The introduction of *Rht-1* genes into modern wheats reduced the plant's ability to respond to gibberellic acid (GA) [28] and allowed the selection of semi-dwarf wheat lines with much higher yields. The *Rht1* genes are functionally equivalent to the *Arabidopsis GA1* gene, which encodes a transcription factor integral to the plant's response to GA. *GA1* carries a characteristic phosphotyrosine signaling domain, and a rice EST with a high degree of similarity at the amino terminus was identified by Peng *et al.* [29]. The rice EST was subsequently used to identify the orthologous genes in wheat [29], which are located on the short arms of chromosomes 4B and 4D (near the centromere).

The *Rht-1* genes differ from the wild-type alleles found in most modern wheat by a single base-pair change that creates a TAG stop codon near the translation start site [29]. The polyploid nature of wheat means that each of the three genomes carries closely related sequences, however. In the case of the *Rht-1* genes, the mutants *Rht-B1b* (on chromosome 4B) and *Rht-D1d* (on chromosome 4D) clearly have to be assayed individually. Differentiating between these closely related sequences is typical of the challenges faced in working with wheat, and Ellis *et al.* [30] were able to utilize sequence differences between the B and D genomes in the promoter regions of the genes to develop specific assays for the respective *Rht* genes. Figure 1 illustrates the location of the primers that were optimized to achieve the specific assay of the *Rht-B1b* and *Rht-D1d* mutant genes.

4. ENGINEERING WHEAT IMPROVEMENT

Many marker-trait combinations are currently being implemented and tested in wheat-breeding programs. Advances in developing gene-specific markers from EST databases, with model systems, and with conventional DNA markers, in particular, are currently leading to strategies for identifying novel alleles in landraces or wild species that can be introgressed to improve wheat varieties. New alleles of complex traits can be identified in wild species using molecular markers [31] and then introduced into elite cultivars. Applying such strategies will broaden the gene pool for wheat improvement and maximize the genetic gain of desirable alleles from the parents in a common background during marker-assisted breeding. The relationship between increasing the number of parent alleles and maximizing genetic gain in a conventional and marker-assisted breeding is shown in Figure 2. Selection based on phenotyping is a major limitation of screening large allelic combinations in a conventional breeding program, so increases in genetic gain are limited. Additional aspects of breeding to which molecular markers can contribute include recurrent parent selection in a back-crossing program, selection of parents for crossing, validation of F1s to ensure crosses have been made and exploitation of primitive germplasm with reference to reducing linkage drag by removing all unwanted chromosome segments more efficiently.

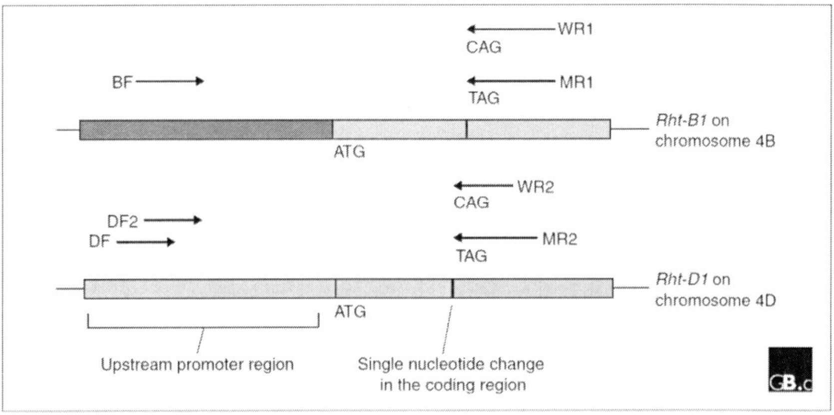

Figure 1 Diagram of the PCR primers used to distinguish the *Rht-B1b* and *Rht-D1b* genes [30]. The primers on the right differentiated the wild-type (WR1 and WR2) from the mutant (MR1 and MR2) genes by a single nucleotide change in the coding region. The primers cannot, however, distinguish between the genes on homeologous chromosomes 4B and 4D. The primers on the left-hand side differentiated between the 4B and 4D genomes because the upstream promoter regions have undergone sufficient evolutionary change to be distinguishable from each other. The figure is based on Ellis *et al.* [30] emphasizing the challenges faced when studying multiple gene copies in the wheat genome.

High-throughput screening capabilities now offer us the opportunity to select larger allelic combinations from wild and domesticated sources during marker-assisted breeding of wheat. Selecting progeny on the basis of genotyping at different stages of the breeding program will result in the accumulation of individuals with desirable allelic combinations and ultimately with a larger combination of desirable alleles at the culmination of the breeding program. Thus, understanding the complexities of the wheat genome leads to optimization of the technologies for combining desirable traits for improved wheat production.

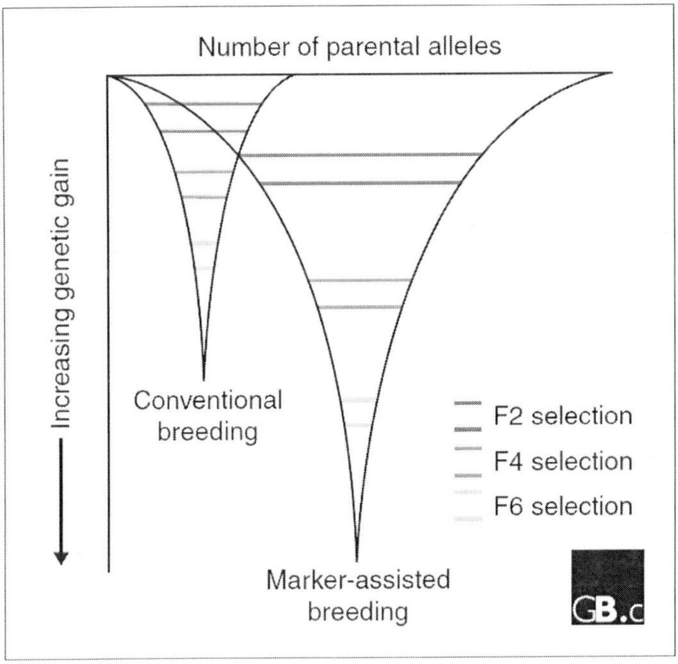

Figure 2 The effect on the genetic gain in elite lines of conventional and marker-assisted breeding. The effect of increasing the number of parental allele combinations, with selection at F2, F4 and F6 generations, highlights the value of a strategic use of molecular markers in breeding [32,33]. The larger base of the inverted triangle implies that more genetic variation is sampled for inclusion into the smaller numbers of lines examined in the F6 generation.

REFERENCES

1. Arumuganathan K, Earle ED: **Nuclear DNA content of some important plant species.** *Plant Mol Biol Rep* 1991, **9:**208-218.
2. Endo TR, Gill BS: **The deletion stocks of common wheat.** *J Hered* 1996, **87:**295-307.
3. Sears ER: **Nullisomic-tetrasomic combinations in hexaploid wheat.** In: *Chromosome Manipulation and Plant Genetics*. Edited by Lewis DR. London: Oliver and Boyd; 29-47.
4. Sears ER: **Chromosome engineering in wheat.** *Stadler Symp* 1972, **4:**23-38.
5. Baum M, Appels R: **The cytogenetic and molecular architecture of chromosome 1R - one of the most widely utilized sources of alien chromatin in wheat varieties.** *Chromosoma* 1991, **101:**1-10.

6. Berzonsky WA, Francki MG: **Biochemical, molecular and cytogenetic technologies for characterizing 1RS in wheat: a review.** *Euphytica* 1999, **108:**1-19.
7. Ozkan H, Levy AA, Feldman M: **Allopolyploidy-induced rapid genome evolution in the wheat (*Aegilops-Triticum*) group.** *Plant Cell* 2001, **13:**1735-1747.
8. Francki MG, Berzonsky WA, Ohm HW, Anderson JM: **Physical location of a *HSP70* gene homologue on the centromere of chromosome 1B of wheat (*Triticum aestivum* L.).** *Theor Appl Genet* 2002, **104:**184-191.
9. Dubcovsky J, Ramakrishna W, SanMiguel PJ, Busso CS,Yan L, Shiloff BA, Bennetzen JL: **Comparative sequence analysis of collinear barley and rice bacterial artificial chromosomes.** *Plant Physiol* 2001, **125:**1342-1353.
10. San Miguel P, Tikhonov A, Jin Y-K, Motchoulskaia N, Zakharov D, Melake-Berhan A, Springer PS, Edwards K, Lee M, Avramova Z, Bennetzen JL: **Nested retrotransposons in the intergenic regions of the maize genome.** *Science* 1996, **274:**765-768.
11. San Miguel PJ, Ramakrishna W, Bennetzen JL, Brusjo CS, Dubcovsky J: **Transposable elements, genes and recombination in a 215-kb contig from wheat chromosome 5Am.** *Funct Integr Genomics* 2002, DOI 10.1007/s10142-002-0056-4.
12. Hirochika H: **Retrotransposons of rice as a tool for forward and reverse genetics.** In *Molecular Biology of Rice*. Edited by Shimamoto K. Tokyo: Springer; 1999: 43-58.
13. Agrawal GK, Yamazaki M, Kobayashi M, Hirochika R, Miyao A, Hirochika H: **Screening of the rice viviparous mutants generated by endogenous retrotransposon *Tos17* insertion.** *Plant Physiol* 2001, **125:**1248-1257.
14. Appels R, Gustafson JP, O'Brien L: **Wheat breeding in the new century: applying molecular genetic analyses of key quality and agronomic traits.** *Aust J Agric Res* 2001, **52:**1043-1423.
15. **International Triticeae Mapping Initiative** [http://www.scri.sari.ac.uk/ITMI/]
16. Langridge P, Lagudah ES, Holton TA, Appels R, Sharp PJ, Chalmers KJ: **Trends in genetic and genome analyses in wheat: a review.** *Aust J Agric Res* 2001, **52:**1043-1077.
17. **U.S. wheat genome project** [http://wheat.pw.usda.gov/NSF/]
18. **GrainGenes: Expressed Sequence Tags (GG-ESTs)** [http://wheat.pw.usda.gov/wEST/binmaps]
19. Clarke BC, Hobbs M, Skylas D, Appels R: **Genes active in developing wheat endosperm.** *Funct Integr Genom* 2000, **1:**44-55.
20. Clarke BC, Larroque OR, Bekes F, Somers D, Appels R: **The frequent classes of expressed genes in wheat endosperm tissue as possible sources of genetic markers.** *Aust J Agric Res* 2002, **53:**1181-1193.
21. Anderson OD, Hsia CC, Adalsteins AE, Lew EJ-L, Kasarda DD: **Identification of several new classes of low molecular weight wheat gliadin related proteins and genes.** *Theor Appl Genet* 2001, **103:**307-315.
22. Salcedo G, Prada J, Aragoncillo C, Low MW: **Gliadin-like proteins from wheat endosperm.** *Phytochemistry*, 1979 **18:**725-727.
23. Rocher A, Calero M, Soriano F, Mendez E: **Identification of major rye secalins as coeliac immunoreactive proteins.** *Biochim Biophys Acta* 1996, **1295:**13-22.
24. **The *Arabidopsis* information resource** [http://www.arabidopsis.org]
25. Li Z, Mouille G, Kosar-Hashemi B, Rahmen S, Clarke B, Gale KR, Appels R, Morrell MK: **The structure and expression of the wheat starch synthase III gene. Motifs in the expressed gene define the lineage of the starch synthase III gene family.** *Plant Physiol* 2000, **123:**613-624.
26. Devos KM, Gale MD: **Genome relationships: the grass model in current research.** *Plant Cell* 2000, **12:**637-646.
27. Lagudah ES, Dubcovsky J, Powell W: **Wheat genomics.** *Plant Physiol Biochem* 2001, **39:**3335-344.
28. Gale MD, Gregory RS: **A rapid method for early generation selection of dwarf genotypes in wheat.** *Euphytica* 1977, **26:**733-738.
29. Peng J, Richards DE, Hartley NM, Murphy GP, Devos KM, Flintham JE, Beales J, Fish LJ, Worland AJ, Pelica F, *et al.*: **'Green revolution' genes encode mutant gibberellin response modulators.** *Nature* 1999, **400:**256-261.

30. Ellis MH, Spielmeyer W, Gale KR, Rebetzke GJ, Richards RA: **'Perfect' markers for the Rht-B1b and Rht1-D1b dwarfing genes in wheat.** *Theor Appl Genet*, in press.
31. Xiao J, Grandillo LJ, Ahn SN, Yuan L, Tanksley SD, McCouch SR: **Identification of trait-improving quantitative trait loci from a wild rice relative, *Oryza rufipogon*.** *Genetics* 1998, **150:**899-909.
32. Ribaut J-M, Hoisington D: **Marker-assisted selection: new tools and strategy.** *Trends Plant Sci* 1998, 3:236-239.
33. Barr AR, Jefferies SP, Warner P, Moody DB, Chalmers KJ, Langridge P: **Marker-assisted selection theory and practice.** In *Barley Genetics VIII - Proceedings of the 8th International Barley Symposium*. Edited by Logue S. Adelaide, Australia: Dept Plant Science, Adelaide University; 2000: 167-178.

CHAPTER 8

Maize Landraces and Adaptation to Climate Change in Mexico

Jon hellin[1], Mauricio R. Bellon[2], and Sarah J. Hearne[1]

[1]*International Maize and Wheat Improvement Center (CIMMYT), Texcoco, Mexico*

[2]*Bioversity International, Rome, Italy*

ABSTRACT

Mexico is the primary center of origin and diversity for maize (*Zea mays* L.). Farmers grow the crop largely under rain-fed conditions. Mexico is at considerable risk from climate change because of predicted rising temperatures, declining rainfall, and an increase in extreme weather events. Small-scale maize farmers are particularly vulnerable because of their geographical location as well as their limited adaptive capacity. Recommended climate change adaptation strategies include farmers' increased use of heat and drought stress-tolerant maize. Farmer adoption of improved germplasm has been disappointing because of inefficient seed input chains and farmers' preference for landraces for culinary, agronomic, and cultural reasons. Scientists have tended to overlook the fact that maize landraces have a critical role to play in climate change adaptation. Landraces may already exist that are appropriate for predicted climates. Furthermore, within the primary gene pool of maize and its wild relatives there exists unexploited genetic diversity for novel traits and alleles that can be used for breeding new high yielding and stress-tolerant cultivars. The breeding component of adaptation strategies should focus more on improving farmers' landraces. The desired result would be a segmented maize seed sector characterized by both (improved) landraces and improved maize varieties. The public and private sector could continue to provide farmers with improved maize varieties and different actors, including farmers themselves, would generate seed of improved landraces for sale and/or exchange.

Keywords: climate change adaptation, seed systems, small-scale farmers, landraces, wild relatives

1. IMPACT OF CLIMATE CHANGE ON AGRICULTURE

Climate change is likely to lead to increased water scarcity in the coming decades (Hendrix et al. 2007; Lobell et al. 2008) and changes in patterns of precipitation. Climate change is also likely to lead to an increase in temperature. Climate models show a high probability (>90%) that by the end of the 21st century, growing season temperatures will exceed the most extreme seasonal temperatures recorded in the past century (Battisti and Naylor 2009). While an increase in temperature of a few degrees is likely to increase crop yields in temperate areas, in many tropical areas, even minimal increases in temperature may be detrimental to food production (Lobell and Burke 2008). High temperatures reduce crop yields by affecting an array of physiological, biochemical, and molecular processes.

While it is true that farmers have a long record of adapting to the impacts of climate variability, predicted climate change represents an enormous challenge that will test farmers' ability to adapt and improve their livelihoods (Adger et al. 2007). Scientific research points to the negative impacts of climate change on small-scale farmers. This is especially the case in developing countries (Jones and Thornton 2003; Fischer et al. 2005; Morton 2007). A scenario of rising temperatures, declining rainfall, increase in extreme weather events, and shifting pest and disease patterns will lead to more short-term crop failures and long-term production declines (Lobell et al. 2011; Kang and Banga 2013; Chauhan et al. 2014). Until recently, the impacts of climate change had largely been regarded as a problem for the future; however, a key finding of The Fifth Assessment Report (AR5) of the Intergovernmental Panel on Climate Change (IPCC) is that climate change impacts on food security are happening now (Vermeulen 2014).

The resulting decline in global per capita food production will threaten future food security (Brown and Funk 2008). There are also gloomy predictions of how environmental crises will affect global security (e.g.,

Raleigh and Urdal 2007). Through direct effects on livelihoods, climate change may under certain circumstances increase the risk of violent conflict. The environmental problems associated with climate change could also force a greater number of people to migrate to more favorable areas (Adger et al. 2003; Feng et al. 2010), leading in some cases to conflict in receiving areas; the arrival of "environmental migrants or refugees" can burden the economic and resource base of the receiving area, promoting native-migrant competition over limited resources, such as cropland and freshwater (Warner 2010).

Mexico is expected to be among the most negatively affected countries. Small-scale maize farmers are particularly vulnerable because of their geographical location as well as their limited adaptive capacity. Climate change models suggest a drying and warming trend for many parts of Mexico during the main maize season (May-October), with this trend predicted to strengthen with time. Predicted trends include: i) a drying of lowland environments, ii) an expansion of lowland and mid-altitude environments up the elevation gradient, and iii) a substantial reduction of the highland environment (through displacement by warmer mid-altitude type environments and an expansion of areas too dry for optimal maize production). A large percentage of poor rural communities is located in the environments that may experience the trends described above (Bellon et al. 2005), highlighting the challenges that the rural poor—many of whom depend on maize for their livelihoods—face with climate change. Agricultural output in Mexico could decrease by 25.7% by 2080 as a result of climate change (Cline 2007) and one of the most affected crops will be the basic staple, maize.

Crop varieties with increased tolerance to heat and drought stress and resistance to pests and diseases are critical for managing current climatic variability and for adaptation to progressive climate change. The challenge is that improved maize germplasm (and other climate-smart technologies) has not been readily adopted by Mexican smallholder farmers. In this paper, we argue that in the case of Mexico, researchers and development practitioners have tended to overlook the importance of farmers' maize landraces (local varieties) in climate change adaptation initiatives. We suggest that a more strategic approach would entail giving greater prominence to maize landraces. This has implications for policy makers, researchers, and development practitioners, as they engage in climate change adaptation, the purpose of which is to manage effectively potential climate risks during the coming decades as climate changes (Howden et al. 2007).

The rest of this paper is structured as follows. In the next section we present and discuss the theme of climate-smart agricultural technologies. This is followed by Section 3, which details how farmer adoption of these technologies is often less than anticipated. In Section 4, we argue for a broadening of the focus of climate-adaptation strategies by directing more resources at farmers' landraces. Using maize in Mexico as an example, we demonstrate the largely untapped potential that maize landraces have in helping maize adapt to climate change. We outline new crop breeding initiatives that are taking advantage of this potential. In Section 5, we emphasize that efforts also need to be directed at enhancing adaptive capacity rather than the promotion of specific adaptation options *per se.* In Section 6 we conclude.

2. CLIMATE-SMART AGRICULTURAL TECHNOLOGIES

Climate-smart agricultural technologies contribute to: i) an increase in global food security, ii) an enhancement of farmers' ability to adapt to a changing climate, and iii) the mitigation of emissions of greenhouse gases. Key technologies and practices include climate-resilient germplasm and sustainable land-management practices.

Improved crop varieties are a key output of agricultural research and have contributed to significant increases in agricultural production and productivity (Pingali 2012). Scientific crop breeding will continue to play a critical role in meeting the challenge of increasing food production in the face of climate change. The development and dissemination of improved germplasm have the potential to offset some of the yield losses linked to climate change. Crop varieties with increased tolerance to abiotic stresses, including heat and drought stress, will play an important role in managing current climatic variability and adapting to progressive climate change (Cooper et al. 2008; Fedoroff et al 2010). Research is required to identify traits associated with combined heat and drought tolerance, and the development of improved germplasm for high-temperature, water-limited environments. The development of climate-resilient germplasm is possible through a combination of conventional, molecular and, in some cases, transgenic breeding approaches (Cairns et al. 2013).

Climate change will be especially detrimental to crop production in cropping systems where soils have been degraded to an extent that they no longer provide adequate water-holding capacity to buffer crops against drought and heat stress. These effects will be most severe if irrigation is not available to compensate for decreased rainfall or to mitigate the effects of elevated temperatures. Improving genetic adaptation to heat or drought stress alone will not address these problems; there is also a need for complementary agronomic interventions (Hobbs and Govaerts 2010).

Agronomic practices involving reduced or zero tillage, enhanced surface retention of crop residues (mulches), and economically viable crop rotations and diversification, e.g., conservation agriculture can contribute to cropping environments that maximize expression of crop genetic potential, buffer crops against erratic weather, and contribute to climate change mitigation. At the same time, trade-offs exist, for example because of the use of crop residues for feed or fuel (Hellin et al. 2013). These trade-offs need to be addressed and, if necessary, appropriate context-specific solutions must be developed to be not only environmentally but also socially and economically sustainable.

3. FARMER (NON)ADOPTION OF RECOMMENDED TECHNOLOGIES

Climate-smart agricultural technologies are just the latest in a long list of technologies that researchers and development practitioners have developed during the past decades and subsequently promoted in rural areas. The salutary truth is that the benefits from these technological innovations have often not reached the majority of resource-poor farmers cultivating marginal lands because farmers have not readily adopted them. The reasons behind farmer (non) adoption of technologies are complex, but previous research on farmers' reluctance to adopt soil and water technologies in the 1980s (see Hudson 1991). and more recently conservation agriculture (Giller et al. 2009; Erenstein et al. 2012; Andersson and D'Souza 2014) sheds much light on this complexity.

An important factor in the non-adoption of climate-smart technologies has been rural labor shortages (Zimmerer 1993). Many farmers depend both upon production from their land and upon off-farm income-generating activities. This is very much the case in many parts of Mexico. This has farreaching implications for the availability of labor at different times of year and can determine farmers' acceptance of practices such as conservation agricultural systems, especially if farmers are unable to purchase labor-saving technologies, such as herbicides to control weeds (Giller et al. 2009).

A major challenge to farmer acceptance of these technologies is that they are knowledge-intensive. While agricultural extension, education, and training can help many farmers maximize the potential of their productive assets through adoption of these technologies, their promotion has coincided with deep cuts to publicly funded extension services in the developing world (Ajieh et al. 2008). The breakdown of classical publicly funded agricultural research and extension services means that these services are now unable to address the needs of farmers living in marginal environments. In the majority of cases, the private sector has proven incapable of replacing previous state services because of high transaction costs, dispersed clientele, and low (or non-existent) profits (Muyanga and Jayne 2008). The absence of relevant and competent extension provision leads to low technology adoption rates.

As for crop breeding, efforts have tended to focus on producing highyielding germplasm. Despite decades of maize breeding and the promotion of improved maize varieties, however, the majority of smallholder maize farmers in rain-fed areas of Mexico continue to use local maize varieties (Barkin 2002). It is widely accepted that farmers maintain crop diversity for social, economic, or cultural purposes, and/or when local varieties show an agronomic performance superior to that of improved varieties (Bellon 2004). Market-related issues in both input and output chains can influence farmers' propensity to adopt improved maize varieties. On the input side, bottlenecks exist in the value chains, which impede farmers' access to seed. On the output side, quality and scale-related barriers also exist, inhibiting the acceptance of farmers' maize in industrial maize markets (Keleman et al. 2013).

Farmers may prefer local maize varieties for culinary and cultural reasons (Brush and Perales 2007; Bellon and Hellin 2011). Perales et al. (2005) documented the importance of culture in determining maize diversity in Mexico. The authors found a significant association between linguistic differentiation and morphological differentiation of maize varieties among Maya-speaking communities in the highlands of the southern Mexican state of Chiapas. There are also differences in preferences between women and men on account of their reproductive and productive roles; women set priorities toward food security and thus tend to favor varieties that are palatable, nutritious, and meet processing and storage requirements. Women can also generate income from the artisanal processing and sale of traditional maize products (Keleman and Hellin 2009).

Adoption studies related to smallholder production systems have shown risk to be an important component in farmers' decision making. From a farmer's perspective, it may well be too risky to purchase seed of an unknown variety if there is a chance that drought and/or frosts are going to have an adverse impact on crop yields. The result is that farmers often trust local varieties, considering farmer-saved seed to be a "known quantity" while fearing that unfamiliar seed would perform in unexpected ways (Arellano Hernández and Arraiga Jordán 2001; Badstue et al. 2007). This may well be a very rational decision; improved maize varieties, often developed on research stations, do not necessarily perform well under farmers' conditions. For example, in Mexico, a study that examined the relationship between maize research and rural poverty showed that maize production was largely coincident with areas of rural poverty; however, it also found a disparity in the locations of trials used for testing improved maize germplasm and where the poor lived, suggesting that the potential for spillovers from germplasm tested in those sites to poor farmers may be limited (Bellon et al. 2005). Improved varieties that crop breeders identify as superior to landraces under experimental conditions may actually yield substantially less under farmers' conditions because of genotype-by-environment interactions that remain undetected in the data from experimental plots (Keleman et al. 2013).

4. BROADENING THE FOCUS OF CLIMATE CHANGE-ADAPTATION STRATEGIES

4.1 Maize Landraces

Landraces have been defined as dynamic populations of a cultivated plant with a historical origin, distinct identity, often genetically diverse and locally adapted, and associated with a set of farmers' practices of seed selection and field management (Camacho Villa et al. 2005). The structure and dynamics of these landraces are the result of both natural and human selection (Bellon and van Etten 2013). Observers have posited that farmers' varieties will be lost in the face of technological change, with modern varieties eventually replacing landraces wholesale (c.f. Harlan 1975). Detailed, ground-level studies (Brush et al. 1992; Chambers et al. 2007; Bellon and Hellin 2011) have both highlighted the threats to farmers' local varieties and also their resilience. In Mexico, 59 well-defined races (e.g., morphologically and genetically distinct sub-populations) of maize have been described (Sanchez et al. 2000; Ureta et al. 2012) and countless varieties exist at the farmer level.

Farmers' maintenance of local maize varieties may have a very important role to play in climate change adaptation in Mexico (Bellon et al. 2011). In many parts of Mexico, small-scale maize farmers recycle seed either by saving their seed from the previous harvest and/or obtaining it from fellow farmers (Badstue et al. 2007; Dyer and Taylor 2008). The vast majority of these farmers operate under rain-fed conditions, and climate is one of the most important risk factors in their agricultural systems. Maize landraces in Mexico show remarkable diversity and climatic adaptability, growing from arid to humid environments and from temperate to tropical environments. Ruiz Corral et al. (2008) found a very high level of variation among and within 41 Mexican maize races for climate adaptation and ecological descriptors. The general overall climatic ranges for maize included 0 to 2900 m altitude, 11.3° to 26.6°C annual mean temperature, 12.0° to 29.1°C growing season mean temperature, and 426 to 4245 mm annual rainfall (Ruiz Corral et al. 2008).

Bioclimatic modeling of the impact of climate change on the distribution of maize races in the future, while showing, on the aggregate, significant reductions in their potential distribution, also point to contrasting responses by race, showing no major shifts in areas with high race richness, highlighting the important role of race diversity in buffering the effects of climate change (Ureta et al. 2012). Providing farmers with access to more diversity of landraces can, hence, be an important strategy to mitigate the negative impacts of climate change on their livelihoods.

In some parts of Mexico, there may already exist crop germplasm in the form of farmers' local maize varieties that are appropriate for predicted climates (Bellon et al. 2011; Mercer et al. 2012). Farmers' local maize varieties should be part of the arsenal of climate-smart technologies and practices. In some cases traditional seed systems may be able to provide farmers with landraces suitable for agro-ecological conditions under predicted climate change, after all traditional seed systems are not closed or static but open and dynamic with seed coming in and out of the systems, with farmers experimenting and incorporating new seeds (Bellon and van Etten 2013). However, if suitable germplasm does exist, those farmers who are going to need it may not currently source it, suggesting that farmers are likely to need to source seed from outside their traditional geographical ranges. This may entail the development of new social networks (Bellon et al. 2011).

In practice, broadening the geographical reach of farmers' seed networks could be achieved through exchange visits; linking farmer groups in different locations; fostering the exchange of germplasm, knowledge, and practices among them; and encouraging cross-community experimentation with local and introduced crop varieties (Bellon and van Etten 2013). Landrace populations may also be able to 'keep up' with climate change because of farmer selection of climate-adapted traits among landraces (Mercer and Perales 2010; Mercer et al. 2012). Furthermore, maize's wild relatives, *Tripsacum* species and *Teosinte* species and subspecies, are expected to respond differently to climate change because of different environments that they currently inhabit (Ureta et al. 2012).

Furthermore, within the primary gene pool of maize and its wild relatives there exists unexploited genetic diversity for novel traits and alleles (Ortiz et al. 2009) that can be used for breeding new high-yielding and stress-tolerant cultivars (Jarvis et al. 2008; Ruiz Corral et al. 2008; Chauhan et al. 2014). Formal breeding efforts as part of climate change-adaptation strategies will likely depend heavily on the diverse crop genetic resources that farmers have helped develop across centuries (Burke et al. 2009).

Hence, along with farmers' use of maize landraces, many farmers may be able to acquire improved germplasm with new adaptive traits. Thus, formal crop breeding still has a very important role to play in the development and diffusion of improved varieties with appropriate traits to cope with climate change. As Bellon (2009) notes, having a diversity of 'winning' (adaptive) combinations of genes and traits that are constantly being updated in response to changing situations and new knowledge should allow us to cope and adapt better to change. This is precisely the idea of the option value of the evolutionary services that on-farm conservation delivers (Bellon and van Etten 2013) and is likely to be critical to climate change-adaptation strategies.

Small-scale maize farmers' adoption of improved germplasm has been minimal and, hence, public and private sector interventions are required to foster a well-functioning seed system that links formal and informal seed systems to enable farmers to access climate-adapted seed. The desired result would be a segmented maize seed sector characterized by both (improved) landraces and improved maize varieties. The public and private sector would provide the latter and different actors, including farmers themselves, would generate seed of landraces for sale and/or exchange.

4.2 Landraces as a Breeding Resource

In 1998, when the Food and Agriculture Organization (FAO) of the United Nations published the State of the World Plant Genetic Resources, there were 261,584 accessions of maize held in gene banks around the world. Of these, the number of unique landraces held in major collections worldwide ranged from 58,000 to 80,000 (Taba and Goodman 2007). The international gene bank at the International Maize and Wheat Improvement Center (CIMMYT) in Mexico holds in excess of 27,500 accessions of maize, covering primarily landraces (24,463; 97% of these from the Americas and Caribbean) but also breeding lines, teosintes, *Tripsacum* sp., and some pools and pre-breeding populations. Within Mexico alone, approximately 18,000 further Mexican maize accessions are held by regional gene banks, universities, and at the new National Genetic Resource Centre within Mexico. These resources are often called an "ark" representing a broad swath of maize genetic diversity, which may offer a reserve of native alleles and genes of importance to crop production now and in the future.

Even under the more moderate climate change scenarios, predicted climate shifts within Mexico are significant; during the past 100 years Mexico has experienced between 0.5 to 0.9°C of warming (Verhulst et al. 2012), which represents 100 generations of maize or 0.009°C per generation. The moderate predicted changes in the next 50 years in Mexico equate to a 0.08°C warming per generation in some maize-growing regions; nine times that previously experienced.

Compared with hybrids, landrace populations have a wider plasticity in the range of environments a particular population can successfully grow in and still maintain yield. This is one of the reasons they are so valued by farmers. This adaptive feature is a reflection of the heterogeneous nature of the populations. However, the ability of landraces to adapt to the rapidly changing climate, as suggested by climate models, is a point of concern. The ex-situ landrace collections contain many materials that have evolved in environments that have a propensity for stress, such as drought, heat, or frost, during the growing season. For example, in the international maize collection at CIMMYT, there are in excess of 3,900 landrace accessions (more than half of which come from Mexico) that originate from locations where the monthly average daily maximum temperature during the flowering period (based on 1901–2009 data) exceeds 30°C. Such landraces offer a unique source of germplasm, genes, and alleles for "climate change tolerance," which may be of benefit to crop improvement activities.

Existing breeding pools contain some genetic variation that can be and is being harnessed. The continuing genetic gain seen in relatively closed private-sector breeding programs is testament to this. However, many other useful genetic characters present in landraces will not have passed through the bottleneck of genetic selection. With increasing environmental variability, heterogeneity in traits deployed for adaptation is important to ensure durability in the long term. The use of non-elite germplasm in plant breeding is limited for very valid reasons, including issues such as linkage drag—desirable traits may be associated with undesirable ones in this type of germplasm; difficulty in making crosses; general adaptation difficulties—non-elite germplasm may be locally adapted and may not perform well outside very specific environmental niches; funding limitations—pre-breeding, the process of making non-elite germplasm suitable for use in breeding, can be costly and takes time; and the relatively short-term goals, from a breeding perspective, of many breeding projects. Indeed, many races of maize have been under-investigated as sources of beneficial genes with a focus being placed on races more representative of an "ideal" plant type, e.g., the *Tuxpeño* race in the case of maize.

Concerted efforts are now focusing on the characterization of landraces at both genetic and phenotypic levels to better inform and enable selection and use of these important resources in plant breeding. Examples include the Seeds of Discovery Initiative funded by Mexico's Ministry of Agriculture, Livestock, Rural Development, Fisheries and Food (SAGARPA) (http://

seedsofdiscovery.org/seed/about/), and the French Government and industry-funded Amaizing Project (http://www.amaizing.fr/ index. php).

While the most direct and simple approach to providing more climate - resilient materials to resource-poor farmers may appear to be simply a re-distribution of those landraces that may be tolerant to changing climates, this may not be productive because of broad and local adaptation issues (Mercer et al. 2012) and local culinary preferences. Indeed, it is not only farmers cultivating landraces who face the challenge of increasing climate variability; improvement of all germplasm, including hybrids, open-pollinated varieties (OPVs), and landraces, for climate resilience is required as a tool in a suite of climate-adaptation strategies.

Using conventional crossing, coupled with modern techniques, such as genomics-assisted breeding and doubled-haploid technology, it is possible to incorporate some important novel traits more rapidly, and, in a more targeted manner (reducing linkage drag) from landraces into more breeder-friendly germplasm. This process, termed pre-breeding, does not seek to produce a finished product but rather a suite of easy-to-use donor materials with associated information that can be deployed in the improvement of all germplasm types from landraces to elite breeding lines. Multiple genetic sources for adaptation characteristics of interest should be explored to enable the development of landrace varieties (and panels of elite varieties) that have (within landraces and across elite varieties) high degrees of heterozygosity and heterogeneity for climate-resilience traits. Such materials better enable buffering capacity in times of stress as demonstrated in studies of the climate-durable crop pearl millet in Africa (Haussmann et al. 2012).

For those organizations involved in crop breeding, the challenge remains of ensuring that beneficial genes and alleles are incorporated into farmer-appropriate genetic backgrounds and that the resulting varieties are evaluated in field trials that best represent farmer conditions and management practices. Resource-poor maize farmers in Mexico tend to grow landraces for a number of reasons, including cultural and culinary preference, and limited access to improved materials. Bellon et al. (2005) documented that in Mexico where formal CIMMYT maize trials had been carried out, only 7 of the 158 sites used were actually in high probability rural poverty areas and only 16 of the 158 were within extremely poor municipalities. The disparity in the locations of the trials largely represents the focus on managed stress trials, the ease of travel to trial locations, and most importantly the evaluation of lines, improved hybrids, and OPVs, which are the mainstay of international breeding efforts at CIMMYT. The importance of improving the genetic resilience of landraces has, however, been recognized and many national breeders in Mexico have components of their breeding programs that look at landraces. Given the range and diversity of landraces and farming niches, however, it is going to be a challenge to scale out improved landraces in a similar way to what has been achieved with OPVs and hybrids.

The development of suitable trait donor germplasm is an essential step towards facilitating the development of climate-resilient germplasm across the whole spectrum of variety types. Strategies, such as participatory plant breeding, need to be discussed and formulated to identify the most effective way to enhance the incorporation of useful genetic variation into the landraces currently used by farmers. These will involve trained plant breeders and active farmer participation. Farmers have been practicing selection for millennia and as long as they are provided suitable donor germplasm, they should be quite capable of making effective selections in their own fields for desired traits, including tolerance to climatic variability.

Considering more formal breeding approaches, while in many arenas the notion is unpopular, some prioritization of landraces will need to be considered to ensure that some landraces are genetically protected against future climatic variability. Irrespective of strategies developed for landrace improvement, it remains that, as a follow up to the development of donor germplasm and some improved varieties, breeders and extension agents should consider re-emphasis on the evaluation of farmer-appropriate technologies and technology packages in locations where target farmers live and work to capitalize on potential for spillovers from germplasm tested in those sites.

4.3 Adaptation Capacity

There also remains the challenge of enhancing adaptive capacity to climate change. Eakin and Lemos (2006) posit that the high uncertainties in climate-change scenarios meant that there was growing interest in improving adaptive capacity as an alternative focus of policy efforts, rather than the promotion of specific adaptation options *per se*. There is an expectation that nation-states will improve their capacity and that of their citizens to adapt to climate change (Eakin and Lemos 2006). Hence, while specific adaptation technologies (such as climate-adapted germplasm) and practices are critical, there is a need to direct more attention at the institutional changes that empower states to design and implement policy to increase adaptive capacity of different actors, including farmers. As Thornton et al. (2009) write, in place of defining large development domains for identifying and implementing adaptation options, what is needed are localized, community-based efforts to increase local adaptive capacity.

A systems approach is needed in which innovation is the result of a process of networking, interactive learning, and negotiation among a heterogeneous set of actors (Klerkx et al. 2009). This very much applies to climate-change adaptation because "the effectiveness of these adaptations for mitigating future sensitivity to climatic risk will be strongly influenced by the ways in which policy enables or inhibits households' capacity to address climatic challenges" (Eakin 2005). This is largely because a households' management of climatic risk is a function of numerous factors, including education, wealth, natural resources, social organization, and institutional relationships (Eakin 2005). This calls for increased farmer participation in stakeholder design of climate-adaptation strategies (Eakin et al. 2007) and an increase in various dimensions of social capital, including peer groups, networks, and collective action (Meinke et al. 2006). Strengthening the social relations of maize production and seed exchange among farmers are fundamental to a successful adaptation strategy (Mercer et al. 2012).

Public- and private-supported extension programs can play a key role in the design of more appropriate adaptation strategies by transferring technology, facilitating interaction, building capacity among farmers, and encouraging farmers to form their own networks. Extension services that specifically address climate-change adaptation include disseminating drought-resistant crop varieties; teaching improved management systems; and gathering information to facilitate national research work. The breeding and agronomic research work needs to be supported by other factors, including complementary investments in climate-responsive information and input-delivery systems; and strengthening of institutions to coordinate grain marketing with seed, fertilizer, and credit delivery. The development of reliable seasonal weather forecast, record of reliable weather, and strengthening of early warning system are also crucial for facilitating adaption to climate change.

The above can best be achieved via a judicious mix of public and private service provision in the agricultural sector that also addresses multiple market and government failures in the delivery of technologies, inputs, and services (Cooper et al. 2008). This requires new institutional arrangements and policy instruments to enhance local capacity and stimulate the adoption of improved technologies for adaptation, managing risks, and protection of vulnerable livelihoods. This requires novel, flexible research and extension approaches that differ from those more commonly used by policy makers, donors, researchers, and extension agents (Ekboir et al. 2009). Enhancing the productivity and profitability in marginal areas will require approaches that promote the translation of innovations in plant science into concrete benefits for poor farmers and in ways that support the emergence of agricultural innovation systems as well as respecting and supporting farmers' preferences for landraces; there is much emphasis on the importance of breeding crops for future climatic conditions even though much of the world's farming population still relies on landrace populations rather than the products of formal breeding networks (Mercer and Perales 2010). Production is clearly important, but food distribution and exchange also determine food availability while access to food and food utilization are other important components of food security (Ingram et al. 2008).

5. CONCLUSIONS

Climate change threatens current agricultural output and, hence, there is a greater need to enhance agricultural yields and resilience of agro-ecosystems as well as to improve the livelihoods of farmers. Despite some uncertainties on the spatially differentiated impact of climate change on agricultural production, there is little doubt that germplasm, more suited to future climates, is critical along with improved agronomic and crop-management practices. Mesoamerica is the center of origin and diversity for maize, and landraces may already exist that possess traits that enable adaptation to predicted climates.

Formal crop breeding using landrace resources has a key role to play in climate change-adaption strategies. Complementary adaptation strategies would include farmers' increased use of climate-adapted maize varieties (with improved tolerance to heat stress, and combined heat and drought stress), coupled with effective soil moisture-conservation techniques. The development and dissemination of climate-responsive germplasm may take several years because the process consists of several steps, including breeding, on-farm testing, release of varieties, and germplasm dissemination. Research efforts need to be directed at facilitating the development of pre-breeding germplasm, germplasm (landrace, OPV, and hybrid) enhancement and varietal release, training seed entrepreneurs, and increasing the provision of foundation seed. Furthermore, widespread farmer uptake of improved climate-adapted maize varieties relies on a functioning and efficient seed sector.

ACKNOWLEDGEMENTS

The authors are grateful to an anonymous reviewer and the Reviewing Editor for invaluable comments on an earlier version of this paper. The views expressed in this paper do not necessarily reflect the views of the authors' institution or the donors.

FUNDING

We acknowledge support provided by the CGIAR Research Program on Climate Change, Agriculture and Food Security (CCAFS), and the Mexican Government's *Secretaría de Agricultura, Ganadería, Desarrollo Rural, Pesca y Alimentación* (SAGARPA) via the Sustainable Modernization of Traditional Agriculture (MasAgro) initiative.

REFERENCES

1. Adger, W. N., S. Agrawala, M. Mirza, C. Conde, K. O'Brien, J. Pulhin, R. Pulwarty, B. Smit, and K. Takahashi. 2007. Assessment of adaptation practices, options, constraints, and capacity. InClimate Change 2007: Impacts, Adaptation and Vulnerability (Contribution of Working Group II to the Fourth Assessment Report of the Intergovernmental Panel on Climate Change), edited by M. L. Parry, O. F. Canziani, J. P. Palutikof, P. J. van der Linden, and C. E. Hanson, 717–743. Cambridge, UK: Cambridge University Press.
2. Adger, W. N., S. Huq, K. Brown, D. Conway, and M. Hulme. 2003. Adaptation to climate change in the developing world. *Prog. Develop. Stud.* 3(3): 179–195.
3. Ajieh, P. C., A. E. Agwu, and A. C. Anyanwu. 2008. Constraints to privatization and commercialization of agricultural extension services as perceived by extension professionals and farmers. *Afr. J. Agric. Res.* 3(5): 343–347.
4. Andersson, J. A., and S. D'Souza. 2014. From adoption claims to understanding farmers and contexts: A literature review of Conservation Agriculture (CA) adoption among smallholder farmers in southern Africa. *Agric. Ecosyst. Environ.* 187:116–132.
5. Arellano Hernández, A., and C. Arriaga Jordán. 2001. Why improved maize (Zea Mays) varieties are utopias in the highlands of central Mexico. *Convergencia* 8(25): 255–276.
6. Barkin, D. 2002. The reconstruction of a modern Mexican peasantry. *J. Peasant Stud.* 30(1): 73–90.
7. Battisti, D. S., and R. L. Naylor. 2009. Historical warnings of future food insecurity with unprecedented seasonal heat. *Science* 323:240–244.
8. Badstue, L. B., M. R. Bellon, J. Berthaud, A. Ramírez, D. Flores, and X. Juarez. 2007. The dynamics of seed flow among small-scale maize farmers in the Central Valleys of Oaxaca, Mexico. *World Develop.* 35(9): 1,579–1,593.
9. Bellon, M. R., and J. Hellin. 2011. Planting hybrids, keeping landraces: agricultural modernization and tradition among small-scale maize farmers in Chiapas, Mexico. World Develop. 39(8): 1,434–1,443.
10. Bellon, M. R., and J. van Etten. 2013. Climate change and on-farm conservation of crop landraces in centres of diversity. In Plant genetic resources and climate change, edited by M. Jackson, B. Ford-Lloyd, M. L. Parry. New York, NY: CABI Publishing.
11. Bellon, M. R., D. Hodson, and J. Hellin. 2011. Assessing the vulnerability of traditional maize seed systems in Mexico to climate change. *Proc. Nat. Acad. Sci. USA.* 108(33): 13,432–13,437.
12. Bellon, M. R., D. Hodson, D. Bergvinson, D. Beck, E. Martinez-Romero, and Y. Montoya. 2005. Targeting agricultural research to benefit poor farmers: Relating poverty mapping to maize environments in Mexico. *Food Policy* 30(5–6): 476–492.
13. Bellon, M. R. 2009. Do we need crop landraces for the future? Realizing the global option value of in situ conservation. In Agrobiodiversity and economic development, edited by A. Kontoleon, U. Pascual, and M. Smale, 51–61. London, UK: Routledge.
14. Bellon, M. R. 2004. Conceptualizing interventions to support on-farm genetic resource conservation. *World Develop.* 32(1): 159–172.

15. Brown, M. E., and C. C. Funk. 2008. Food security under climate change. *Science* 319(5863): 580–581.
16. Brush, S. B., and H. R. Perales. 2007. A maize landscape: Ethnicity and agro-biodiversity in Chiapas Mexico. *Agric. Ecosyst. Environ.* 121(3): 211–221.
17. Brush, S. B., J. E. Taylor, and M. R. Bellon. 1992. Technology adoption and biological diversity in Andean potato agriculture. *J. Develop. Econ.* 39:365–387.
18. Burke, M. B., D. B. Lobell, and L. Guarino. 2009. Shifts in African crop climates by 2050, and the implications for crop improvement and genetic resources conservation. *Global Environ. Change* 19(3): 317–325.
19. Cairns, J. E., J. Hellin, K. Sonder, J. L. Araus, J. F. MacRobert, C. Thierfelder, and B. M. Prasanna. 2013. Adapting maize to climate change in sub-Saharan Africa. *Food Security* 5:345–360.
20. Camacho Villa, T. C., N. Maxted, M. A. Scholten, and B. V. Ford-Lloyd. 2005. Defining and identifying crop landraces. *Plant Genet. Res.* 3:373–384.
21. Chambers, K. J., S. B. Brush, M. N. Grote, and P. Gepts. 2007. Describing maize (*Zea mays* L.) landrace persistence in the Bajío of Mexico: A survey of 1940s and 1950s collection Locations. *Econ. Bot.* 61(1): 60–72.
22. Chauhan, B. S., Prabhjyot-Kaur, G. Mahajan, R. K. Randhawa, H. Singh, and M. S. Kang. 2014. Global warming and its possible impact on agriculture in India. In Advances in agronomy, vol. 123, edited by D. L. Sparks, 65–121. Burlington, VT: Academic Press.
23. Cline, W. R. 2007. Global warming and agriculture: impact estimates by country. Washington, D.C., USA: Center for Global Development and Peterson Institute for International Economics.
24. Cooper, P. J. M., J. Dimes, K. P. C. Rao, B. Shapiro, B. Shiferaw, and S. Twomlow. 2008. Coping better with current climatic variability in the rain-fed farming systems of sub-Saharan Africa: An essential first step in adapting to future climate change? *Agric. Ecosyst. Environ.* 126(1–2): 24–35.
25. Dyer, G. A., and J. E. Taylor. 2008. A crop population perspective on maize seed systems in Mexico. *Proc. Nat. Acad. Sci. USA.* 105(2): 470–475.
26. Eakin, H. 2005. Institutional change, climate risk, and rural vulnerability: Cases from Central Mexico. *World Develop.* 33:1,923–1,938.
27. Eakin, H., and M. C. Lemos. 2006. Adaptation and the state: Latin America and the challenge of capacity-building under globalization. *Global Environ. Change* 16(1): 7–18.
28. Eakin, H., V. Magaña, J. Smith, J. L. Moreno, J. M. Martínez, and O. Landavazo. 2007. A stakeholder driven process to reduce vulnerability to climate change in Hermosillo, Sonora, Mexico. *Mitig. Adapt. Strategies Global Change* 12(5): 935–955.
29. Ekboir, J. M., Dutrénit, G., Martínez, G., Torres Vargas, A., and A. Vera-Cruz, A. 2009. Successful organizational learning in the management of agricultural research and innovation: The Mexican Produce Foundations. IFPRI Research Report No. 162. Washington, DC: International Food Policy Research Institute.
30. Erenstein, O., K. Sayre, P. Wall, J. Hellin, and J. Dixon. 2012. Conservation agriculture in maize and wheat based systems in the (sub) tropics: Lessons from adaptation initiatives in South Asia, Mexico and Southern Africa. *J. Sustainable Agric.* 36(2): 180–206.
31. Fedoroff, N. V., D. S. Battisti, R. N. Beachy, P. J. M. Cooper, D. A. Fischhoff, C. N. Hodges, V. C. Knauf, et al. 2010. Radically rethinking agriculture for the 21st Century. *Science* 327:833–834.
32. Feng, S., A. B. Krueger, and M. Oppenheimer. 2011. Linkages among climate change, crop yields, and Mexico–US cross-border migration. *Proc. Nat. Acad. Sci. USA.* 107(32): 14,257–14,262.
33. Fischer, G., M. Shah, F. N. Tubiello, and H. van Velhuizen. 2005. Socio-economic and climate change impacts on agriculture: an integrated assessment 1990-2080. *Philos. Trans. R. Soc. London, Series B* 360(1463): 2067–2083.
34. Giller, K. E., E. Witter, M. Corbeels, and P. Tittonell. 2009. Conservation agriculture and smallholder farming in Africa: The heretics' view. *Field Crops Res.* 114:23–34.
35. Harlan J. 1975. Our vanishing genetic resources. *Science* 188:618–621.

36. Haussmann, B. I. G., H. Fred Rattunde, E. Weltzien -Rattunde, P. S. C. Traoré, K. vom Brocke, H. K. Parzies. 2012. Breeding strategies for adaptation of pearl millet and sorghum to climate variability and change in West Africa. *J. Agron. Crop Sci.* 198:327–339.
37. Hellin, J., O. Erenstein, T. Beuchelt, C. Camacho, and D. Flores. 2013. Maize stover use and sustainable crop production in mixed crop–livestock systems in Mexico. *Field Crops Res.* 153:12–21.
38. Hendrix, C. S, and S. M. Glaser. 2007. Trends and triggers: Climate, climate change and civil conflict in Sub-Saharan Africa. *Pol. Geog.* 26:695–715.
39. Hobbs, P. R., and B. Govaerts. 2010. How conservation agriculture can contribute to buffering climate change. In *Climate Change and Crop Production*, edited by M. P. Reynolds, 151–176. CABI Series in Climate Change. New York, NY: CABI International.
40. Howden, S. M., Soussana, J.-F., Tubiello, F. N., Chhetri, N., Dunlop, M., Meinke, H. 2007). Adapting agriculture to climate change. *Proc. Nat. Acad. Sci. USA.* 104(50): 19691–6.
41. Hudson, N.W. 1991. A study of the reasons for success and failure of soil conservation projects. Soils Bulletin 64. Rome, Italy: Food and Agriculture Organization of the United Nations.
42. Ingram, J. S. I., P. J. Gregory, and A. M. Izac. 2008. The role of agronomic research in climate change and food security policy. *Agric. Ecosyst. Environ.* 126(1–2): 4–12.
43. Jarvis, A., A. Lane, and R. J. Hijmans. 2008. The effect of climate change on crop wild relatives. *Agric. Ecosyst. Environ.* 126(1–2): 13–23.
44. Jones, P., and P. Thornton. 2003. The potential impacts of climate change on maize production in Africa and Latin America in 2055. *Global Environ. Change* 13(1): 51–59.
45. Kang, M. S., and S. S. Banga. 2013. Global agriculture and climate change. *J. Crop Improv.* 27(6): 667–692. doi:10.1080/15427528.2013.845051.
46. Keleman, A., and J. Hellin. 2009. Specialty Maize Varieties in Mexico: A Case Study in Market-Driven Agro-Biodiversity Conservation. *J. Latin Am. Geog.* 8(2):147–174.
47. Keleman, A., J. Hellin, and D. Flores. 2013. Diverse varieties and diverse markets: scale-related maize "profitability crossover" in the central Mexican Highlands. *Human Ecol.* 41:683–705.
48. Klerkx, L., A. Hall, and C. Leeuwis. 2009. Strengthening agricultural innovation capacity: are innovation brokers the answer? *Int. J. Agric. Res. Gov. Ecol.* 8(5/6): 409–438.
49. Lobell, D. B., M. Bänziger, C. Magorokosho, and B. Vivek. 2011. Nonlinear heat effects on African maize as evidenced by historical yield trials. *Nature Clim. Change* 1:42–45.
50. Lobell, D. B., and M. B. Burke. 2008. Why are agricultural impacts of climate change so uncertain? The importance of temperature relative to precipitation. *Environ. Res. Lett.* 3(3): 1–8.
51. Meinke, H., R. Nelson, P. Kokic, R. Stone, and R. Selvaraju. 2006. Actionable climate knowledge: from analysis to synthesis. *Climate Res.* 33:101–110.
52. Mercer, K. L., and H. R. Perales. 2010. Evolutionary response of landraces to climate change in centers of crop diversity. *Evolut. Appl.* 3(5–6): 480–493.
53. Mercer, K. L., H. R. Perales, and J. D. Wainwright. 2012. Climate change and the transgenic adaptation strategy: Smallholder livelihoods, climate justice, and maize landraces in Mexico. *Global Environ. Change* 22(2): 495–504.
54. Morton, J. F. 2007. The impact of climate change on smallholder and subsistence agriculture. *Proc. Nat. Acad. Sci. USA.* 104:19,680–19,685.
55. Muyanga, M., and T. S. Jayne. 2008. Private agricultural extension system in Kenya: Practice and policy lessons. *J. Agric. Edu. Exten.* 14(2): 111–124.
56. Ortiz, R., S. Taba, V. H. C. Tovar, M. Mezzalama, Y. Xu, J. Yan, and J. H. Crouch. 2009. Conserving and enhancing maize genetic resources as global public goods: A perspective from CIMMYT. *Crop Sci.* 50:13–28.
57. Perales, H. R., B. F. Benz, and S. B. Brush. 2005. Maize diversity and ethno-linguistic diversity in Chiapas, Mexico. *Proc. Nat. Acad. Sci. USA.* 102(3): 949–954.
58. Pingali, P. L. 2012. Green Revolution: Impacts, limits, and the path ahead. *Proc. Nat. Acad. Sci. USA.* 109(31): 12,302–12,308.
59. Raleigh, C., and H. Urdal. 2007. Climate change, environmental degradation and armed conflict. *Poly. Geogr.* 26(2007): 674–694

60. Ruiz Corral, J. A., N. Durán Puga, J. D. J. Sánchez González, J. Ron Parra, D. R. González Eguiarte, J. B. Holland, and G. Medina García. 2008. Climatic adaptation and ecological descriptors of 42 Mexican maize races. *Crop Sci.* 48(4): 1502–1512.
61. Sanchez, J. J. G., M. M. Goodman, and C. W. Stuber. 2000. Isoenzymatic and morphological diversity in the races of maize in Mexico. *Econ. Bot.* 54(1): 43–59.
62. Taba, S., and M. Goodman. 2007. Background on the development of the "Global Strategy for the Ex Situ Conservation and Utilization of Maize Germplasm." Available at http://www.croptrust.org/documents/cropstrategies/maize.pdf
63. Thornton, P. K., P. G. Jones, G. Alagarswamy, and J. Andresen. 2009. Spatial variation of crop yield response to climate change in East Africa. *Global Environ. Change* 19(1): 54–65.
64. Ureta, C., E. Martínez-Meyer, H. R. Perales, amd E. R. Álvarez-Buylla. 2012. Projecting the effects of climate change on the distribution of maize races and their wild relatives in Mexico. *Global Change Biol.* 18(3): 1,073–1,082.
65. Verhulst, N., B. Govaerts, K. D. Sayre, K. Sonder, R. Romero-Perezgrovas, M. Mezzalama, and L. Dendoovenc. 2012. Conservation agriculture as a means to mitigate and adapt to climate change, a case study from Mexico. In Climate change mitigation and agriculture, edited by E. Wollenberg, A. Nihart, M.-L. Tapio-Biström, and M. Grieg-Gran, 287–300. Oxford, UK: Earthscan.
66. Vermeulen, S. J. 2014. Climate change, food security and small-scale producers. CCAFS InfoBrief. Copenhagen, Denmark: CGIAR Research Program on Climate Change, Agriculture and Food Security (CCAFS).
67. Warner, K. 2010. Global environmental change and migration: Governance challenges. *Global Environ. Change* 20(3): 402–413.
68. Zimmerer, K. 1993. Soil erosion and labor shortages in the Andes with special reference to Bolivia 1953-91: Implications for 'conservation-with-development.' *World Develop.* 21(10): 1,659–1,675.

CHAPTER 9

Crop Management as an Agricultural Adaptation to Climate Change in Early Modern Era: A Comparative Study of Eastern and Western Europe

Qing Pei [1,*], David D. Zhang [2], Harry F. Lee [2] and Guodong Li [3]

[1] *Department of Social Sciences, The Education University of Hong Kong, Hong Kong*

[2] *Department of Geography and International Centre for China Development Study, The University of Hong Kong, Hong Kong*

[3] *Department of Statistics and Actuarial Science, The University of Hong Kong, Hong Kong*

AUTHOR CONTRIBUTIONS

Q.P. conceived and designed the experiments; Q.P. performed the experiments; Q.P., D.D.Z. and H.F.L. analyzed the data; Q.P., D.D.Z. and G.D.L. contributed reagents/materials/analysis tools; Q.P. and H.F.L. wrote the paper.

ABSTRACT

Effective adaptation determines agricultural vulnerability to climate change, especially in the pre-industrial era. Crop management as an agricultural adaptation to climate change in recent human history, however, has rarely been systematically evaluated. Using Europe as our study area, we statistically compared yield ratio of wheat, rye, barley, and oats (an important performance indicator of an agrarian economy) between Eastern and Western Europe in AD 1500–1800. In particular, a statistical comparison was made of crop yield ratio in the two regions during the warm agricultural recovery period AD 1700–1800. The general trend of crop yield in Eastern and Western Europe basically followed the alternation of climatic epochs, in which the extreme cooling period in AD 1560–1660 drastically reduced the crop yield ratio. The yield ratio of rye in Eastern and Western Europe was very similar throughout the entire study period. However, the yield ratio of wheat, barley, and oats showed different patterns in the two regions and increased drastically in Western Europe in the warm agricultural recovery period, which might have contributed to rapid socio-economic development in Western Europe and eventually the East–West Divide in Europe in the following centuries.

Keywords: climate change; crop yield ratio; non-parametric analysis; Eastern and Western Europe; early modern era

1. INTRODUCTION

Global climate change has exerted a significant influence on human life, covering almost every aspect of our societies [1]. Agricultural production is regarded as the sector most sensitive to climate change because it is highly dependent on climatic conditions; this was especially true in the early modern era when the technological level was low [2]. The interaction between climate change and agricultural production has long been conceived as an arena where only a minimum level of technology is available to overcome natural impacts. Deteriorating climate could drastically reduce agricultural production within the context of the early modern period. Given that the agricultural sector was the basis of agrarian economies in the pre-industrial era, climate-induced agricultural shrinkage often translated into various socio-economic crises during the time, an observation supported by empirical evidence [3,4].

Crop yield ratio has been highlighted in historical studies and employed in examining agricultural productivity [5]. In agrarian societies, crop yield ratio not only reflects their social production capacity, but also indicates their macro- and micro-economic performance [6]. As the implication of analyzing crop yield ratio has been increasingly realized, a growing number of studies have examined the socio-economic impact of climate change in pre-industrial Europe via the use of crop yield ratio, and those studies are often conducted in a quantitative manner [7,8,9]. In addition, crop yield ratio has also been employed to explore the effectiveness of agricultural adaptation in pre-industrial Europe [10].

Nevertheless, the analysis of crop yield ratio has multiple constraints:

First, although crop yield ratio has been employed to investigate the effectiveness of labor and land input in boosting agricultural production [10], various types of crops (i.e., wheat, rye, barley, and oats) are treated as an aggregate, while their combination has not been considered. Wheat, rye, barley and oats are categorized as "small grain" crops, and between 80 and 95 percent of the early modern population directly depended on them [11,12]. In recent human history, bumper crops contributed positively to the agricultural sector and facilitated economic recovery, while people chose bumper crops according to their biological characteristics [13]. In this regard, historical combination/selection among these four crops deserves extra attention. Even in the present day, altering crop combination is considered to be an important adaption to climate change [14]. Therefore, historical choices of crops should be further investigated.

Second, pre-industrial Europe is often taken as a spatial aggregate when examining the link between climate change and agricultural production [7,8,10,15]. However, agricultural adaption to climate change at the sub-continental level is insufficiently explored, though there are some studies on the social impacts of climate change at the sub-continental level [16,17]. Hence, the regional dynamics of crop management in response to climate change in early modern Europe are not well understood.

Third, previous studies focused on the disastrous impact of extreme cooling on agricultural production in Europe during the Little Ice Age [18,19], while recovery of agricultural production after the cooling was neglected. As part of a complete economic cycle, the recovery stage of an agrarian economy deserves as much attention as its crisis stage.

In this study, we sought to address the above issues. Fine-grained paleo-climate and historical crop yield data together with statistical methods were employed. We focused on the assessment of agricultural production efficiency under major climate fluctuations in early modern Europe via the scrutiny of various crop yield ratios. Through regional comparisons, the differences of crop management efficiency and agricultural recovery capacity were revealed.

2. MATERIALS AND METHODS

We delineated our study area as Europe as per current administrative boundaries. Europe is one of the original major grain centers in the world and has a long history of crop cultivation, especially for "small grain" crops [20]. In addition, Europe has detailed records on climate history [21]. Data of crop yield ratios are also available, which have been repeatedly employed in previous studies [7,8,9,10].

Scale is both a methodological issue inherent to observation and an objective characteristic of complex interactions within and among social and natural processes [22]. This study was mainly conducted at the sub-continental scale. We followed the Köppen Climate Classification to divide Europe into two macro regions, namely Eastern Europe and Western Europe (Figure 1) [23,24]. The classification is based on the concept that native vegetation is the best expression of climate [25]. Boundaries of climate zone are delineated according to the types of vegetation cover. This study focuses on the climate-agriculture nexus. As the classification considers the connection between climate and vegetation, we followed it to boil down our study area into sub-regions (i.e., Eastern and Western Europe).

A major methodological difficulty, however, persists. Early history is less precise than recent history, and written records are minimal and insufficient. Records in earlier eras are more fragmented than those in modern times. Nowadays, the agrarian economy is not only influenced by agricultural production, but also by industrial goods and international trade [26]. Thus, studying the pure relationship between climate fluctuations and the agrarian economy is difficult. Consequently, the study has been confined to AD 1500–1800. Our study time span overlaps with the Little Ice Age (between the sixteenth and mid-nineteenth centuries), as dated in most publications [27]. In addition, it covers the "General Crisis of the Seventeenth Century in Europe," which is described as a period of widespread social disorder characterized by economic distress, political unrest, and population decline.

During this period, agricultural shrinkage is one of the most important aspects of the General Crisis. Widespread agricultural shrinkage has been reported, covering a number of countries such as England [19], Poland [28], Germany [29], France, and Russia [30]. In addition, social instability triggered and fueled by economic (agricultural) failure, happened across Europe [31]. Many conflicts and wars occurred in Europe during the crisis period, including the Thirty Years War in Germany, the Wars of the Three Kingdoms, the Protectorate and the Glorious Revolution in Britain and Ireland, internal revolts in the Ottoman Empire, the climax of the

Dutch Revolt and so on. Briefly, "General Crisis of the Seventeenth Century in Europe" is characterized by widespread socio-economic and political chaos.

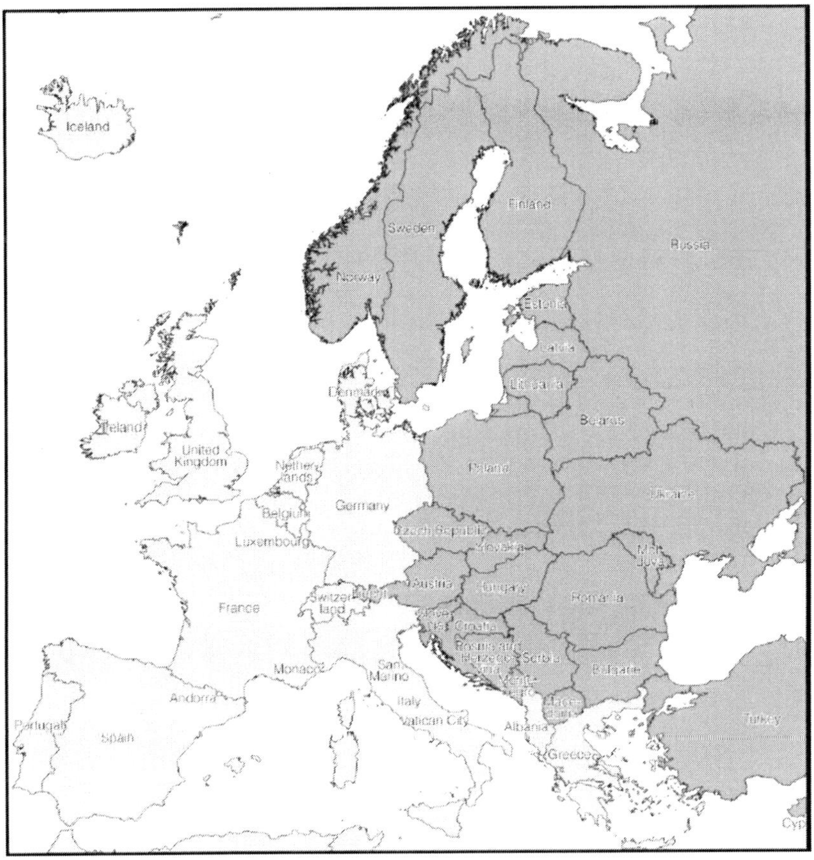

Figure 1. Map of Europe divided by Köppen Climate Classification. **Blue** part—Eastern Europe; **Green** part—Western Europe.

The major time span of the crisis period is generally considered as AD 1560–1660, and it officially ended at the start of the eighteen century (AD 1700) as stated by contemporary historians [32,33,34]. This study thus focuses particularly on the agricultural recovery period in Europe in the warm eighteenth century (AD 1700–1800).

Our goal was to compare historical trends of crop management as the adaptation approach to major climate fluctuations at the centennial and sub-continental scales. Thus, attention was paid neither to individual incidents that could temporarily distort agricultural production nor to exceptional cases. We believed that this rather broad-brush approach, although not without limitations, suits the scope of the present study.

2.1. Data Source

Our datasets were primarily obtained from our previous study on pre-industrial Europe [9]. The data from this previous publication have been validated. In the following sections, the source of data will be illustrated more in detail. In this study, quantitative methods were primary and qualitative methods were supplementary in verifying the effectiveness of crop management. Details of our datasets and methods are stated below.

2.1.1. Temperature and the Delineation of Climatic Phases

Temperature has been proven to be more important than precipitation in determining the agrarian economies in pre-industrial Europe at the large spatial scale [7,15]. Moreover, temperature is also the appropriate indicator of climate change at the large spatial scale [35] and the most important factor of plant cultivation [36]. Thus, we adopted temperature to represent climatic conditions in this study within the macro-scale geography.

The temperature anomaly series was derived from six authoritative annual temperature reconstructions of Europe, namely the annual temperature reconstruction for European land areas by Luterbacher et al. [37]; the summer temperatures since Roman times by Luterbacher et al. [38]; the annual temperature dataset by Osborn and Briffa [27], who combined the regional temperature series nested within Europe; the temperature reconstruction based on tree ring chronologies in central Europe by Büntgen et al. [39]; the temperature reconstruction based on documentary records in central Europe by Dobrovolný et al. [40]; and spring–summer temperature reconstructions in Europe by Guiot et al. [41]. Because different proxies and methods were employed in the above reconstructions, we first normalized them to homogenize their original variability. These normalized temperature series were then arithmetically averaged to generate the European temperature composite we adopted in this study (Figure 2). Based on this temperature composite, two major epochs were delineated. First, the period AD 1560–1660 was delineated as "extreme cooling period" during the Little Ice Age, which is also coincident with the major time span of the General Crisis of the Seventeenth Century in European history. Second, the warm AD 1700–1800 period was delineated as "agricultural recovery period".

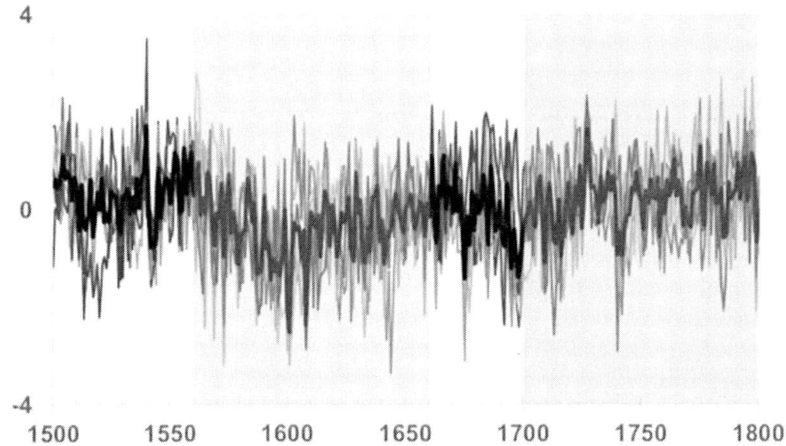

Figure 2. Temperature anomalies in Europe in AD 1500–1800. Note: The **orange** line is the normalized temperature reconstruction of Luterbacher et al. [37]; the **green** line is of Luterbacher et al. [38]; the **blue** line is of Osborn and Briffa [27]; the **grey** line is of Büntgen et al. [39]; the **purple** line is of Dobrovolný et al. [40]; and the **red** line is of Guiot et al. [41]. The bold **black** line is the arithmetic average of the six normalized temperature reconstructions, which was employed in this study. **Blue** shade indicates the extreme cooling period during the Little Ice Age (AD 1560–1660), while **red** shade indicates the warm agricultural recovery period (AD 1700–1800).

2.1.2. Crop Yield Ratio

Our analyses were primarily based on crop yield ratio of wheat, rye, barley, and oats to assess agricultural productivity and scrutinize the effectiveness of crop management in our study area [5]. There are few good substitutes for wheat, rye, barley, and oats, particularly in the pre-industrial period [42,43,44]. The crop yield ratio data used in this study were derived from the dataset of van Bath [45]. Crop yield ratio is defined as grain yield ratio in relation to seed. Yield ratio was calculated as follows:

$$Yield\ ratio = \frac{Harvest\ amount}{Seed\ amount} \quad (1)$$

Crop yield ratio is taken as one of the most important indicators reflecting the general picture of agricultural productivity. Yield ratio could be considered as an indicator of past agricultural production efficiency and has been adopted for a number of reasons, mainly on account of the widely varying usage by medieval stewards of local customary measures, which often precludes any accurate means of yield measurement in terms of grain output per unit area [5]. Thus, an impression of the vulnerability of the surpluses upon which the elaborate civilization of Europe rested can be derived by examining the harvest yield ratios [46].

Van Bath's dataset contains the crop yield ratio in 18 European countries [45]. The crop yield ratios of those countries were arithmetically averaged to obtain the one for the whole of Europe, as well as for Eastern Europe and Western Europe, respectively. Aggregating the data from different countries will help eliminate or neutralize a specialty in some local regions and provide a more accurate impression over a large area. Data aggregation should often be treated with caution, especially if data are from different sources and in different units, as in the case of Olmstead and Rhode [47]. However, in our study, the yield ratio data are from a single source (van Bath [45]), and the data across those European countries are in uniform units. Given the above circumstances, it is believed that the risk of our data aggregation is within acceptable range.

2.2. Formatting of Mathematical Components

Although the data employed in the study are published in international refereed journals, with their reliability verified, the data quality issue still cannot be totally ignored.

In this study, to further minimize the possible risk brought by data imperfection, we thus adopt a long term and macro perspective to investigate the link between major climate fluctuations and agricultural adaption. We also recognize that some records in some specific years may be not accurate enough. However, if we seek to obtain a big picture by including a large number of cases or records, the associated findings are reliable to draw an accurate historical account [48], similar to a questionnaire study [49]. In the study, two different kinds of statistical methods have been applied, which are explained in detail below.

2.2.1. Non-Parametric Analysis

The records of crop yield ratio in van Bath's [45] dataset, which is the original source of yield ratio in Zhang et al.'s research [9], were derived from different sources all over Europe. Against this background, we applied non-parametric analysis to extract the general trends of crop yield ratio. The non-parametric method makes fewer assumptions on the nature of data quality and distribution. The results of the non-parametric method would be particularly useful to the assessment of performance from a general perspective [50]. This method is operationalized by the statistical software R. The model of non-parametric analysis is described as:

$$y_i = f(x_i) + \varepsilon_i \tag{2}$$

$$f(x_0) = \frac{1}{nh}\sum_{i=1}^{n}\{W(\frac{x_0 - x_i}{h}) \times x_i\} \tag{3}$$

where h is the bandwidth of the local regression. W is the weighted function to each value x_i around x_0. The fitted value at x_0, that is, the estimated height of the regression curve, is $\hat{y}_0=a$. Therefore, in a very small neighborhood, the y_i can be written as the following simple function:

$$y_i = a + e_i \tag{4}$$

The above function is correct only in a small neighborhood. In the model, we use average weight to calculate the trend. If the x_i within the small neighborhood of x_0, weight value is 0.5; otherwise, it is 0. Because the weight in total should be 1, that is, $\sum Wi=1$; therefore, the weight value is 0.5:

$$\begin{array}{lll} W(z) = 0.5 & for & |z| < 1 \\ W(z) = 0 & for & |z| \geqslant 1 \end{array} \tag{5}$$

In the above context, $z_i = (x_i - x_0)/h$. It is typical to adjust h so that each local regression includes a fixed proportion s of the data. s is called the span of the local-regression smoother. The larger the span, the smoother the result is. We built on the general trends of crop yield ratio, which were obtained from the above method, to compare the agricultural production efficiency between Eastern and Western Europe.

2.2.2. Analysis of Variance (ANOVA)

One-way analysis of variance (ANOVA) was employed to statistically compare the recovery capacity of the agrarian economy (in terms of crop yield ratio) between Eastern and Western Europe in the warm AD 1700–1800 period (i.e., the agricultural recovery period).

3. RESULTS

Figure 3 shows the general trends of crop yield ratio in Eastern and Western Europe that were obtained from the non-parametric analysis. The overall condition of crop yield ratio was calculated by averaging the yield ratio of the four types of crops (i.e., wheat, rye, barley, and oats). The averaged crop yield ratio in both Eastern and Western Europe (Figure 3) exhibited similar fluctuations with regard to the change of temperature (Figure 2). Beginning in AD 1500, the crop yield ratio dropped in both of the two regions. The drop was drastic in AD 1560–1660, the extreme cooling period occurring during the Little Ice Age. Only after AD 1700 did the crop yield ratio begin to rise above the bottom. Despite the above similarities, the differences between Eastern (Figure 3A) and Western (Figure 3B) Europe are also apparent. During the extreme cooling period, the decrease of crop yield ratio in Eastern Europe was more drastic than that in Western Europe. Since AD 1650, the grain yield in Western Europe has been higher than that in Eastern Europe. In addition, the crop yield ratio in the eighteenth century was higher than that in the sixteenth century in Western Europe but that was not the case in Eastern Europe. Furthermore, only Western Europe showed an elevated level of crop yield ratio throughout the study period.

Following the analysis of the general pattern of averaged crop yield ratios in the two macro regions, the four types of crops were separated for further analysis. In Eastern Europe, the patterns of four types of crops in Figure 4 are similar to the general pattern of crop yield change as shown in Figure 3A, except for rye. Furthermore, as shown in Figure 5, the fluctuation of rye yield ratio exhibits a different pattern in Western Europe, while the other three types of crops are more or less similar to the general trend of crop yield ratio as shown in Figure 3B.

Fixing on the regional variation of crop yield ratio changes, it was observed that the yield ratio of rye in both Eastern and Western Europe was very similar during the entire study period. Furthermore, the rye yield ratio bounced back first after the extreme cooling of the Little Ice Age, in both Eastern and Western Europe. Nevertheless, the regional divergences of crop yield became more pronounced afterwards. For instance, in Eastern Europe, only rye showed signs of recovery after the extreme cooling, while the yield of the other three types of crops in the eighteenth century was lower than that in the sixteenth century. In Western Europe, however, the yield of all of the crops in the eighteenth century exceeded their previous levels in the sixteenth century except rye. Furthermore, in Western Europe, the yield ratio of wheat, barley, and oats increased remarkably and reached a new height in the eighteen century.

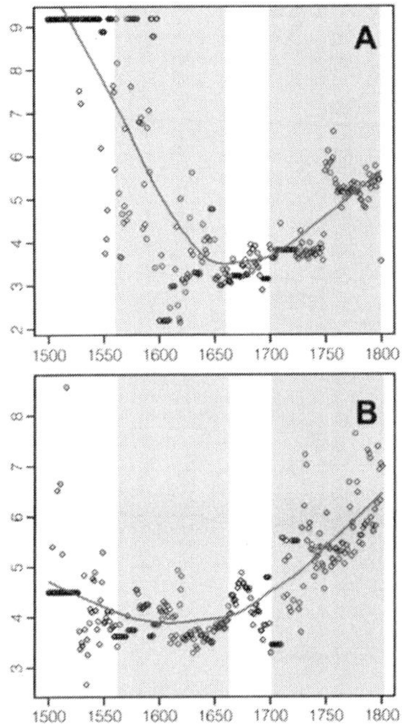

Figure 3. Average crop yield ratio in Eastern and Western Europe. (**A**) Average crop yield ratio in Eastern Europe; (**B**) average crop yield ratio in Western Europe. Note: Scatter plots in the figure are the raw data of crop yield ratio. **Red** line represents the general trend of crop yield ratio that is derived from non-parametric analysis. **Blue** shade indicates the extreme cooling period during the Little Ice Age (AD 1560–1660), while **red** shade indicates the warm agricultural recovery period (AD 1700–1800).

In the comparison of crop yield ratio between Eastern and Western Europe in the agricultural recovery period AD 1700–1800, rye differentiated from the other three types of crops, as its yield ratio between Eastern and Western Europe was not statistically significant. However, the crop yield ratios of wheat, barley, and oats in Western Europe were all higher than those in Eastern Europe. As the crop yield ratio is a chief indicator of agricultural productivity, the higher crop yield ratios of wheat, barley, and oats in Western Europe imply that, in general, agricultural productivity in Western Europe was higher than that in Eastern Europe in the eighteenth century. Given that the period is after the extreme cooling in the Little Ice Age, the higher crop yield ratio in Western Europe in the eighteenth century also implies a more effective agricultural recovery mechanism there.

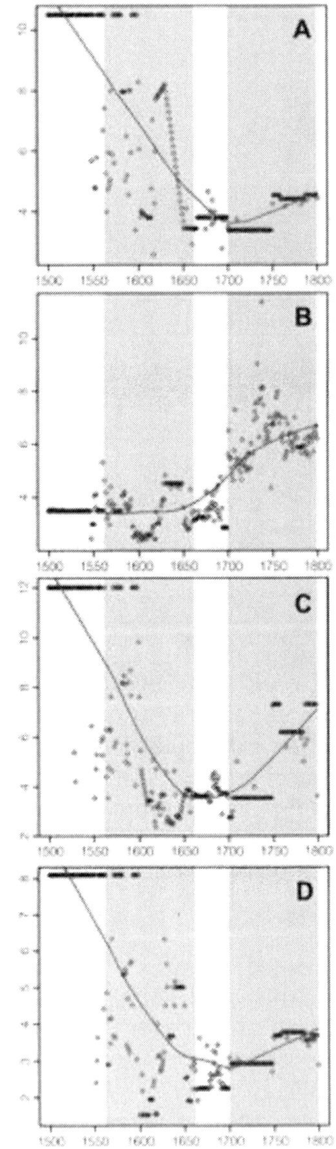

Figure 4. Yield ratio of various crop types in Eastern Europe. (**A**) wheat; (**B**) rye; (**C**) barley; (**D**) oats. Note: Scatter plots in the figure are the raw data of crop yield ratio. **Red** line represents the general trend of crop yield ratio that is derived from non-parametric analysis. **Blue** shade indicates the extreme cooling period during the Little Ice Age (AD 1560–1660), while **red** shade indicates the warm agricultural recovery period (AD 1700–1800).

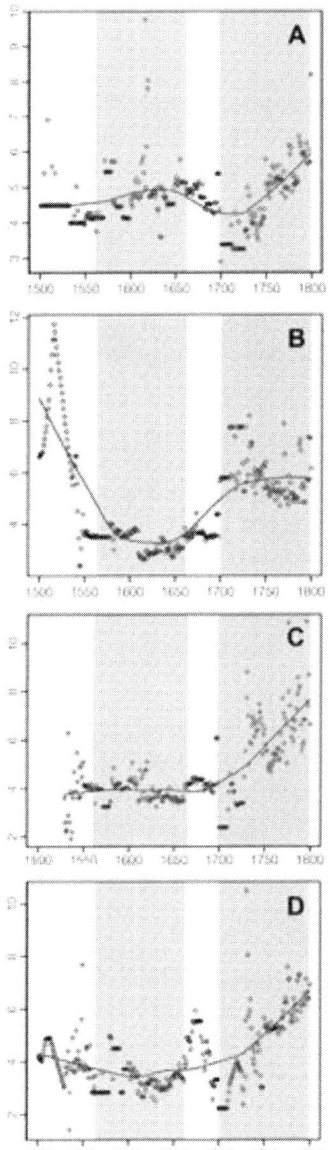

Figure 5. Yield ratio of various crop types in Western Europe. (**A**) wheat; (**B**) rye; (**C**) barley; (**D**) oats. Note: Scatter plots in the figure are the raw data of crop yield ratio. **Red** line represents the general trend of crop yield ratio that is derived from non-parametric analysis. **Blue** shade indicates the extreme cooling period during the Little Ice Age (AD 1560–1660), while **red** shade indicates the warm agricultural recovery period (AD 1700–1800).

4. DISCUSSION

The general trend of crop yield ratio in Eastern and Western Europe basically followed the alternation of climatic epochs, and the extreme cooling period of AD 1560–1660 resulted in the drastic reduction of crop yield ratio. Additionally, the yield ratio of rye in both Eastern and Western Europe was very similar throughout the entire study period. On the other hand, the yield ratio of wheat, barley, and oats was higher in Western Europe, especially in the agricultural recovery period in AD 1700–1800. In addition, the yield of wheat, barley, and oats increased drastically in Western Europe during the time.

Within the context of the agricultural sector, simple adaptation to climate change usually refers to substituting current vulnerable cultivars with climate-resistant ones [51], including any changes in crops, cultivars, sowing dates, sowing rates, tillage practices, use of inputs and so on. In the past and in modern days, some cases occur wherein current varieties (species) become unfeasible choices under climate change [52,53]. Farmers find it relatively easy to cope with change through the planting of new crop species [54]. This also applies to past agricultural production, which was characterized by a lower technological level.

Notably, new crops such as maize and potatoes were introduced to meet subsistence needs. Both of these could be good substitutes for grains to significantly augment the food-producing capacity of the land [55]. However, in the early modern era, farmers had very few alternatives. The introduction of new food was a prolonged process, though maize was introduced into Europe beginning in the mid-1500s [56]. The acceptance of potatoes and maize as food crops in Europe was closely related to rapid population growth after AD 1750 [57]. Potato was an uncommon crop before the second half of the eighteenth century, and most farmers were prejudiced against maize as food [58], leading them to plod along the same paths that their ancestors had trodden [59,60]. Wheat, rye, barley, and oats were the only acceptable options for farmers adapting to climate change during the study period. Therefore, in early modern Europe, selection of bumper or cold-resistant crops, which is a feasible means of adaptation, was largely limited to wheat, rye, barley, and oats. In this regard, a brief background of these four crops is discussed below for additional information.

4.1. Relative Importance of Wheat, Rye, Barley, and Oats in Pre-Industrial Europe

Rye was the dominant crop in Europe in our study period. However, it was not initially intentionally sown. Before its cold-resistant characteristics were widely recognized, rye was sown as a separate crop [61]. After the Renaissance, rye was more prominent than wheat and became the predominant bread grain. It was gradually replaced by wheat in the nineteenth century [62]. During the Middle Ages in Europe, rye was more important than wheat [63,64]. As agriculture extended to marginal areas to cope with the negative climatic impacts on grain harvest during the Little Ice Age [10], rye was generally the most popular food crop in Europe in our study period [65].

The prevalence of wheat in the human diet is relatively recent. The use of wheat as a bread grain grew rapidly between the late nineteenth century and the first half of the twentieth century [66]. In the past, most people in Europe relied more on coarse crops [62]. Furthermore, wheat was only cultivated to pay for dues and rent, or for export as an economic commodity [67]. Global expansion of wheat production mainly occurred in the twentieth century, followed by an increase in wheat planting area and yield [68].

Barley was grown more extensively than wheat, and apparently was more highly regarded as human food [69]. Although barley was the most important crop in the Roman era, it was replaced by rye during the Middle Ages [64]. In addition, barley is often valued for its forage quality of hay and for its relative earliness, providing harvestable grain when other crops are still growing.

Oats were always marginal in status. Oats were regarded as an uneconomical and poor-land crop, useful for planting in poorer soils [70]. They were only grown when population pressure forced encroachment into marginal areas. Oats were usually employed as a nurse crop for other agricultural use [71], and only in climates with lower temperatures were they considered as food [72].

Briefly, rye was the major food crop on the European continent in our study period. This may explain why the yield ratio of rye in Eastern and Western Europe was very similar in the entire study period (Table 1 and Figure 4 and Figure 5). On the other hand, wheat, barley, and oats were marginal crops, particularly in Eastern Europe, where their yield ratios in the eighteenth century were lower than those in the sixteenth century. The marginal status of wheat, barley, and oats implies that their cultivation was more determined by regional context, which may account for the variation of their yield ratios in Eastern and Western Europe.

Table 1. ANOVA for various crop yield ratio between Eastern and Western Europe in AD 1700–1800.

	Mean	Eastern Europe	Western Europe	Comparison	F	p
Wheat	4.243	3.885	4.601	Eastern<Western	37.499	0.00 **
Rye	6.013	6.065	5.961	-	0.598	0.44
Barley	5.324	5.058	5.591	Eastern<Western	4.674	0.03 *
Oats	4.075	3.291	4.859	Eastern<Western	88.612	0.00 **

** Significant at the 0.01 level ($p < 0.01$); * Significant at the 0.05 level ($p < 0.05$).

4.2. Crop Selection as a Means of Adaptation to Climate Change

Given that rye was the major crop and that wheat was not popular in Europe in our study time span, only barley and oats could serve as additional choices during the time. Those crops were usually cultivated in newly opened farmland. Barley could be cultivated at temperate regions in spring and summer [73], but it is not particularly cold-tolerant [74]. Oats could adapt to a wide range of soil conditions, but it is also not cold-tolerant [62,75]. When the cooling prevailed, barley and oats could not serve as bumper crops.

Since rye is the most cold-resistant crop among the four types of crops, in a cold climate, rye should be widely cultivated to maintain the food supply. In a previous study, it was found that increasing land input was the most effective method to bump up harvest and relieve climatic impacts among past societies [10]. Barley and oats could serve as nurse crops (particularly oats) because of their biological characteristics [71]; they were planted in marginal areas to increase arable land area and, eventually, the food supply in pre-industrial Europe.

Indeed, there were limits for increasing land input to increase harvest. Although Europe was an agrarian hub, there had been little opportunity for simple expansion of cultivable land. Only Eastern Europe had unused lands that could be agriculturally productive [76]. In Eastern Europe, arable land existed in relatively unlimited supply, and there was nothing to prevent people from inhabiting these new lands [77]. However, this was not the case in Western Europe. For instance, shrinkage of wheat production in Spain in the seventeenth century was primarily caused by the exhaustion of cultivable land [78].

Overall, crop management was obviously employed as a means of adaptation to major climate fluctuations in early modern Europe. Rye was chosen as major crop in our study period, probably to relieve the impact of low temperature during the Little Ice Age. However, the efficiency and effectiveness of crop management were often determined by many other factors. Agriculture was sensitive to price changes and other market factors in early modern Europe [79]. Crop (particularly bumper crop) cultivation was determined by market factors [80]. For instance, oats were regarded as "uneconomical," except for planting in poorer soils, due to their lower yield compared to wheat, rye, and barley [70]. The prevalence of wheat resulted primarily from the improvement of bread-making techniques [63]. Furthermore, government played an essential role in crop management [81]. In addition, making the best choice at precisely the right time was always a challenge. Most importantly, the selection of cold-resistant crops might have played a positive role in maintaining harvest in some small regions or short time spans, but it may not have been effective at long temporal and large spatial scales [7,9]. This phenomenon can be revealed by the drastic reduction of the yield ratio of rye in both Eastern and Western Europe during the extreme cooling period in the Little Ice Age (Figure 3, Figure 4 and Figure 5).

Notably, in the study, we have emphasized that the analysis is implemented at a large spatial and long term scale. Each scale brings new problems and information coinciding with a new nature or social interface [82,83,84]. Besides being a research question at the research level, scale is also an issue of recognizing how our living planet has been regarded in history [85]. At a large scale, the complex interactions between nature and human society are more obvious [86].

In terms of adaption, the institutional or technological capacity of past societies should not be neglected. However, in the pre-industrial period, the general low level of technology is the common image of the early modern world [87,88]. Even though humans could relieve the climatic effect in the short run by adaptation, institutional and social buffering mechanisms ultimately would be exhausted by the recurrent subsistence crises caused by long-term cooling [89]. Moreover, differences in institutional or technological capacity at the national scale would therein be neutralized, if this study takes Europe as the research region [90]. To clarify, at the scale in the study, we could identify the significant influences of climate change on agricultural activities. If the scale is narrowed down or shortened, other socio-economic factors should be considered, and their influences will possibly exceed those of climate change [91]. Furthermore, in individual countries, some advancement in technology could also be identified, such as in England [92,93] and the Netherlands [94]. In addition, some agricultural techniques like the development of heavy plows might have contributed to agricultural development [95], through more intensive and extensive cultivation, which was facilitated by the improvement of farm tools, also resulting in severe environmental degradation in return [96].

4.3. Agricultural Divide between Eastern and Western Europe

As mentioned in the introduction, most of the studies of early modern Europe focus on the climatic impacts on agricultural production in the Little Ice Age. However, the period of agricultural recovery (the eighteenth century) has not been examined in detail. This study demonstrates that crop yield ratio, which is a good indicator of agricultural productivity, is closely related to temperature. Thus, the change of crop yield ratio after the extreme cooling in the Little Ice Age could indicate the recovery capacity of the agrarian economy.

Table 1 shows that rye is the only crop whose yield ratio did not demonstrate significant differences between Eastern and Western Europe. On the other hand, wheat, barley, and oats in Western Europe showed higher yields than those in Eastern Europe. In agrarian societies, yield was the essential performance indicator of the economy. As revealed by the crop yield ratio, economic disparity of Eastern and Western Europe increased in the warm eighteenth century. The bumper crops positively affected the agrarian sector, as well as other sectors, which initiated economic recovery [13]. The yield of wheat, barley and oats increased drastically in Western Europe, but not in Eastern Europe, in the eighteenth century, which may suggest that the speed of recovery in Western Europe was faster than that in Eastern Europe.

In recent studies, although the East–West Divide in Europe since the last century has attracted great research attention [97], it has been emphasized that such divide originated from a long historical tradition, as well as from differences in culture and civilization [98]. The East–West Divide is reflected in almost every aspect of life, and the course of agrarian history in Eastern Europe is in notable contrast with that in Western Europe [99]. Economic conditions have been highlighted to explain the East–West Divide, as well as religion, philosophy, culture and other factors [100]. However, there are very few studies to explain the divide from the perspective of agricultural productivity. The above mentioned crop yield ratio upsurge in Western Europe in AD 1700–1800 significantly contributed to the regional economy.

In the agricultural sector, increasing land input was more feasible in Eastern Europe, while increasing labor input was more feasible in Western Europe [46,101]. Labor input could have been the momentum needed to increase crop yield ratio and agricultural production in the early modern period, in spite of its limited effect [10]. As highlighted in previous studies, production output is usually estimated by analyzing the labor input [102], especially in past agrarian societies with a low technological level. In the pre-industrial era, agricultural production was labor-intensive; as such, the availability of farm labor was significant [103]. Labor input also has special advantages because it is more elastic in supply compared to land [104]. Thus, labor is generally regarded as the abundant and principal input in production [105]. In this regard, in Western Europe, better crop yield ratio in the warm eighteenth century should have triggered the transfer of labor from agriculture to other sectors, as it did in England [88]. Human resources are fundamental to national development; for example, the advancement in technology in England was due to the increasing supply of unskilled workers in cities

[106]. With the help of large labor inputs, industrial sectors started to gain the momentum for development, which might have contributed to the rapid social and economic development in Western Europe in the following centuries.

5. CONCLUSIONS

In this study, we based our analysis on crop yield ratio, which is the essential indicator of the economy in agrarian societies, to assess agricultural production efficiency under major climate fluctuations in early modern Europe. Via regional comparison, the yield ratio of rye in Eastern and Western Europe was found to be very similar. On the other hand, the yield ratio of wheat, barley, and oats was higher in Western Europe. As revealed by the changes of crop yield ratio in different climatic epochs, the disparity of the agrarian economies in Eastern and Western Europe was enlarged in the warm eighteenth century. Bumper crops positively contributed to agriculture, as well as other sectors, which initiated economic recovery [13]. The yield of wheat, barley and oats increased drastically in Western Europe in the warm eighteenth century, which may imply that the speed of economic recovery in Western Europe was more rapid than that in Eastern Europe during the time. In Western Europe, larger crop output facilitated the transfer of labor from agriculture to other sectors. Surplus human resources that are fundamental to national development would possibly have supported other sectors to drive social development. This might have contributed to rapid socio-economic development in Western Europe, and eventually to the East–West Divide in the following centuries. We do not deny the importance of other factors in establishing the divide in Europe. However, we offer a new perspective to explain the divide, which is grounded in the quantitative analysis of agricultural productivity in early modern periods.

Currently, the issue of global warming has attracted much scholarly attention. Warming has been unprecedented over the last two millennia, which is different from the warm eighteenth century mentioned in this study. Potential impacts of climate change on human societies are still under debate. Further research is needed to assess the climate–agriculture link under the global warming trend, especially in some under-developed regions or countries that are still heavily dependent on agricultural production. The findings in the study not only evaluated the effectiveness of crop management to relieve climatic impact at a low technological level in past society, but also investigated the potential to recover after climatic shocks. At a low technological level, the primary adaptation approach, like the simple crop management strategy described in the study, is not enough. Thus, no doubt the development of technology is the key step to addressing social vulnerability to climate change. In addition, after climatic shocks, we also may accelerate recovery speed if we have more choices in terms of socio-economic development, as illustrated in the study: Western Europe maintained a stronger pace to heal the damages caused by the cooling impacts than did Eastern Europe in the recovery period. Perhaps our empirical analysis of pre-industrial Europe could serve as a reference to attract more attention to the recovery capacity of society after the impact of climate change in the future.

ACKNOWLEDGMENTS

The research grants are generously provided by a post-doctoral fellowship from The University of Hong Kong, Hong Kong, Hui Oi-Chow Trust Fund (201502172003 and 201602172006), and the Research Grants Council of The Government of the Hong Kong Special Administrative Region of the People's Republic of China (HKU745113H and HKU17610715). The funders had no role in study design, data collection and analysis, decision to publish, or preparation of the manuscript.

REFERENCES

1. IPCC. Climate Change 2007: Impacts, Adaptation and Vulnerability. Contribution of Working Group II to the Fourth Assessment Report of the Intergovernmental Panel on Climate Change; Cambridge University Press: Cambridge, UK, 2008.
2. Galloway, P.R. Long-term fluctuations in climate and population in the preindustrial era. Popul. Dev. Rev. **1986**, 12, 1–24.
3. Lee, H.F.; Zhang, D.D.; Pei, Q.; Fei, J. Downscaling and disaggregating NAO-conflict nexus in pre-industrial Europe. Chin. Geogr. Sci. **2016**, 1–14.
4. Zhang, D.D.; Lee, H.F.; Wang, C.; Li, B.; Zhang, J.; Pei, Q.; Chen, J. Climate change and large-scale human population collapses in the pre-industrial era. Glob. Ecol. Biogeogr. **2011**, 20, 520–531.
5. Brandon, P.F. Cereal yields on Sussex estate of Battle Abbey during the later middle ages. Econ. Hist. Rev. **1972**, 24, 404–412.
6. Van Bath, B.H.S. The rise of intensive husbandary in the Low Countries. In Agrarian Conditions in Modern European History; Warner, C.K., Ed.; The Macmillan Company: New York, NY, USA, 1966.
7. Pei, Q.; Zhang, D.D.; Lee, H.F.; Li, G. Climate change and macro-economic cycles in pre-industrial Europe. PLoS ONE**2014**, 9, e88155.
8. Pei, Q.; Zhang, D.D.; Li, G.; Lee, H.F. Climate change and the macroeconomic structure in pre-industrial Europe: New evidence from wavelet analysis. PLoS ONE **2015**, 10, e0126480.

9. Zhang, D.D.; Lee, H.F.; Wang, C.; Li, B.; Pei, Q.; Zhang, J.; An, Y. The causality analysis of climate change and large-scale human crisis. Proc. Natl. Acad. Sci. USA **2011**, 108, 17296–17301.
10. Pei, Q.; Zhang, D.D.; Lee, H.F. Evaluating the effectiveness of agricultural adaptation to climate change in preindustrial society. Asian Geogr. **2015**, 32, 85–98.
11. Parker, G.; Smith, L.M. Introduction. In The General Crisis of the Seventeenth Century, 2nd ed.; Parker, G., Smith, L.M., Eds.; Routledge: London, UK, 1997.
12. Zohary, D.; Hopf, M. Domestication of Plants in the Old World; Oxford University Press: New York, NY, USA, 2000.
13. Appleby, A.B. Grain prices and subsistence crises in England and France. J. Econ. Hist. Rev. **1979**, 39, 865–887.
14. Howden, S.M.; Soussana, J.-F.; Tubiello, F.N.; Chhetri, N.; Dunlop, M.; Meinke, H. Adapting agriculture to climate change. Proc. Natl. Acad. Sci. USA **2007**, 104, 19691–19696.
15. Pei, Q.; Zhang, D.D.; Li, G.; Lee, H.F. Short and long term impacts of climate variations on the agrarian economy in pre-industrial Europe. Clim. Res. **2013**, 56, 169–180.
16. Lee, H.F.; Zhang, D.D.; Brecke, P.; Pei, Q. Regional geographic factors mediate the climate-war relationship in Europe. Br. J. Interdiscip. Stud. **2015**, 2, 1–28.
17. Lee, H.F.; Zhang, D.D.; Brecke, P.; Fei, J. Positive correlation between the North Atlantic Oscillation and violent conflicts in Europe. Clim. Res. **2013**, 56, 1–10.
18. Brenner, R. Agrarian class structure and economic development in pre-industrial Europe. Past Present **1976**, 70, 30–75.
19. Campell, B.M.S. Nature as historical protagonist: Environment and society in pre-industrial England. Econ. Hist. Rev.**2010**, 63, 281–314.
20. Hartmann, H.T.; Kofranek, A.M.; Rubatzky, V.E.; Flocker, W.J. Plant Science: Growth, Development, and Utilization of Cultivated Plants; Prentice-Hall Inc.: Englewood Cliffs, NJ, USA, 1981.
21. Jones, P.D.; Bradley, R.S. Climatic variations over the last 500 years. In Climate Since AD 1500, Revised ed.; Bradley, R.S., Jones, P.D., Eds.; Routledge: London, UK; New York, NY, USA, 1995.
22. Sayre, N.F. Ecological and geographical scale: Parallels and potential for integration. Prog. Hum. Geogr. **2005**, 29, 276–290.
23. Gerstengarbe, F.-W.; Werner, P.C. A short update on Koppen climate shifts in Europe between 1901 and 2003. Clim. Chang. **2008**, 92, 99–107.
24. Peel, M.C.; Finlayson, B.L.; McMahon, T.A. Updated world map of the Köppen-Geiger climate classification. Hydrol. Earth Syst. Sci. **2007**, 11, 1633–1644.
25. McKnight, T.L.; Hess, D. Climate Zones and Types: The Köppen System. In Physical Geography: A Landscape Appreciation; Prentice Hall: Upper Saddle River, NJ, USA, 2000.
26. Persson, K.G. The seven lean years, elasticity traps, and intervention in grain markets in pre-industrial Europe. Econ. Hist. Rev. **1996**, 49, 692–714.
27. Osborn, T.J.; Briffa, K.R. The spatial extent of 20th-century warmth in the context of the past 1200 years. Science **2006**, 311, 841–844.
28. Stone, D. The Polish-Lithuanian State, 1386–1795; University of Washington Press: Seattle, Washington, DC, USA, 2001.
29. Fulbrook, M. A Concise History of Germany; Cambridge University Press: Cambridge, UK, 2004.
30. Pennington, D.H. Europe in the Seventeenth Century; Longman: London, UK, 1989.
31. Hsiang, S.M.; Burke, M.; Miguel, E. Quantifying the influence of climate on human conflict. Science **2013**, 341, 1235367.
32. Fischer, D.H. The Great Wave: Price Revolutions and the Rhythm of History; Oxford University Press: New York, NY, USA, 1996.
33. Parker, G.; Smith, L.M. Introduction. In The General Crisis of the Seventeenth Century; Parker, G., Smith, L.M., Eds.; Routledge & Kegan Paul: London, UK, 1978.
34. Aston, T. Crisis in Europe: 1560–1660; Routledge & Kegan Paul: London, UK, 1996.
35. Jones, P.D.; Bradley, R.S. Climatic variations in the longest instrumental records. In Climate Since AD 1500; Bradley, R.S., Jones, P.D., Eds.; Routledge: London, UK, 1992.

36. Mathias, R.J. Factors Affecting the Establishment of Callus Cultures in Wheat. In Wheat; Bajaj, Y.P.S., Ed.; Springer: Heidelberg, Germany, 1990.
37. Luterbacher, J.; Dietrich, D.; Xoplaki, E.; Grosjean, M.; Wanner, H. European seasonal and annual temperature variability, trends, and extremes since 1500. Science **2004**, 303, 1499–1503.
38. Luterbacher, J.; Werner, J.P.; Smerdon, J.E.; Fernández-Donado, L.; González-Rouco, F.J.; Barriopedro, D.; Ljungqvist, F.C.; Büntgen, U.; Zorita, E.; Wagner, S.; et al. European summer temperatures since Roman times. Environ. Res. Lett.**2016**, 11, 024001.
39. Büntgen, U.; Tegel, W.; Nicolussi, K.; McCormick, M.; Frank, D.; Trouet, V.; Kaplan, J.O.; Herzig, F.; Heussner, K.-U.; Wanner, H.; et al. 2500 years of European climate variability and human susceptibility. Science **2011**, 331, 578–582.
40. Dobrovolný, P.; Moberg, A.; Brázdil, R.; Pfister, C.; Glaser, R.; Wilson, R.; Engelen, A.v.; Limanówka, D.; Kiss, A.; Halíčková, M.; et al. Monthly, seasonal and annual temperature reconstructions for Central Europe derived from documentary evidence and instrumental records since AD 1500. Clim. Chang. **2010**, 101, 69–107.
41. Guiot, J.; Corona, C.; members, E. Growing season temperatures in Europe and climate forcings over the past 1400 years. PLoS ONE **2010**, 5, e9972.
42. Van Bath, B.H.S. The yields of different crops (mainly cereal) in relation to the seed c. 810–1820. Acta Hist. Neederlandica **1963**, 2, 26–106.
43. McConnell, C.R.; Bruce, S.L. Economics: Principles, Problems and Policies; McGraw-Hill/Irwin: Boston, MA, USA, 2002.
44. Spencer, M.H.; Orley, M.; Amos, J. Contemporary Economics, 8th ed.; Worth Publishers: New York, NY, USA, 1993.
45. Van Bath, B.H.S. Yield Ratios, 810–1820; A.A.G. Bijdragen, No. 10; Afdeling Agrarische Geschiedenis, Landbouwhogeschool: Wageningen, Pays-Bas; p. 963.
46. Vries, J. Economy of Europe in an Age of Crisis; Cambridge University Press: New York, NY, USA, 1976.
47. Olmstead, A.L.; Rhode, P.W. Adapting North American wheat production to climatic challenges, 1839–2009. Proc. Natl. Acad. Sci. USA **2011**, 108, 480–485.
48. Bunge, M. Causality and Modern Science, 4th ed.; Transaction Publishers: New Brunswick, NB, Canaada, 2009.
49. Babbie, E. The Practice of Social Research; Wadsworth Cengage Learning: Belmont, CA, USA, 2013.
50. Gibbons, J.D.; Chakraborti, S. Nonparametric Statistical Inference, 4th ed.; Marcel Dekker: New York, NY, USA, 2003.
51. Smit, B.; Wandel, J. Adaptation, adaptive capacity and vulnerability. Glob. Environ. Chang. **2006**, 16, 282–292.
52. Meza, F.J.; Silva, D. Dynamic adaptation of maize and wheat production to climate change. Clim. Chang. **2009**, 94, 143–156.
53. Iglesias, A.; Erda, L.; Rosenzweig, C. Climate change in Asia: A review of the vulnerability and adaptation of crop production. Water Air Soil Pollut. **1996**, 92, 13–27.
54. Rosenzweig, C.; Parry, M.L. Potential impact of climate change on world food supply. Nature **1994**, 367, 133–138.
55. McEvedy, C.; Jones, R. Atlas of World Population History; Allen Lane: London, UK, 1978.
56. Zirkle, C. How maize get to Europe. J. Hered. **1952**, 43, 116.
57. Langer, W.L. American foods and Europe's population growth 1750–1850. J. Soc. Hist. **1975**, 8, 51–66.
58. Van Bath, B.H.S. The Agrarian History of Western Europe A.D. 500–1850; Bulter & Tanner Ltd: Frome/London, UK, 1963.
59. Coleman, D.C. Technology and economic history, 1500–1750. Econ. Hist. Rev. **1959**, 11, 506–514.
60. Fussell, G.E. Farming Technique from Prehistoric to Modern Times; Pergamon Press: Edinburgh/London, UK, 1966.
61. Martin, J.H.; Leonard, W.H.; Stamp, D.L. Principles of Field Crop Production; Macmilan Publishing Co., Inc.: New York, NY, USA, 1976.
62. Leonard, W.H.; Martin, J.H. Cereal Crops; The Macmilam Company: London, UK, 1970.

63. Orth, R.A.; Shellenberger, J.A. Origin, production, and utilization of wheat. In Wheat: Chemistry and Technology, 3rd ed.; Pomeranz, Y., Ed.; American Association of Cereal Chemists, Inc.: Saint Paul, MN, USA, 1988; Volume 1.
64. Hugh, J.C.; Hoveling, A.W. Wheat: Chemistry and Utilization; Technomic Publishing Co., Inc.: Lancaster, PA, USA, 1998.
65. Küster, H. Rye. In The Cambridge World History of Food; Kiple, K.F., Ornelas, K.C., Eds.; Cambridge University Press: Cambridge, UK, 2000; Volume I.
66. Duff, B.; Rasmussen, P.E.; Smiley, R.W. Wheat/fallow systems in semi-arid regions of the Pacific NW America. In Agriculture Sustainability: Economic, Environmental and Statistical Considerations; Barnett, V., Payne, R., Steiner, R., Eds.; Jhon Wiley & Sons: West Sussex, UK, 1995.
67. Lough, J. An Introduction to Seventeenth Century France; Orient Longmans Private Ltd.: New York, NY, USA, 1957.
68. Bell, G.D.H. The history of wheat cultivation. In Wheat Breeding: Its Scientific Basis; Lupton, F.G.H., Ed.; Chapman and Hall: London, UK, 1987.
69. Takahashi, R. The origin and evolution of cultivated barley. Adv. Genet. **1955**, 7, 227–266.
70. Moore-Colyer, R.J. Oats and oat production in history and prehistory. In The Oat Crop: Production and Utilization; Welch, R.W., Ed.; Chapman & Hall: London, UK, 1995.
71. Schrickel, D.J. Oats production, value, and use. In Oats: Chemistry and Technology; Webster, F.H., Ed.; American Association of Cereal Chemists, Inc.: Saint Paul, MN, USA, 1986.
72. Albala, K. Food in Early Modern Europe; Greenwood Press: London, UK, 2003.
73. Shands, H.G.; Dickson, A.D. Barley-botany, production, harvesting, processing, utilization, and economics. Econ. Bot.**1953**, 8, 3–26.
74. Tarjoc, F.C.; Tabără, V. Issues concerning the impact of foliar fertilization on the production of barley in Timisoara. Res. J. Agric. Sci. **2011**, 43, 184–189.
75. Nilan, R.A.; Ullrich, S.E. Barley: Taxonomy, origin, distibution, production, genetics, and breeding. In Barley: Chemistry and Technology; MacGregor, A.W., Bhatty, R.S., Eds.; American Association of Cereal Chemists, Inc.: Saint Paul, MN, USA, 1993.
76. Jan, D.V. Economy of European in an Age of Crisis, 1600–1750; Cambridge University Press: New York, NY, USA, 1976.
77. Gunst, P. Agrarian Development and Social Change in Eastern Europe, 14th–19th Centuries; Variorum: Hampshire, UK, 1996.
78. Weisser, M.R. The agrarian depression in seventeenth-century Spain. J. Econ. Hist. **1982**, 42, 149–154.
79. Romano, R. Between the sixteenth and seventeenth centuries: The economic crisis of 1619–1622. In The General Crisis of the Seventeenth Century, 2nd ed.; Parker, G., Smith, L.M., Eds.; Routledge: London, UK, 1997.
80. Welch, R.W. The Oat Crop: Production and Utilization; Chapman & Hall: London, UK, 1995.
81. Karl, G.P. Grain Markets in Europe, 1500–1900: Integration and Deregulation; Cambridge University Press: Cambridge, UK, 1999.
82. Dolidon, H. La multiplicité des échelles dans l'analyse d'un phénomène d'interface nature/société.L'exemple des feux de brousse en afrique de l'ouest. Cybergeo: European J. Geography **2007**, 363, 20.
83. Pei, Q.; Zhang, D.D.; Li, G.; Forêt, P.; Lee, H.F. Temperature and precipitation effects on agrarian economy in late imperial china. Environ. Res. Lett. **2016**, 11, 064008.
84. Pei, Q.; Zhang, D.D.; Lee, H.F. Contextualizing human migration in different agro-ecological zones in ancient China. Quat. Int. **2016**.
85. Chaplin, J.E. Earthsickness: Circumnavigation and the terrestrial human body, 1520–1800. Bull. Hist. Med. **2012**, 86, 515–542.
86. MacDonald, K.I. Push and shove: Spatial history and the construction of a Portering economy in northern Pakistan. Comp. Stud. Soc. Hist. **1998**, 40, 287–317.
87. Clark, G. A Farewell to Alms: A Brief Economic History of the World; Princeton University Press: Princeton, NJ, USA; Oxford, UK, 2007.

88. Allen, R.C. Economic structure and agricultural productivity in Europe, 1300–1800. Eur. Rev. Econ. Hist. **2000**, 3, 1–25.
89. Orlove, B. Human adaptation to climate change: A review of three historical cases and some general perspectives. Environ. Sci. Policy **2005**, 8, 589–600.
90. Zhang, D.D.; Brecke, P.; Lee, H.F.; He, Y.Q.; Zhang, J. Global climate change, war, and population decline in recent human history. Proc. Natl. Acad. Sci USA **2007**, 104, 19214–19219.
91. Gibson, C.C.; Ostrom, E.; Ahn, T.K. The concept of scale and the human dimensions of global change: A survey. Ecol. Econ. **2000**, 32, 217–239.
92. Overton, M. Agricultural Revolution in England: The Transformation of the Agrarian Economy 1500–1850; Cambridge University Press: Cambridge, UK; New York, NY, USA; Melbourne, Australia, 1996.
93. Pretty, J.N. Farmers' extension practice and technology adaptation: Agricultural revolution in 17–19th century Britain. Agric. Hum. Values **1991**, 8, 132–148.
94. Bieleman, J. Five Centuries of Farming: A Short History of Dutch Agriculture 1500–2000; Wageningen Academic Publishers: Wageningen, The Netherlands, 2010.
95. Andersen, T.B.; Jensen, P.S.; Skovsgaard, C.V. The heavy plough and the agricultural revolution in medieval Europe. J. Dev. Econ. **2016**, 118, 133–149.
96. Fraser, E.D.G. Can economic, land use and climatic stresses lead to famine, disease, warfare and death? Using Europe's calamitous 14th century as a parable for the modern age. Ecol. Econ. **2011**, 70, 1269–1279.
97. Challand, B. 1989, contested memories and the shifting cognitive maps of Europe. Eur. J. Soc. Theor. **2009**, 12, 397–408.
98. Stråth, B. A European identity: To the historical limits of a concept. Eur. J. Soc. Theor. **2002**, 5, 387–401.
99. Warner, C.K. Introduction. In Agrarian Conditions in Modern European History; Warner, C.K., Ed.; The Macmillan Company: New York, NY, USA, 1966.
100. Young, C. Religion and economic growth in Western Europe: 1500–2000. In Manuscript; Stanford University: Stanford, CA, USA, 2009.
101. Clark, G. Productivity growth without technical change in European agriculture before 1850. J. Econ. Hist. **1987**, 47, 419–432.
102. Douglas, P.H. The cobb-Douglas production function once again: Its history, its testing, and some new empirical values. J. Political Econ. **1976**, 84, 903–916.
103. Symons, L. Agricultural Geography, Revised ed.; Westview Press: Boulder, CO, USA, 1979.
104. Floyd, J.E. The effects of farm price supports on the returns to land and labor in agriculture. J. Political Econ. **1965**, 73, 148–158.
105. Cornia, G.A. Farm size, land yields and the agricultural production function: An analysis for fifteen developing countries. World Dev. **1985**, 13, 513–534.
106. Acemoglu, D. Technical change, inequality and the labor market. J. Econ. Lit. **2002**, 40, 7–72.

CHAPTER 10

A new mutant genetic resource for tomato crop improvement by TILLING technology

Silvia Minoia[1,3], Angelo Petrozza[1], Olimpia D'Onofrio[1], Florence Piron[2], Giuseppina Mosca[1], Giovanni Sozio[1], Francesco Cellini[1], Abdelhafid Bendahmane[2], Filomena Carriero[1*]

[1]*Metapontum Agrobios, SS Jonica 106 Km 448.2, 75010 Metaponto (MT), Italy.*

[2]*Unité de Recherche en Génomique Végétale, UMR INRA-CNRS, Rue Gaston Crémieux, 91057 Evry Cedex, France.*

[3]*ENEA, Casaccia Research Center, PO Box 2400 Roma 00100AD, Italy.*

AUTHORS' CONTRIBUTIONS

FCa planned and headed the development of the mutant populations. AP and GS took care and visually phenotyped the mutant populations; OD set up the LycoTILL database; SM extracted the DNA; SM, FP and GM did TILLING screens and analysis; AB supervised the TILLING platform set up in Evry; FCe and FCa co-directed the TILLING project in M. Agrobios; FCa and SM were responsible for drafting and revising the manuscript with contributions from co-authors.

ABSTRACT

Background: In the last decade, the availability of gene sequences of many plant species, including tomato, has encouraged the development of strategies that do not rely on genetic transformation techniques (GMOs) for imparting desired traits in crops. One of these new emerging technology is TILLING (Targeting Induced Local Lesions In Genomes), a reverse genetics tool, which is proving to be very valuable in creating new traits in different crop species.

Results: To apply TILLING to tomato, a new mutant collection was generated in the genetic background of the processing tomato cultivar Red Setter by treating seeds with two different ethylemethane sulfonate doses (0.7% and 1%). An associated phenotype database, LycoTILL, was developed and a TILLING platform was also established. The interactive and evolving database is available online to the community for phenotypic alteration inquiries. To validate the Red Setter TILLING platform, induced point mutations were searched in 7 tomato genes with the mismatch-specific ENDO1 nuclease. In total 9.5 kb of tomato genome were screened and 66 nucleotide substitutions were identified. The overall mutation density was estimated and it resulted to be 1/322 kb and 1/574 kb for the 1% EMS and 0.7% EMS treatment respectively.

Conclusions: The mutation density estimated in our collection and its comparison with other TILLING populations demonstrate that the Red Setter genetic resource is suitable for use in high-throughput mutation discovery. The Red Setter TILLING platform is open to the research community and is publicly available via web for requesting mutation screening services.

1. BACKGROUND

Tomato (*Solanum lycopersicum*) is one of the most important vegetable plants in the world. Its fruits are end products both for the fresh market and food processing industry. Tomato presents a relatively small genome highly syntenic to others economically important Solanaceae species and was selected as a reference species for sequencing a Solanaceae genome. In addition to the availability of a number of genomic resources, including transcriptome [1, 2, 3] and metabolome [4], large collections of genetic resources are available to dissect the biochemical and the metabolic pathways in tomato [5]. Large EMS and fast neutron mutant collections, in the background of M82 tomato cultivar, have been generated and more then 3,000 phenotype alterations catalogued [6]. An EMS-induced mutation library of the miniature dwarf tomato cultivar Micro-Tom has also been produced and this constitutes another resource for tomato genetic studies [7].

In recent years, the genome sequencing program of many plant species [8, 9, 10], including tomato [11] has led to the availability of a large number of gene sequences in public databases which subsequently has encouraged the development of reverse genetics approaches. T-DNA and transposon insertional mutagenesis have been exploited to inactivate genes in tomato [12, 13]. However, unless a high-throughput tomato transformation protocol is developed, systematic functional analysis of tomato genes with these approaches is not realistic. In recent years, TILLING (Targeting Induced Local Lesions IN Genomes) [14, 15] a new emerging technology that doesn't rely on genetic transformation techniques, allows systematic functional genomic studies. The only prerequisite for its application is the knowledge of the gene nucleotide sequences. TILLING is a reverse genetic strategy that utilises chemical mutagenesis for inducing variability and sensitive molecular screenings to identify point mutations responsible for phenotype alteration. The strength and potency of this reverse genetic strategy has been validated by its successful application in both plants (Arabidopsis thaliana [16, 17, 18], pea [19], wheat [20], rice [21, 22], barley [23], maize [24], soybean [25], Lotus japonicus [26], sorghum [27], tomato [28, 29]) and animals (zebrafish [30], drosophila [31]).

In the present paper we report the construction of a high-quality tomato genetic mutant reference collection which could be used for both forward and reverse genetic studies. We have developed such a population by mutagenizing the processing tomato variety Red Setter with EMS and establishing an associated phenotype database, LycoTILL, and a TILLING platform. The database also serves as a portal for users to request materials or TILLING experiments.

2. RESULTS

2.1 Generation of the mutant collection

Red Setter is a processing tomato variety that completes its reproductive cycle within 110 days, it is a high productive variety and its architecture permits mechanical harvesting. In order to optimize the EMS mutagenesis, we first conducted a 'kill-curve' analysis, using a range of doses from 0.3-1.5% EMS. Two EMS doses were then chosen to generate the mutant collection. The first mutagen treatment was performed by incubating about 11,000 seeds with 0.7% EMS that caused 20% reduction in seed germination (LD_{20}) with respect to untreated control seeds. The second mutant tomato population was produced by treating 12,000 seeds with 1% EMS (LD_{49}). Out of the 23,000 treated seeds, 13,000 seedlings were grown to fruit maturity in controlled conditions and M2 seeds were collected from individual M1 plants from different plant internodes. In total we collected 6,667 distinct M2 seed stocks, among which 4,741 and 1,926 M2 seed stocks were obtained with 0.7% and 1% EMS treatment respectively. For the production of M3 seed stocks, three seeds per M2 family were sown in nursery, grown to fruit maturity in open field and M3 seeds harvested from individual M2 plants. In total we collected 5,508 M3 seed stocks (Table 1) as 1,159 M2 families out of 6,667 M2 families didn't produce M3 seeds. Specifically there were 585 M2 families (12.33%) generated from the 0.7% mutagenesis experiment and 574 M2 families (29.8%) generated from the 1% EMS treatment.

Table 1 Summary of the Red Setter tomato mutant collection development

EMS Concentration	Mutagenized Seeds (No)	Transplanted M1 plants	M2 seed Families	M3 seed Families
0.7% (LD 20)	11,000	8,500	4,741	4,156
1% (LD 49)	12,000	4,500	1,926	1,352
Total	**23,000**	**13,000**	**6,667**	**5,508**

EMS concentrations, number of seeds treated with the mutagen, number of M1 plants and collected M2 and M3 seed families are shown.

2.2 M2 plant phenotyping

Three plants per M2 family of Red Setter mutant population were scored for visual phenotype alteration at key developmental stages, from germination until fruit maturation. The data collected from individual plants were organized in 17 classes and 51 subclasses of phenotypes. The vocabulary used to describe the phenotypes was derived from the plant phenotype ontology and from previous investigation of systematic phenotyping of the mutant tomato collection [6]. We also introduced three new classes of phenotype alterations, the cotyledon, the fruit number and the seed germination into fruit. The cotyledon class describes mutants showing alterations in the number, color and morphology of the cotyledons. The fruit number class describes mutants affected in the fruit yield and it contains three subclasses, few, many or absence of fruits. By adding the fruit number class and the subclass "absent" we distinguished the plant sterility due to the absence of fruits from those caused by the seedless fruits. The class of phenotype seed germination into fruit describes mutants having pre-germinated seeds still in the fruit. This phenotype is presumed to result from an altered fruit flesh pH or by a hormonal imbalance [32, 33].

The complete list of the vocabulary used and the number of lines found in each major phenotype category are shown in Table 2.

Table 2 List of phenotype classes and subclasses

	Class	Subclass	No. of plants
1	Seed	No germination	3,904
		Seedling lethality	1,674
2	Cotyledons	Colour	264
		Number	82
		Morphology	17
		Size	2
		Other cotyledon development	3
3	Plant size	Small plant	303
		Large plant	4
4	Plant habit	Aborted growth	81
		Branching	109
		Internode length	16
		Other plant habit	174
5	Leaf morphology	Leaf complexity	27
		Leaf size	98
		Leaf texture	16
		Leaf width	12
		Other leaf development	188
6	Leaf colour	Dark green leaf	19
		Dull green/grey leaf	19
		Purple leaf	17
		Variegation	20
		White leaf	4
		Yellow leaf	28
		Yellow green leaf	79

7	Flowering	Late flowering	142
8	Inflorescence	Inflorescence structure	28
9	Flower morphology	Flower homeotic mutation	5
		Flower organ size	10
		Flower organ width	12
		Other flower morphology	4
10	Flower colour	Pale yellow flower	13
		White flower	5
11	Fruit size	Large fruit	212
		Small fruit	192
12	Fruit morphology	Long fruit	43
		Other fruit morphology	2
		Rounded fruit	4
13	Fruit colour	Dark red fruit	0
		Green fruit	0
		Orange fruit	0
		Yellow fruit	3
14	Fruit number	Absent	1,073
		Few	1,393
		Many	64
15	Sterility	Partial sterility	2,125
		Total sterility	576
16	Seed germination into fruit	Seed germination into fruit	208
17	Disease and stress response	Necrosis	48
		Wilting	80
		Other disease response	0

Number of plants bearing a specific mutant phenotype. Since a single plant may also be recorded more than once, if it was scored for more than one phenotype, the numbers are not additive.

The data reported in this table refer to the phase of M2 population development.

39% of tomato M2 plants showed at least one visual mutant trait and among these lines 37% displayed multiple phenotypes that fall into more than one major class of phenotypes. The most commonly observed phenotypes are related to the cotyledons (368), the leaf morphology (341), the habit (380) and the plant size (307) classes. In Figure 1 examples of tomato mutant traits are shown.

2.3 LycoTILL database

In order to manage and integrate the recorded phenotypic data, we implemented the database LycoTILL. LycoTILL was developed according to a relational database system, interconnecting three main modules: lines, class and subclass of phenotypes. The database interrogation can be done according to the phenotypic catalog, previously reported, or by plant code number (plant name) or family name. The result displays all the collected phenotypic information as well as photos of the mutant lines. LycoTILL, that is an evolving database, is publicly accessible through the web interface [34].

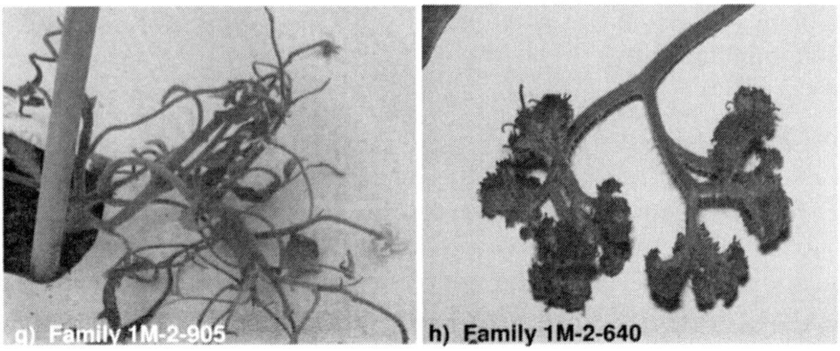

Figure 1 Examples of tomato mutant phenotypes. Mutations affecting: **a**) fruit morphology, **b**) fruit colour, **c**) plant habit, **d**) leaf morphology, **e**) cotyledon number, **f**) flower morphology, **g**) plant habit, **h**) inflorescence structure.

2.4 Mutation discovery in Red Setter mutant collection

To set up the tomato TILLING platform, DNA samples were prepared from 5,221 M3 families corresponding to 3,924 and 1,297 families obtained by treatment with 0.7% and 1% EMS respectively. The selection of M3 families was based on the M3 seed abundance, 287 families were discarded due to their low seed set. DNA samples were organized in pools of 8 M3 families. To validate the 0.7% and the 1% EMS Red Setter TILLING platforms and to estimate the mutation density of the populations, we chose seven genes involved in fruit quality traits. In particular we analysed ripening-inhibitor (*RIN*) and green ripe (*Gr*) genes involved in the ripening of tomato fruit, rab11 GTPase (*Rab11a*) and expansin 1 (*Exp1*) genes related to the tomato softening control, polygalacturonase (*PG*) gene involved in the cell wall hydrolysis, and lycopene beta cyclase (*Lcy-b*) and lycopene epsilon cyclase (*Lcy-e*) involved in the carotenoid biosynthesis pathway.

The CODDLE program (Codons Optimized to Discover Deleterious Lesions [35, 36] combined with the PRIMER3 tool [37] was used to define the best amplicon for TILLING analysis. Mutations were detected in the amplified targets using the mismatch-specific endonuclease ENDO1 as previously described [19, 38].

In total 9.5 kb of tomato genome were screened and 66 induced point mutations were identified (Table 3) of which 41 and 25 mutations were derived from the 0.7% and 1% EMS treated populations respectively. As expected for EMS mutagenesis, single nucleotide substitutions were identified both in coding and non-coding regions [17].

Table 3 Mutation density in 0.7% EMS and 1% EMS Red Setter populations

Target gene		No. of screened M3 families		No. of identified mutations		Overall mutation density	
Name	Amplicon size (kb)	0.7% EMS	1% EMS	0.7% EMS	1% EMS	0.7% EMS	1% EMS
Rab11a	0.407	1,373	713	1	3	1/559 kb	1/97 kb
PG	2.587	2,791	963	7	2	1/1031 kb	1/1246 kb
Exp1	1.025	3,885	1,284	14	6	1/284 kb	1/219 kb
RIN	1.331	3,885	1,284	4	8	1/1293 kb	1/214 kb
Gr	1.409	3,885	1,284	5	3	1/1095 kb	1/603 kb
Lcy-b	1.274	3,801	1,252	4	3	1/1211 kb	1/532 kb
Lcy-e	1.414	3,630	1,185	6	0	1/855 kb	-
Total/mean	9.447			41	25	1/574 kb	1/322 kb

The accession numbers of the analyzed seven target genes are the following: *Rab11a* [GenBank:AJ245570], *PG* [GenBank:M37304], *Exp1* [GenBank:AF548376], *RIN* [GenBank:AF448522], *Gr* [GenBank:DQ372897], *Lcy-b* [GenBank:CQ788383], *Lcy-e* [GenBank:Y14387]. The number of screened M3 families, the number of identified mutations and the overall mutation density, estimated as described in Methods, are reported both for 0.7% and 1% EMS Red Setter populations.

Missense mutations while no stop codon type of mutations was found. Using the SIFT programme (Sorting Intolerant From Tolerant [39, 40]), we analysed the putative impact of the missense mutations on the function of the tilled genes and 57.14% of the missense mutations were predicted deleterious for the protein's activity.

We calculated the mutation density in the seven targeted genes (Table 3) according to Dalmais et al. [19]and Greene et al. [17]. We estimated the mutation density at 1 mutation/322 kb in the 1% EMS and 1 mutation/574 kb in the 0.7% EMS Red Setter population.

3. DISCUSSION

The optimization of mutagenesis is a critical parameter in establishing a good mutant collection for forward and reverse genetic studies. In order to balance maximum mutation density with an acceptable plant survival rate we decided to utilise two different doses of EMS, 0.7% and 1%. A strict correlation was observed between the EMS doses and the toxicity, the mutation density obtained and the frequency of phenotype alterations. At 1% EMS the plant fertility rate was 41% less than the plant treated with 0.7%. In contrast, the 1% EMS yielded 1.78 fold more mutations per genome than 0.7% treated plant. At the phenotype level, 60% of the mutant phenotypes scored in the M2 generation were derived from the 1% EMS treated seeds.

In the TILLING screens we analysed seven genes and discovered a total of 66 induced point mutations. The spectrum of expected mutations in an EMS-treated population is essentially GC/AT transition because of the frequent alkylation of guanine residues by EMS [41]. In Arabidopsis, maize, wheat and pea, more than 99% of identified mutations are GC/AT transition [17, 19, 20, 24]. In our mutant Red Setter populations the percentage of observed GC/AT transition was 60% in the 0.7% EMS population and only 28.6% in the 1% EMS population. We also identified GC/TA, AT/TA, AT/CG, GC/CG and AT/GC transversions (Table 4). The spectrum of observed nucleotide changes in the 0.7% EMS-treated tomato population is similar to the spectrum of mutations observed in the reverse TILLING screens of rice and barley for which transitions in the range of 70% are reported [21, 23]. In contrast, the mutational spectrum of the 1% EMS population is different (AT/CG, GC/CG, see Table 4).

Table 4 Spectrum of mutations identified in Red Setter populations and their comparison to other organisms

Mutation		Tomato Red Setter		Barley [23]	Rice [21]
Type	Change	0.7% EMS	1% EMS		
Transition	GC/AT	60.0	28.6	70.0	70.0
Transversion	GC/TA	6.7	14.3	10.0	4.0
	AT/TA	20.0	14.3	10.0	15.0
	AT/GC	13.3	14.3	10.0	11.0
	AT/CG	0	14.3	0	0
	GC/CG	0	14.3	0	0
Total (%)		100	100	100	100

Distribution of the identified mutations in the different classes of nucleotide changes. In addition to the Red Setter mutant population data, reported as percentage values, the mutation spectrum of barley and rice are also shown.

In order to rule out the probability that natural polymorphisms, introduced through pollen or seed contamination, could be responsible for the non-GC/AT changes observed in our mutant populations, we analysed the natural sequence variation of the tilled genes using BLAST analysis [Additional file 1: Supplemental Figure S1] and EcoTILLING [42] of 150 tomato varieties among which 45 were Italian varieties (unpublished data). These analyses revealed that the nucleotide changes identified by TILLING were present neither in the available gene bank sequences nor in the screened tomato varieties. Based on this, we concluded that the non-GC/AT changes discovered in the TILLING screens do not result from cross pollination, but are new allelic variants generated by the mutagen action. This conclusion is also consistent with the non recovery of non-GC/AT changes in multiple genes in the same individual as reported for the Seattle Arabidopsis population, where rare contaminants were observed to introduce polymorphisms in more than one gene in the same plant [17].

Based on this we speculate that tomato might differ from other plant species in its mutagenic response to EMS doses. Moreover, we think that the choice of ENDO1 enzyme was fundamental in the detection of all types of changes that we observed in our mutant populations. For its high specificity in recognizing mismatches at the same rate [38] we could identify mutations never found in other plant species and with a higher frequency.

The TILLING screening performed on seven tomato genes permitted the calculation of the mutation density in the two mutant Red Setter populations. We estimated the mutation density at 1 mutation/322 kb in the 1% EMS and 1 mutation/574 kb in the 0.7% EMS Red Setter population.

The mutation densities calculated in the 1% and 0.7% EMS Red Setter populations are 2.3 and 1.2 times respectively higher than one mutation every 737 kb reported by Gady et al[29] in the 1% EMS TPAADASU tomato population.

This comparison demonstrate that our populations have a higher number of mutations respect to those so far available and published for tomato. The high mutation density of our populations, especially for the 1% EMS one, increases the size of allelic series that can be obtain and reduces the population size that needs to be screened.

Comparing the mutation densities estimated in the 1% and 0.7% EMS Red Setter populations with those described in other plant species results that they are 1.9 and 3.4 times respectively lower than one mutation per 170 kb reported previously for Arabidopsis [17] but their average (1/448 kb) is similar to those reported for maize (1/500 kb) [24] and rice (1/500 kb) [22] and 2.2 fold higher than one mutation per Mb found in barley by Caldwell et al. [23].

So far higher mutation densities were observed only in tetraploid wheat (1/40 kb) and hexaploid wheat (1/24 kb) [20]. It's likely that the polyploidy nature of their genomes helps in withstanding the mutagen action and consequently higher mutation frequencies can be obtained.

A mutant population is considered saturated with at least a single "hit" in every gene [6]. In the Red Setter TILLING platform more than one mutation was identified per gene analysed. We can therefore conclude that our mutant populations are sufficiently saturated. Furthermore by comparing other plant species used in public TILLING projects we can also affirm that our populations are suitable for use in high-throughput mutation discovery.

4. CONCLUSIONS

We have developed a new genetic resource in the tomato Red Setter genetic background by means of EMS mutagenesis. The mutant collection is organized as such that it could be used for both forward (EMS saturated mutant collection and the associated phenotypic database) and reverse (high-throughput TILLING platform) genetics in tomato, for both basic science or crop improvement.

The Red Setter TILLING platform is open to the scientific community to request TILLING screenings in genes of interest and to obtain material.

These services can be requested via database that also serves as portal for user need. In addition to our platform, at present, other tomato TILLING platforms are publicly accessible via web for requesting TILLING services.

All the available tomato TILLING platforms, including the Red Setter one, utilise mutant collections generated in different genetic backgrounds and with different EMS doses which increase the chance of obtaining a larger spectrum of alleles. Thus, it is of interest for the scientific community to have different tomato TILLING resources for the possibility of identifying a greater number of mutations of interest.

5. METHODS

5.1 EMS mutagenesis

Tomato seeds (cv Red Setter) were treated with two different concentrations (0.7% and 1%) of the chemical mutagen EMS (ethylmethane sulfonate) for 18 h at RT with gentle shaking. The seeds were then thoroughly washed, dried and sown in compost in 96 well seed trays which allowed an accurate determination of germination frequency.

Control seeds, those not exposed to EMS treatment, were treated in the same manner.

5.2 Plant material

M2 seeds: for the M2 seed production, M1 plants were grown according to standard tomato agronomic practice and at the end of the fruit-ripening phase, M2 seeds were collected from individual M1 plants and kept separate.

M3 seeds: 3 seeds belonging to each mutant M2 family were sown in 96 well seed trays and the corresponding seedlings transplanted in open field. M3 seeds were collected from single M2 plants.

5.3 M2 plant phenotyping and data collection tools

Phenotype scoring was performed at different developmental stages from seed germination through fruit ripening and seed harvest. Each mutant candidate was characterized according to 17 classes and 51 subclasses which are reported in Table 2. The selection of classes and subclasses was for the most part carried out on the basis of the phenotypic catalog reported by Menda et al [6].

Data were collected using a hand-held Asus MyPal 730w while pictures were taken by using the Nikon Coolpix 4500 digital camera.

5.4 Database construction

The phenotype database was developed using MySQL [43] as a relational database system.

5.5 DNA extraction and sample pooling

For each M3 family, the genomic DNA was extracted from four young leaves collected from four different plants of the same family. The leaf samples were collected in 96-well plates and the DNA was isolated by using DNeasy 96 Plant Kit (Quiagen, Hilden, Germany). The quantification of extracted DNA was carried out on a 0.8% agarose gel using λ DNA (Invitrogen, Carlsbad, CA, USA) as a concentration reference. Genomic DNA samples were then diluted tenfold and pooled eightfold to obtain the working material.

5.6 PCR amplification, mutation detection and validation

PCR amplification was based on nested-PCR and was carried out using two couples of target-specific primers. 4 ng of pooled genomic DNA was used for the first PCR and forward-strand primers and reverse-strand primers 5'-end labelled with IRDye 700 and IRDye 800 dye (LI-COR®, Lincoln, NE, USA) respectively were employed for the second PCR [19].

Mutation detection was performed as previously described [38]. Electrophoresis were performed on a LI-COR 4300 (LI-COR®, Lincoln, NE, USA) and gel images were analysed using Adobe Photoshop software (Adobe Systems Inc., San José, CA, USA).

After discovery, mutations were validated by sequence analysis.

The mutation frequency for each amplicon was calculated as previously described [19]. For the average mutation frequency we have summed the sizes of all amplicons and we have divided by the total number of identified mutants. The data were multiplied for 0.75 in order to eliminate the missing evaluation due to the presence of one-fourth wild-type alleles in the 1:2:1 Mendelian segregation in M3 generation [17].

ACKNOWLEDGEMENTS

This work was mainly funded by the Italian Ministry of University and Research (MIUR, ITALYCO Project, DD n° 603/RIC) and by the European project EU-SOL (Contract number: FOOD-CT-2006-016214) through the 6th framework programme.

This work was also supported by a Marie Curie fellowship. The authors thank J.J. Giovannoni for providing the Rin gene sequence and Stephan Summere for reviewing the English. The authors wish to thank all colleagues of Metapontum Agrobios for helping through all phases of the project.

REFERENCES

1. Tomato Functional Genomics Database. [http://ted.bti.cornell.edu/]
2. Fei Z, Tang X, Alba RM, White JA, Ronning CM, Martin GB, Tanksley SD, Giovannoni JJ: Comprehensive EST analysis of tomato and comparative genomics of fruit ripening. Plant J. 2004, 40 (1): 47-59. 10.1111/j.1365-313X.2004.02188.x.
3. Chiusano ML, D'Agostino N, Traini A, Licciardiello C, Raimondo E, Aversano M, Frusciante L, Monti L: ISOL@: an Italian SOLAnaceae genomics resource. BMC Bioinformatics. 2007, 9 (Suppl): S7-10.1186/1471-2105-9-S2-S7.
4. Moco S, Bino RJ, Vorst O, Verhoeven HA, de Groot J, van Beek TA, Vervoort J, de Vos CH: A liquid chromatography-mass spectrometry-based metabolome database for tomato. Plant Physiol. 2006, 141 (4): 1205-1218. 10.1104/pp.106.078428.
5. TGRC: Tomato Genetic Resourse Center. [http://tgrc.ucdavis.edu/]
6. Menda N, Semel Y, Peled D, Eshed Y, Zamir D: In silico screening of a saturated mutation library of tomato. Plant Journal. 2004, 38: 861-872. 10.1111/j.1365-313X.2004.02088.x.
7. Watanabe S, Mizoguchi T, Aoki K, Kubo Y, Mori H, Imanishi S, Yamazaki Y, Shibata D, Ezura H: Ethylmethansulfonate (EMS) mutagenesis of Solanum lycopersicon cv. Micro-Tom for large-scale mutant screens. Plant Biotechnololy. 2007, 24: 33-38.
8. Analysis of the genome sequence of the flowering plant Arabidopsis thaliana. Nature. 2000, 408 (6814): 796-815. 10.1038/35048692.
9. The map-based sequence of the rice genome. Nature. 2005, 436 (7052): 793-800. 10.1038/nature03895.
10. Jaillon O, Aury JM, Noel B, Policriti A, Clepet C, Casagrande A, Choisne N, Aubourg S, Vitulo N, Jubin C: The grapevine genome sequence suggests ancestral hexaploidization in major angiosperm phyla. Nature. 2007, 449 (7161): 463-467. 10.1038/nature06148.
11. International Tomato Genome Sequencing Project. [http://www.sgn.cornell.edu/about/tomato_sequencing.pl]
12. Mathews H, Clendennen SK, Caldwell CG, Liu XL, Connors K, Matheis N, Schuster DK, Menasco DJ, Wagoner W, Lightner J: Activation tagging in tomato identifies a transcriptional regulator of anthocyanin biosynthesis, modification, and transport. Plant Cell. 2003, 15 (8): 1689-1703. 10.1105/tpc.012963.
13. Cooley MB, Goldsbrough AP, Still DW, Yoder JI: Site-selected insertional mutagenesis of tomato with maize Ac and Ds elements. Mol Gen Genet. 1996, 252 (1-2): 184-194. 10.1007/BF02173219.
14. McCallum CM, Comai L, Greene EA, Henikoff S: Targeting Induced Local Lesions IN Genomes (TILLING) for plant functional genomics. Plant Physiol. 2000, 123 (2): 439-442. 10.1104/pp.123.2.439.
15. Colbert T, Till BJ, Tompa R, Reynolds SH, Steine MN, Yeung AT, McCallum CM, Comai L, Henikoff S: High-throughput screening for induced point mutations. Plant Physiol. 2001, 126: 480-484. 10.1104/pp.126.2.480.
16. Till BJ, Colbert T, Codomo C, Enns L, Johnson J, Reynolds SH, Henikoff JG, Greene EA, Steine MN, Comai L: High-throughput TILLING for Arabidopsis. Methods Mol Biol. 2006, 323: 127-135.

17. Greene EA, Codomo CA, Taylor NE, Henikoff JG, Till BJ, Reynolds SH, Enns LC, Burtner C, Johnson JE, Odden AR: Spectrum of chemically induced mutations from a large-scale reverse-genetic screen in Arabidopsis. Genetics. 2003, 164 (2): 731-740.
18. Martin B, Ramiro M, Martinez-Zapater JM, Alonso-Blanco C: A high-density collection of EMS-induced mutations for TILLING in Landsberg erecta genetic background of Arabidopsis. BMC Plant Biol. 2009, 9: 147-10.1186/1471-2229-9-147.
19. Dalmais M, Schmidt J, Le Signor C, Moussy F, Burstin J, Savois V, Aubert G, Brunaud V, de Oliveira Y, Guichard C: UTILLdb, a Pisum sativum in silico forward and reverse genetics tool. Genome Biol. 2008, 9 (2): R43-10.1186/gb-2008-9-2-r43.
20. Slade AJ, Fuerstenberg SI, Loeffler D, Steine MN, Facciotti D: A reverse genetic, nontransgenic approach to wheat crop improvement by TILLING. Nat Biotechnol. 2005, 23 (1): 75-81. 10.1038/nbt1043.
21. Till BJ, Cooper J, Tai TH, Colowit P, Greene EA, Henikoff S, Comai L: Discovery of chemically induced mutations in rice by TILLING. BMC Plant Biol. 2007, 7: 19-10.1186/1471-2229-7-19.
22. Wu JL, Wu C, Lei C, Baraoidan M, Bordeos A, Madamba MR, Ramos-Pamplona M, Mauleon R, Portugal A, Ulat VJ: Chemical- and irradiation-induced mutants of indica rice IR64 for forward and reverse genetics. Plant Mol Biol. 2005, 59 (1): 85-97. 10.1007/s11103-004-5112-0.
23. Caldwell DG, McCallum N, Shaw P, Muehlbauer GJ, Marshall DF, Waugh R: A structured mutant population for forward and reverse genetics in Barley (Hordeum vulgare L.). Plant Journal. 2004, 40: 143-150. 10.1111/j.1365-313X.2004.02190.x.
24. Till BJ, Reynolds SH, Weil C, Springer N, Burtner C, Young K, Bowers E, Codomo CA, Enns LC, Odden AR: Discovery of induced point mutations in maize genes by TILLING. BMC Plant Biol. 2004, 4: 12-10.1186/1471-2229-4-12.
25. Cooper JL, Till BJ, Laport RG, Darlow MC, Kleffner JM, Jamai A, El-Mellouki T, Liu S, Ritchie R, Nielsen N: TILLING to detect induced mutations in soybean. BMC Plant Biol. 2008, 8: 9-10.1186/1471-2229-8-9.
26. Perry JA, Wang TL, Welham TJ, Gardner S, Pike JM, Yoshida S, Parniske M: A TILLING reverse genetics tool and a web-accessible collection of mutants of the legume Lotus japonicus. Plant Physiol. 2003, 131 (3): 866-871. 10.1104/pp.102.017384.
27. Xin Z, Wang ML, Barkley NA, Burow G, Franks C, Pederson G, Burke J: Applying genotyping (TILLING) and phenotyping analyses to elucidate gene function in a chemically induced sorghum mutant population. BMC Plant Biol. 2008, 8: 103-10.1186/1471-2229-8-103.
28. Rigola D, van Oeveren J, Janssen A, Bonne A, Schneiders H, Poel van der HJ, van Orsouw NJ, Hogers RC, de Both MT, van Eijk MJ: High-throughput detection of induced mutations and natural variation using KeyPoint technology. PLoS One. 2009, 4 (3): e4761-10.1371/journal.pone.0004761.
29. Gady AL, Hermans FW, Wal Van de MH, van Loo EN, Visser RG, Bachem CW: Implementation of two high through-put techniques in a novel application: detecting point mutations in large EMS mutated plant populations. Plant Methods. 2009, 5: 13-10.1186/1746-4811-5-13.
30. Moens CB, Donn TM, Wolf-Saxon ER, Ma TP: Reverse genetics in zebrafish by TILLING. Brief Funct Genomic Proteomic. 2008, 7 (6): 454-459. 10.1093/bfgp/eln046.
31. Winkler S, Schwabedissen A, Backasch D, Bokel C, Seidel C, Bonisch S, Furthauer M, Kuhrs A, Cobreros L, Brand M: Target-selected mutant screen by TILLING in Drosophila. Genome Res. 2005, 15 (5): 718-723. 10.1101/gr.3721805.
32. Matilla AJ: Ethylene in seed formation and germination. Seed Science Research. 2000, 10: 111-126.
33. Kucera B, Cohn MA, Leubner-Metzger G: Plant hormone interactions during seed dormancy release and germination. Seed Science Research. 2005, 15: 281-307. 10.1079/SSR2005218.
34. LycoTILL: Tomato Mutant DB. [http://www.agrobios.it/tilling/index.html]

35. Till BJ, Reynolds SH, Greene EA, Codomo CA, Enns LC, Johnson JE, Burtner C, Odden AR, Young K, Taylor NE: Large-scale discovery of induced point mutations with high-throughput TILLING. Genome Res. 2003, 13 (3): 524-530. 10.1101/gr.977903.
36. CODDLE: Codons Optimized to Discover DeleteriousLEsions. [http://www.proweb.org/coddle]
37. Rozen S, H S: Primer3 on the WWW for general users and for biologist programmers. Bioinformatics Methods and Protocols: Methods in Molecular Biology. Edited by: Krawetz SA, Misener S. 1999, Totowa NJ: Humana Press, 365-386. full_text.
38. Triques K, Sturbois B, Gallais S, Dalmais M, Chauvin S, Clepet C, Aubourg S, Rameau C, Caboche M, Bendahmane A: Characterization of Arabidopsis thaliana mismatch specific endonucleases: application to mutation discovery by TILLING in pea. Plant J. 2007, 51 (6): 1116-1125. 10.1111/j.1365-313X.2007.03201.x.
39. SIFT: Sorting Intolerant From Tolerant. [http://blocks.fhcrc.org/sift/SIFT.html]
40. Ng PC, Henikoff S: SIFT: Predicting amino acid changes that affect protein function. Nucleic Acids Res. 2003, 31 (13): 3812-3814. 10.1093/nar/gkg509.
41. Sega GA: A review of the genetic effects of ethyl methanesulfonate. Mutat Res. 1984, 134: 113-142.
42. Comai L, Young K, Till BJ, Reynolds SH, Greene EA, Codomo CA, Enns LC, Johnson JE, Burtner C, Odden AR: Efficient discovery of DNA polymorphisms in natural populations by Ecotilling. Plant J. 2004, 37 (5): 778-786. 10.1111/j.0960-7412.2003.01999.x.
43. MySQL. [http://www.mysql.com/]

CHAPTER 11

Overview of Organic Cover Crop-Based No-Tillage Technique in Europe: Farmers' Practices and Research Challenges

Laura Vincent-Caboud [1,*], Joséphine Peigné [1], Marion Casagrande [1,2] and Erin M. Silva [3]

[1] *Department of Agroecology and Environment, ISARA-Lyon (member of the University of Lyon), 23 rue Jean Baldassini, F-69364 Lyon Cedex 07, France*

[2] *ITAB, Quartier Marcellas, F-26800 Etoile sur Rhône, France*

[3] *Department of Agronomy, University of Wisconsin-Madison, Madison, 53706 WI, USA*

AUTHOR CONTRIBUTIONS

L. Vincent-Caboud reviewed the literature and wrote the initial draft of the paper with assistance from J. Peigné. L. Vincent-Caboud, J. Peigné, M. Casagrande, and E. M. Silva contributed to revising the manuscript.

ABSTRACT

Cover crop mulch–based no-tillage (MBNT) production is emerging as an innovative alternative production practice in organic farming (OF) to reduce intensive soil tillage. Although European organic farmers are motivated to implement MBNT to improve soil fertility and achieve further management benefits (e.g., labor and costs savings), low MBNT practice is reported in Europe. Thus, this paper aims to understand the challenges of both farmers and researchers limiting the further adoption of MBNT in organic farming in temperate climates. The primary no-tillage (NT) practices of organic European farmers and findings of organic MBNT studies conducted in Europe are reviewed, focusing on living or mulch cover crop-based NT (LBNT or MBNT) for arable crop production. Major conclusions drawn from this review indicate consistent weed control and an establishment of best practices for cover crop management as the two main overarching challenges limiting adoption. In view of substantial gaps of knowledge on these issues, additional research should focus on cover crop selection and management (species, date of sowing) to increase cover crop biomass, particularly in warmer climates. Lastly, further research is needed to optimize cover crop termination to prevent competition for water and nutrients with cash crops, particularly in wetter northern conditions which promote vigorous cover crop growth.

Keywords: no tillage; cover crops; organic farming; roller crimper; conservation tillage

1. INTRODUCTION

Conservation agriculture (CA) has been developed in conventional farming to minimize and prevent soil erosion, reduce labor and energy inputs, and preserve soil fertility [1,2,3,4]. The Food and Agriculture Organization (FAO) has defined conservation agriculture as a range of practices based on three main principles: (1) minimum soil disturbance obtained using reduced-tillage (RT) or no-tillage (NT); (2) a permanent soil cover via living or dead (mulch) cover crops; and (3) diversified crop rotations [2]. NT involves production practices with which a cash crop is sown into soil that has not been tilled since the previous harvest [5] (Figure 1). As highlighted by several authors, NT may also be referred to as direct seeding or direct drilling, terms that are often employed in European publications [3,6], although some North American researchers have stressed differences in soil disturbance between NT and direct drilling [7]. Typically, with the use of NT practices, the soil surface is only disturbed down to three to five centimeters in depth in the seeding row, with significant amounts of crops or cover crop residues is maintained on the soil surface to preserve soil quality and suppress weeds [8].

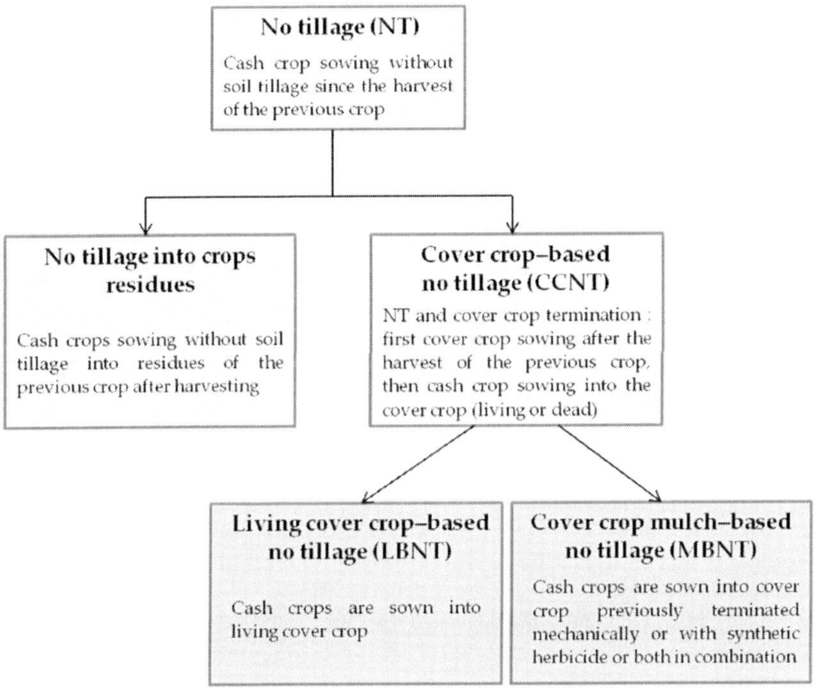

Figure 1. Diagram of different techniques of no-tillage and cover crop management.

The specific integration of cover crops into NT practices is an innovative alternative recently developed to mitigate environmental pollution, such as nitrate leaching, reduce inputs (fertilization, herbicides), and suppress weeds via cover crop competition for nutrient and water resources [9,10] (Figure 1). Cover crop mulch–based NT and living cover crop-based NT have gained increasing attention as methods to further enhance the agroecosystem services of NT systems, particularly in organic agriculture. Depending on the cover crop species used in the system, weed control is achieved through physical suppression preventing access to light, and reducing soil temperatures, competition with nutrient resources, and allelopathic interactions [11]. Cover crop-based NT (CCNT) techniques, i.e., MBNT and LBNT, consist of sowing the main crop into a living cover crop, frost-killed cover crop, or cover crop terminated either mechanically (roller crimper, mower) or with a synthetic herbicide, or both in combination (Figure 1). These strategies, in addition to limiting soil disturbance and suppressing weeds, also provides an opportunity to address another principle of conservation agriculture by improving living soil cover. Although CCNT is employed in conventional agriculture to promote soil fertility, control weeds, increase water infiltration, and preserve soil moisture [12], these systems often integrate the use of herbicides to manage weeds and/or terminate the cover crop, thereby raising environmental sustainability issues [8,13].

Organic farming (OF) has also emerged as an alternative for maintaining soil fertility and reducing the harmful environmental impacts of agriculture. The European standard No. 834/2007 bans the use of chemical synthetics (pesticides and fertilizers) and limits off-farm and synthetic inputs in OF. Current organic practices use crop rotation as a foundational practice to provide nutrient recycling and to control weeds, pests, and diseases [14]. Theoretically, OF offers many advantages in the shift toward sustainable agricultural systems via the increase of soil organic matter and biodiversity. In reality, however, without use of synthetic herbicides, farmers often depend on intensive tillage to manage weeds [15,16]. This can result in soil degradation: decreases in biological activity and biodiversity, increases in soil erosion, losses in organic matter content, and destruction of soil physical structure due to frequent machine traffic and tillage activities on the field [17,18]. Organic NT represents an opportunity to mitigate these drawbacks, improve soil fertility components, and provide further benefits regarding European farmers' working conditions (labor savings, energy consumption reduction, etc.) [13].

Despite the potential benefits of integrating NT practices into organic cropping systems (including, but not limited to, MBNT and LBNT) which have been reported by researchers, and the interest shown by European farmers, NT has not been widely implemented in Europe and few research studies have been conducted investigating this technique [19,20]. The trials that have been carried out in Europe and interviews with farmers using organic conservation tillage including MBNT bring to light the typical practices, motivations, and challenges related to conservation techniques [19,20]. This paper aims to synthesize knowledge and conclusions drawn from previous European research projects to identify the overarching challenges facing research in Europe in the promotion of organic MBNT in arable crops.

Based on a literature review and interviews of European organic farmers, we first explain the methodology selected to address the goal of this paper. Then, we present a broad overview of organic farmers' NT and cover crop practices in Europe and identify their main challenges. Next, we examine, in detail, the knowledge drawn from European research to address the organic farmers' difficulties. Lastly, we discuss the primary future research challenges to improve our understanding of organic MBNT under a temperate climate.

2. METHODOLOGY

Several papers have been published in North America on organic MBNT, summarizing the results of research trials based on cover crop species [21,22,23,24,25,26], cover crop termination methods and timing [24,27, 28,29, 30,31,32,33], weed suppression [34,35,36,37,38,39,40,41], and cash crop sowing (fertilization strategies, row spacing, seeding rate) [24,42,43]. Nevertheless, in Europe, very few studies exist that focus on organic MBNT, raising questions of the effectiveness and appropriateness of this technique to enhance the sustainability of European organic farms [44]. Concerns related to the implementation of organic MBNT in Europe must be articulated to identify future research specifically addressing the implementation of MBNT on European farms. Indeed, despite the robust discussion of the challenges related to MBNT by American researchers [16,35,36,39], these challenges may differ in Europe due to its unique agricultural context (e.g., climate, information or equipment access, organic seed availability, etc.).

In Europe, although a range of studies and review articles have been carried out on NT in conventional farming [8,45,46], these papers did not specifically address MBNT, leaving a dearth of information on MBNT, particularly regarding its unique integration into organic production. In Europe, published papers dealing with tillage issues in OF are focused on RT, and it is difficult to discern from this knowledge about NT [13,15,17,20,44,47,48,49]. While there are some references referring to organic MBNT in arable crops, data analysis on this technique is often presented in the context of a comparison with traditional ploughing and RT [15,20,50,51,52]. This research strategy limits the generation of data which could contribute to the refinement of the MBNT technique and the development of best management practices, while diluting access of MBNT information.

In order to address this resource gap, this review paper focuses solely on organic MBNT and LBNT, identifying challenges specific to these NT techniques on organic European farms. The review is organized into four sections: the first two sections summarize MBNT and LBNT practices and challenges on European organic farms, considering separately two distinct climatic regions in Europe: the northern region (≤45° N) and the southern region (≤45° N). These sections are informed by two surveys, one conducted across Europe and the second focused on France, which aimed to identify conservation practices of organic farmers focusing on NT, RT, and green manure techniques [19,20,53]. Green manure, also called cover crop, is referred to as any crops implemented with the aim of soil fertility preservation and improvement. While certain aspects of these surveys' data have been published elsewhere, we will employ a unique approach focusing solely on the farmer-reported practices and perceptions regarding NT and cover crop management, and place the survey information against the backdrop of the European literature on conservation agricultural practices in both organic and conventional systems, creating a framework upon which to overview knowledge and challenges specific to the organic MBNT in Europe [13,17,18,54,55]. In particular, review papers focused on the implementation of NT practices within the European biological, physical, and chemical contexts are used to describe the pedo-climatic effects on MBNT implementation in Europe [3,8,46,56]. In the third section, we summarize research focused on organic CCNT (i.e., MBNT and LBNT) in Europe, which further informs the fourth section, in order to identifying remaining gaps in the understanding of MBNT in Europe needed to promote farmer adoption and success with techniques combining NT with intensive cover cropping.

3. EUROPEAN ORGANIC FARMERS' NO-TILLAGE AND COVER CROP PRACTICES

Existing published literature describing the results of controlled NT research studies, do not necessarily report on the specific attitudes and practices employed by European farmers' using MBNT or LBNT practices. Within the literature that does describe farmer implementation of MBNT practices, farmers report low adoption of organic MBNT and LBNT techniques, in part due to challenges with consistent termination of cover crops without the use of synthetic herbicides [13,47]. A recent European survey carried out in 2012 with 159 organic farmers who are applying conservation principles (i.e., at least one of the following practices: reduced tillage, cover crop, no tillage) (see [20]) revealed that even within this progressive cohort of farmers utilizing sustainable RT and cover crop practices, only 27% of interviewed farmers used NT techniques [20]. This mirrored the results of an earlier 2010 French survey interviewing 24 producers, which also confirmed that very few farmers had experimented MBNT [53].

3.1. Organic No-Tillage and Cover Crop Practices in Europe

Organic NT approaches vary across Europe, largely due to differences in climate, national cultures, and/or information access [20]. Much of Europe experiences a temperate climate, with consistent humid and cool conditions, as compared to other regions of the globe with higher seasonal rainfall and/or drought periods, as encountered in North America [8]. However, warmer and drier climates can be found in Southern Europe, with these types of environments expected to expand with global warming, increasing the risk of soil degradation [8]. With these changes in climatic conditions, interest in organic NT is expected to increase as additional production strategies are sought to reduce soil erosion and preserve soil moisture in the face of increasing extreme weather events. In this section, we will discuss the main practices and challenges in two European areas: the northern region (>45° N), where 109 organic farmers were interviewed in 2012, and the southern region (≤45° N, with 50 interviewed farmers interviewed in 2012). Conservation practices implemented by these farmers may cover a range of techniques [20], but we focus in this paper on NT (including MBNT and LBNT), in tandem with the cover crop information drawn from these surveys.

As reported in the 2012 aforementioned survey of organic farmers using progressive, sustainable practices, in the northern part of Europe, only 23% of the 109 interviewed farmers used NT, while 93% of farmers implemented a cover crop in their crop rotation to increase soil N supply to the subsequent crop [20]. Maëder and Berner [13] have shown that the humid and cool conditions experienced in northern climates increases the difficulty of implementing NT. Researchers have recognized the role of tillage in managing weeds and accelerating soil warming to foster crop emergence, which otherwise could be delayed by large quantities of mulch left on the soil surface (Table 1). Furthermore, humid conditions enable cover crops to be established rapidly, thus influencing northern farmers' focus on cover crop practices rather than NT [20].

Conversely, in the southern part of Europe, 34% of the 50 interviewed farmers use NT, with only 48% implementing a cover crop [20]. NT is more widespread among southern farmers, likely due to soil degradation and soil water loss caused by intensive tillage under warm and dry conditions where soil is more exposed to water and wind erosion [8]. Moreover, higher temperatures accelerate soil warming, promoting crop emergence and contributing to the success of NT (Table 1). However, while implementation NT practices were greater among the southern farmers, this group demonstrated a decreased use of cover crops as compared with northern European farmers. This illustrates the trade-offs regarding sustainable practices which exist in the south; while cover crops are difficult to establish in the water-limited southern climate, this same condition promotes alternative sustainable practices, the adoption of intensive non-inversion tillage strategies to manage weeds, maintain yields, and incorporate fertilizers to limit ammoniac volatilization [20].

Table 1. Opportunities and challenges for organic MBNT adoption in the European context.

European Regions	Opportunities	Challenges	References
Northern areas (>45° N)	High cover crop biomass potential Large number of farmers with experience of cover crop practices Soil preservation to cope with climate change	Slow soil warming Moisture conditions foster rapid weed development and degradation of cover crop residues Cover crop management/termination difficulties Low NT adoption Fewer experienced farmers in CA	[8,13,19,20]
Southern areas (≤45° N)	Fast soil warming reduces soil erosion and maintains soil moisture in the face of climate change More experienced farmers in NT	Water and nutrient deficiencies for the cash crop (more irrigation required) Less application of cover crops in the crop rotation Cover crop establishment difficulties (under warm and dry conditions) Fewer farmers experienced in cover crop practices	

NT = no-tillage; CA = conservation agriculture.

Due to these trade-offs, the implementation of multiple practices integrating NT principles, i.e., including LBNT and MBNT, represents a real challenge.

3.2. Cover Crop Management

The majority of the 159 interviewed farmers implemented cover crops in their crop rotation strategies, but only 19% of them combined NT with a cover crop [20], likely due to the limited termination methods for cover crops in OF, which relies exclusively either on a mechanical solution or frost [18]. Cover crop regrowth due to inconsistent or incomplete termination may compete with the cash crop for nutrients and water resources. Detailed management descriptions of the winter cropping practices of 68 farmers applying cover crops showed that most terminated the cover crop by undercutting or mowing, but only 2% of them by rolling [20]. Thus, the cover crop often remained a living cover when combined with MBNT. Moreover, organic farmers recognized the critical trade-offs between the management of the cover and cash crops, including the impacts on (i) managing weeds with the use of a false seed-bed or (ii) incorporating a cover crop to provide nitrogen (N) to the subsequent crops [57]. Such practices do not appear to be aligning cover crop management with MBNT principles. Thus, alternatives are needed to help organic farmers manage weeds and allow for the incorporation of N-rich cover crops without tilling.

3.3. Application in Crop Rotations

NT primarily integrates the use of winter cereal crops, such as wheat (*Triticum vulgare* L.), barley (*Hordeum vulgare* L.), or triticale (*x Triticosecale* spp. L.), seeded into a living cover crop or by direct sowing following the harvest of the previous main crop. Only 2.5% of the 40 interviewed farmers using NT in Europe sowed spring crops (e.g., soybean (*Glycine max* L.) or corn (*Zea mays* L.)) using NT practices [20], with the practice only implemented where conditions are ideal (e.g., low weed pressure, good climate conditions, etc.).

The length of crop rotations in which NT crops are included averaged seven years [20]. Diversified crop rotations are typically used, including cereals, legumes, and other species, such as *Brassicacae*. Among the 40 farmers using NT, crop rotations are mainly based on alternating spring and winter crops to manage weeds, and include an average of 25% spring crops, 35% winter crops, 10% other legume crops, and 30% cover crops used to increase soil N supply to the subsequent crop. Tillage is generally not completely absent from crop rotations, and occasional ploughing is performed at some stages (e.g., before spring sowing, or to incorporate cover crops). Farmers limit the application of NT on specific crops in crop rotation (e.g., wheat, barley, etc.) with occasional ploughing at a mean frequency of 0.6 times over 10 years among the interviewed farmers [20].

3.4. Farmers' Experience in No Tillage Production

The European survey showed that organic farmers lack experience in NT, with only five years of practice compared with an average of 16 years with the use of cover crops [20]. Results indicated that only 10% of 40 organic farmers using NT applied this technique without adopting RT practices, with farmers using NT averaging 12 years of experience in RT and 13 years in cover crops. This latter finding suggests that farmers using NT are not the most experienced in CA, confirming what has also been observed in the French survey [53]. Organic farmers' progress in NT techniques occurs through exploring solutions on a farm-by-farm basis, integrating the unique objectives and constraints of their farm organization.

The situation across Northern Europe contrasts with southern farmers who have slightly greater experience in NT, with an average of seven years of practice, as opposed to five years in northern areas among the 40 interviewed farmers applying NT [20]. These differences in farmers' practices and experiences using NT highlight the need for better understanding of the main challenges unique to their regions and farming systems if relevant solutions are to be provided in a broader European context.

This overview of NT and cover crop practices in Europe reveals the obstacles facing organic farmers to combine both techniques in accordance with the pedo-climatic context (Table 1).

4. THE CHALLENGES OF LIVING OR KILLED (MULCH) COVER CROP-BASED NO-TILLAGE IN EUROPEAN ORGANIC FARMING

4.1. Weed Management

European researchers indicated insufficient weed suppression and resulting high weed pressures as the main reason for the low adoption of organic NT practices in Europe [48,49,58,59]. For some farmers, CCNT practices (MBNT and LBNT) offer a means of suppressing weeds without mechanical methods; however, maintaining weed suppression throughout the entire cash crop production season from crop sowing through to harvest was cited as a major difficulty raised by farmers (Figure 2) [19]. Lefèvre et al. [53] also stressed weed suppression as a primary concern of interviewed French farmers. Indeed, the possible need to intensify mechanical weed control during a crop rotation including NT due to inadequate weed management in the NT phase may lead to additional issues at the rotational scale (e.g., soil erosion, increase in labor and investment, etc.). This problem is illustrated by current farmer practices where tillage is not entirely removed from the crop rotation [20]; weed management concerns appear more important for northern farmers who are slower to adopt NT practices (Figure 2). In northern environments, humid conditions promote weed development throughout the cash crop production season, including in CCNT if the residue biomass is inadequate to provide season-long weed suppression; with significant cover crop residue on the soil surface, CCNT limits the ability of farmers to employ mechanical methods of weed management.

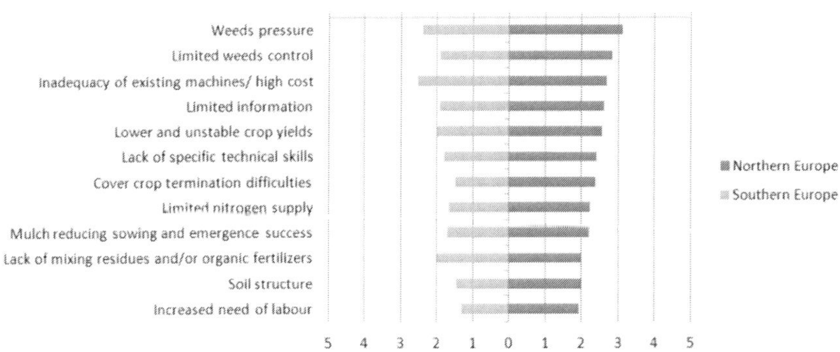

Figure 2. The challenges of no-tillage for 159 European organic farmers (according to Casagrande et al. [19]). The means of Likert scale values of the challenges facing organic farmers in Southern Europe (**left**) and Northern Europe (**right**) among the forty interviewed farmers practicing NT in Europe in 2012 are shown. The Likert scale is based on five values: (1) not important, (2) low importance, (3) moderate importance, (4) very important, (5) highly important (the number of interviewed farmers is, respectively, 25 and 15 in northern and southern areas).

4.2. Inadequacy of Available NT Equipment and Low Technical Skills and Information

European farmers reported an inability to access information related to the organic NT technique to guide their management decision [19]. More specifically, Peigné et al. [20] found that southern farmers have less access to NT information resources than northern farmers, despite the higher number of NT practitioners observed in South Europe. According to Lefèvre et al. [53], the shortage of technical guidelines leads farmers to explore solutions independently on their own farms. For example, lack of knowledge as to the best practices for seeding cash crops into a flattened cover crop is an overarching challenge which is consistent with the low level of NT experience observed in Europe [20].

Additionally, the high cost and the low availability of NT equipment is a limiting factor in the farmers' ability to invest in effective machinery, with the resulting lack of adequate equipment reducing NT success [8]. Indeed, Peigné et al. [18] have demonstrated the importance of specific machines needed for MBNT production (e.g., roller crimper, NT seeder), which remain very scarce on European organic farms with only a few interviewed farmers rolling the cover crop for termination. Equipment must be designed, manufactured, and/or modified to be more adapted to MBNT [15]. European researchers in conventional farming also presented the small quantity of NT equipment manufactured in Europe as a main challenge [8,46]. Above all, according to Derpsch et al. [46] and Freidrich et al. [56], European farmers who want to use specific and efficient machinery in CA must import the equipment from North America. The authors further explain that while efficient equipment exists in Europe, the lack of technical support for farmers could explain some of the failures observed in NT on European farms. Thus, supporting farmers to build and/or modify NT equipment, or invest in the purchase of NT equipment with other farmers, could generate more appropriate machines to specific soil and climate conditions and reduce costs.

4.3. Unstable Crop Yields

Lower yields in NT compared with traditional tillage illustrates the effect of weeds and cover crop competition for water and nutrient resources, which may lead to a lower net return, as evidenced by French farmers [53]. Furthermore, inconsistent crop yields may be exacerbated by climate change, which may be even more profound using NT practices [8]. Under humid and cool conditions, cash crop yields experience greater instability if the cover crop is not successfully managed, leading to insufficient weed control and competition from the cover crop. In southern parts of Europe, the occurrence of summer drought in relation to cash crop planting and establishment may lead to significant yield losses if irrigation is not available, as illustrated by Delate et al. [21] in Iowa. As such, farmers who do implement NT in southern regions tend to use, more frequently, the technique only on winter crops without a cover crop to limit competition with the main crop.

4.4. Cover Crop Termination

MBNT and LBNT are challenging for farmers with respect to issues that arise with cover crop termination without the use of synthetic pesticides, relying instead on mechanical methods or termination of the cover crop at frost [18,19]. Although farmers tend to use frost-sensitive cover crop species (e.g., certain clover species, common vetch (*Vicia sativa*), etc.), a winter-killed cover crop may not allow for adequate maintenance of residue on the soil surface after the winter and through the cash crop growing season, which is essential to ensure weed suppression until crop harvest, with biomass limited by either fall growth of the cover crop or decomposition of residues. Cover crop termination remains an even greater concern for northern farmers, as wet and cool spring conditions can create an environment fostering strong cover crop growth while soil conditions limit field operations, creating difficulties with management (Figure 2). Additionally, the French survey also indicated additional issues with cover crop management, including the high cost of seed purchasing, greater labor needs, and difficulties related to cover crop establishment (unevenly spread, lack of N availability) [53].

4.5. Crop Emergence

In the case of MBNT, technical and mechanical issues related to direct seeding into a flattened cover crop may limit the seed-to-soil contact and contribute to farmers' concerns regarding crop emergence (Figure 2). This is of particular concern for northern farmers, who must grapple with the challenge of planting through thick cover crop biomass [15]. Moreover, Morris et al. [60] and Soane et al. [8] have highlighted that, in conventional farming systems, soil temperatures are lower under large amounts of residue on the soil surface, delaying crop emergence. Additionally, concerns exist related to limited N supply to the subsequent cash crops, with slower mineralization rates resulting from cooler soil temperatures and significant soil carbon additions. The challenge regarding N availability, also noted by Cooper et al. [44], is of even greater importance for northern farmers experiencing lower spring temperatures, thus aggravating the delay in N supply.

4.6. Soil Structure

Improving soil properties is often a primary motivator for farmers interested in adopting CA [8,19,61]. Maintaining soil structure, however, may represent a challenge in NT due to the heavy required equipment (e.g., direct seeder, roller crimper) [19]. In particular, topsoil compaction may begin as quickly as five years after the implementation of NT practices in conventional agriculture and seems strongly dependent on soil type and climate [6,8,62]. Unstable soil with low organic matter content and greater soil moisture may increase soil compaction, possibly leading to the greater emphasis on soil structure issues cited by northern organic farmers [8]. In Finland, Alakukku et al. [63], cited in Soane et al. [8], observed similar yields in NT soil as a ploughing treatment on silty clay, where lower yields were obtained in NT on clay soil under similar conditions. Soane et al. [8] highlighted that soil compaction is also associated with the previous crop harvest and the lack of tillage prior to sowing subsequent crops. Consequently, the short timeframe between cash crop harvest and cover or cash crop planting in NT does not allow for improved soil structural formation through biological activity.

4.7. Increased Labor Requirements

Potential savings in labor is a primary farmer motivation when deciding to adopt NT as documented by European researchers, as significant labor is required for soil tillage [2,53,64]. However, farmers' concerns about increased labor requirements related to the implementation of a cover crop phase may explain why relatively few famers combine NT and cover crops, as perceptions exist that the time required for cover crop management will negatively impact the expected labor-saving benefits of NT.

5. OVERVIEW OF LIVING OR KILLED (MULCH) COVER CROP-BASED NO-TILLAGE ORGANIC RESEARCH IN EUROPE

A recent meta-analysis on shallow non-inversion tillage in OF showed that impacts of NT management on farming system characteristics are poorly documented in Europe [44]. While a significant body of North American research is concentrated on organic NT, European studies are focused on RT and are mainly carried out in Germany, France,

Switzerland, Greece, Italy, Slovakia, and the Netherlands [13,47,54,59,65,66]. Therefore, relatively little knowledge has been provided on organic MBNT or LBNT, despite the recognition by international researchers that this innovation may help to manage weeds and decrease soil disturbance while maintaining biodiversity [10,17]. The meta-analysis presented by Cooper et al. [44] indicates that while promising prospects exist related to the improvement of soil quality, substantial yield losses due to perennial weed pressure and the delay in N supply remain concerns, as evidenced by European organic farmers' challenges outlined in the preceding sections [19].

In Europe, trials including organic NT have been established in Greece, Switzerland, and France (Table 2). These studies have been carried out on NT without the integration of cover crops [50,67], or on CCNT [15,68,69] (Figure 1). In this latter case, most trials have concentrated on using a living cover crop with winter wheat, flax (*Linum usitatissimum* L.), or quinoa (*Chenopodium quinoa* L.) crops [52,68,69]. Only two studies have been conducted on MBNT in France, in 2005 (corn crop) and 2008 (soybean crop) [15]. Mäder and Berner [13] highlighted that additional NT trials have been led in Europe, such as in Germany, but little feedback on scientific knowledge is easily available within the scientific community because of the different national languages in which these trials were reported.

5.1. Living Cover Crop-Based No-Tillage (LBNT) Production

LBNT assessed in Europe is often compared to other tillage techniques mostly relying on traditional tillage (20–25 cm) and RT (5–10 cm) [51]. These studies have documented soil fertility improvement in LBNT with higher macroporosity, more organic matter, and higher soil N content due to benefits provided by the cover crop (biological activity, soil structure, etc.) [50,51,52,70]. Researchers have also reported the major challenge of identifying the best cover crop species which ensure weed suppression without affecting cash crop yield and quality. For instance, studies conducted in Switzerland on the direct seeding of wheat into a living cover crop demonstrated competition of the cover crop with the main crop [69]. According to Den Hollander et al. [71], clovers, while providing soil fertility benefits, had limited utility for weed control and high competition with the main crop. However, Hiltbrunner et al. [68,69] observed improved weed control with white clover (*Trifolium repens* L.) and submediterranean clovers (*Trifolium lappaceum* L.) as compared to alfalfa (*Medicago sativa* L.), with increased wheat seeding rates potentially improving yield.

Table 2. Published experimental trials, including organic no-tillage techniques in Europe.

Country	Soil Type	Year	Cash Crop	Cover Crop	References
France	Cambisol (silty)	2003	corn	clover (living)	[15,54,73]
	Fluvisol (sandy loam)	2005	corn	alfalfa (mulch *)	
		2008	soybean	rye (mulch *)	
Greece	Tagnic Eutric Cambisol (clay loam)	2007	corn	Common vetch (*Vicia sativa* L.) (living), fababean (*Vicia faba* L.) (living)	[50–52,70]
		2008	winter wheat		
		2010	flax		
		2011	quinoa		
Switzerland	Partial gleyic Cambisol	2002	winter wheat	birdsfoot trefoil (*Lotus corniculatus* L.), white clover (*Trifolium repens* L.), subclover (*Trifolium subterraneum* ssp.), strong-spined medick (*Medicago truncatula Gaertner*) (living)	[67–69]
		2004	winter wheat		
	Orthic Luvisol	2005	winter wheat		

* Direct seeding of cash crop into mulch cover crop terminated with roller crimper.

Across living cover crops trials, overall, researchers have identified white clover as the legume species providing the best compromise between competing with weeds while limiting the competition for light with the wheat crop [69,71,72]. The use of a living cover crop is still debated regarding its potential competition with cash crops and related cash crop yield losses.

5.2. Cover Crop Mulch-Based No-Tillage (MBNT) Techniques

In France, where MBNT was tested (Figure 1), four tillage treatments were compared: (1) traditional ploughing (30 cm depth); (2) shallow ploughing (18 cm depth); (3) RT (12 cm–15 cm depth); and (4) MBNT [15]. For this latter treatment, cash crops were sown into a cover crop terminated with a 1.7 m wide roller crimper with steel blades welded onto a cylinder. This equipment was built following the design model of Brazilian rollers. In 2005, corn was sown into alfalfa mulch and soybean was planted under rye (*Secale cereale* L.) mulch in 2008. Results showed significantly more earthworms in MBNT with the cover crop biomass left on the soil surface providing food resources which foster biological activity [54]. This finding has been confirmed by several authors in conventional farming who underscore the role of high biological activity in generating resilient organic systems [6,58]. Vian et al. [73] showed higher microbial carbon content, organic carbon activity and N mineralization in MBNT from a 0 to 30 cm soil depth. These findings are consistent with conventional studies where more soil C and N content are observed after longer-term trials, although additional fertilization is often required during the first three years of NT adoption [8,9].

Yield losses were also observed in the two French trials, with 25% soybean yield reduction in MBNT treatments in 2008 compared to ploughing, and as much as 75% corn yield reduction under MBNT in 2005 [15]. Crop failures have been attributed to two main factors: (1) weed infestation and (2) cover crop regrowth competition with main crops.

In 2005, alfalfa mulch was unevenly distributed on the soil surface and did not ensure sufficient weed control. More than 100 weeds/m^2 infested the MBNT plots on average, while only one weed/m^2 was observed after ploughing at the corn emergence stage [15]. In North America, studies also indicate contrasting results in maintaining corn yield due to competition by both weeds and the cover crop [21,74]. Indeed, although the Rodale Institute (Pennsylvania, USA) achieved as much as 10.25 t/ha corn yield in 2007 with a hairy vetch (*Vicia villosa* L.) cover crop [24], substantial yield losses have also been reported in Iowa due to drought conditions [21]. While legume cover crops increase N supply to the subsequent corn crop, legume species are also more difficult to terminate through rolling and experience less effective weed control due to their faster residue degradation rate (low C:N rate) and lower biomass production as compared to cereal species. This lack of cover crop biomass on the soil surface throughout corn production season could explain the failure observed in the French trial [15] with alfalfa, as well as the interest of some North American researchers to use cover crop mixtures composed of both legume and cereal species prior to corn MBNT [75]. Cereal species could reduce the amount of N supplied to corn; however, this hypothesis must be further tested to obtain a better understanding of N dynamics in MBNT production.

In France in 2008, cereal rye demonstrated a greater ability to suppress weeds on soybean MBNT, but did not provide enough biomass to ensure weed control throughout the season [15]. Weed density was more abundant at the end of the cash crop season due to unusually wet conditions. Cereal rye has been the most researched cover crop in North America prior to soybean MBNT and has led to the most promising results regarding the range of benefits conferred (e.g., allelopathic effect, high biomass, early flowering, flexibility of sowing date, kill by rolling, etc.) [26,39,42,76]. Researchers also extolled the benefits of other cereal cover crops (e.g., triticale, barley, etc.) related to their longer persistence on the soil surface (high C:N rate) and greater biomass production than legume species. Above all other factors related to cover crop management, according to Mirsky et al. [38], the cover crop must reach more than 8000 kg/ha to lead to satisfactory weed control [77]. This biomass level was likely not obtained in the European studies.

In France, despite weed infestation, Peigné et al. [15] reported greater soybean populations in MBNT compared with ploughing. The researchers explained that seed predation by pigeons occurring on the trial affected more plants on the ploughed plots because of the lack of mulch cover, which protected the soybean seed and seedlings in the MBNT plots. In North America, Delate et al. [21] and Lefèbvre et al. [78] also tended to observe higher soybean populations in MBNT in the United States and in Canada due to efficient weed control at the early vegetative stage of soybean. In addition, the authors indicated that rotary hoeing on ploughed plots likely decreased soybean stands. Based on the French results, we could expect additional benefits from MBNT to ensure crops and soil protection regarding both the risks of pests and climate (storage, drought periods etc.). Substantial work is still required to obtain good weed suppression until crop harvest and maintain similar yields to those obtained in traditional tillage to fully address farmers' concerns [19].

As underlined by a large number of authors, effective cover crop termination is essential to MBNT success [38]. Nevertheless, French researchers have highlighted the difficulty of killing a cover crop with a roller crimper [15]. Indeed, some plots needed to be rolled more than once and cover crop regrowth occurred throughout the crop season, likely impacting cash crop yield [15]. According to North American researchers comparing various stages of cover crop growth and the related efficacy of termination, cover crops must reach at least the flowering growth stage to be successfully controlled [24,31,38,40]. However, the French trial experienced difficulties in delaying the sowing of the cash crop due to climate conditions which led to the cover crop being terminated prior to flowering, contributing to the failure to control the cover crop [15]. In light of these problems, North American researchers have screened different cover crop species and cultivars to allow for earlier cover crop flowering the following spring.

6. EUROPEAN RESEARCH CHALLENGES FOR THE FUTURE

As evidenced by Freidrich et al. [56] in conventional farming, this review of European organic NT trials indicates that projects often compare MBNT or LBNT with other tillage techniques, an experimental framework that does not focus on optimizing MBNT success. Thus, considerable work should be carried out to compare several management systems under various pedo-climatic conditions to find the best strategies leading to the suppression of weeds until crop harvest, yield improvement, and cover crop control.

This paper also indicates the potential soil fertility benefits which could be provided by the further implementation of organic MBNT in Europe [44,54], although there is still a poor understanding of the longer-term effects of MBNT production on soil quality. Furthermore, despite farmers' recognized interest in MBNT, European research has provided very little knowledge to address their concerns (effects on soil structure, NT viability, skills, etc.), thus inhibiting the adoption of MBNT in organic farming. Two primary factors remain poorly researched in Europe, challenging the ability of farmers to maintain crop yields in organic MBNT production: (1) cover crop management (species, cultivars, termination), and (2) reliable, consistent, and effective weed suppression. The following key agronomical issues should be addressed under temperate conditions to eliminate these knowledge gaps:

- **Choice of relevant cover crop species:** In Europe, few cover crop species have been assessed for MBNT production, even though the appropriate choice of cover crop is essential for efficient weed control [38]. Further screening of species and cultivars under temperate climate conditions is needed to identify optimal cover crop/cash crop combinations and support the decision-making process of farmers. Other questions must also be addressed: Can cover crop species that have given promising results in North America (e.g., rye and triticale) lead to similar results on soybean MBNT in Europe? Which cover crop(s) lead to the best trade-off between weed suppression, crop yield and cover crop control?

- **Increasing cover crop biomass:** As indicated by many researchers, adequate cover crop biomass is a key factor to ensure weed control in MBNT; however, doubts have been raised regarding the ability to achieve high levels of cover crop biomass (8000–9000 kg/ha) determined to be optimal by North American researchers in Europe, particularly in southern areas. Thus, the following questions must be addressed: What level of cover crop biomass must be reached to ensure satisfactory weed control? Which management strategies could optimize cover crop biomass and achieve a comparable degree of weed suppression as ploughing or chemical methods? How does cover crop biomass impact weed development and crop yield? What are the weed control implications of earlier cover crop sowing?

- **Improving cover crop termination:** Improving cover crop termination by mechanical methods (e.g., rolling-crimping and mowing) leads to greater success in organic MBNT; however, consistent termination of the cover crop remains challenging using the current practices of European farmers. Cover crop termination remains particularly challenging in Northern Europe, where a humid climate fosters vigorous cover crop development. While considerable efforts must be made to develop appropriate machinery adapted to work with the high biomass remaining on the soil surface, further refinement is needed regarding the optimal cover crop termination practices (number of rollings, date of rolling, etc.), with the aim of improving mulch distribution, prolonging its persistence on the ground until crop harvest, and enhancing the quality of crop sowing (seeds/soil contact). In particular, additional studies are required to quantify the effect of cover crop rolling on the mulch degradation rate until crop harvest under a temperate climate and the effectiveness of the roller crimper on legume species termination.

- **Designing crop rotation:** Integrating MBNT into organic crop rotations in accordance with organic farmers' challenges (e.g., weed management at rotational scale, maintaining crop yields, etc.) while incorporating both production and economic limitations remains a major issue.

ACKNOWLEDGMENTS

We acknowledge Paul Mäder (FiBL), F.Xavier Sans (UB), José Manuel Blanco-Moreno (UB), Daniele Antichi (CiRAA), Paolo Bàrberi (SSSA), Annelies Beeckman (INAGRO, department of organic crop production), Federica Bigongiali (SSSA), Julia Cooper (UNEW, NEFG), Hansueli Dierauer (FiBL), Kate Gascoyne (UNEW, NEFG), Meike Grosse (University of Kassel), Juergen Heß (University of Kassel), Andreas Kranzler (FiBL), Anne Luik (EULS), Elen Peetsmann (EULS), Andreas Surböck (FiBL), and Koen Willekens (ILVO), who carried out the European survey within the frame of TILMAN-ORG (www.tilman-org.net) supported by CORE Organic.

REFERENCES

1. Altieri, M.A.; Lana, M.A.; Bittencourt, H.V.; Kieling, A.S.; Comin, J.J.; Lovato, P.E. Enhancing crop productivity via weed suppression in organic no-till cropping systems in santa catarina, Brazil. *J. Sustain. Agric.* **2011**, *35*, 855–869.
2. Hobbs, P.R.; Sayre, K.; Gupta, R. The role of conservation agriculture in sustainable agriculture. *Philos. Trans. R. Soc. B Biol. Sci.* **2008**, *363*, 543–555.
3. Kassam, A.; Friedrich, T.; Shaxson, F.; Pretty, J. The spread of conservation agriculture: Justification, sustainability and uptake. *Int. J. Agric. Sustain.* **2009**, *7*, 292–320.
4. Masutti, C. Action publique et expertise dans la conservation des ressources agricoles aux états-unis dans les années 1930. *Ruralia* **2007**, *25*, 1–25.
5. Horowitz, J. *No-Till Farming is a Growing Practice*; DIANE Publishing: Collingdale, PA, USA, 2011; p. 28.
6. Cannell, R.Q.; Hawes, J.D. Trends in tillage practices in relation to sustainable crop production with special reference to temperate climates. *Soil Tillage Res.* **1994**, *30*, 245–282.
7. Gohlke, T.; Ingersoll, T.; Roe, R.D.; Oregon, N.; Pullman, W.N. Soil disturbance in no-till and direct seed planting systems. *Nat. Res. Conserv. Serv. Agron. Tech. Note* **2000**, *39*, 1–6.
8. Soane, B.D.; Ball, B.C.; Arvidsson, J.; Basch, G.; Moreno, F.; Roger-Estrade, J. No-till in northern, western and south-western Europe: A review of problems and opportunities for crop production and the environment. *Soil Tillage Res.***2012**, *118*, 66–87.
9. Teasdale, J.R.; Coffman, C.B.; Mangum, R.W. Potential long-term benefits of no-tillage and organic cropping systems for grain production and soil improvement. *Agron. J.* **2007**, *99*, 1297–1305.
10. Triplett, G.B.; Dick, W.A. No-tillage crop production: A revolution in agriculture! *Agron. J.* **2008**, *100*, 153–165.
11. Favarato, L.; Galvão, J.; Souza, J.; Guarçoni, R.; Souza, C.; Cunha, D. Population density and weed infestation in organic no-tillage corn cropping system under different soil covers. *Planta Daninha* **2014**, *32*, 739–746.
12. Carrera, L.M.; Abdul-Baki, A.A.; Teasdale, J.R. Cover crop management and weed suppression in no-tillage sweet corn production. *HortScience* **2004**, *39*, 1262–1266.

13. Mäder, P.; Berner, A. Development of reduced tillage systems in organic farming in Europe. *Renew. Agric. Food Syst.***2012**, *27*, 7–11.
14. Watson, C.A.; Atkinson, D.; Gosling, P.; Jackson, L.R.; Rayns, F.W. Managing soil fertility in organic farming systems. *Soil Use Manag.* **2002**, *18*, 239–247
15. Peigné, J.; Lefevre, V.; Vian, J.F.; Fleury, P. Conservation agriculture in organic farming: Experiences, challenges and opportunities in europe. In *Conservation Agriculture*; Farooq, M., Siddique, K.H.M., Eds.; Springer International Publishing: Cham, Switzerland, 2015; pp. 559–578.
16. Shirtliffe, S.J.; Johnson, E.N. Progress towards no-till organic weed control in western Canada. *Renew. Agric. Food Syst.* **2012**, *27*, 60–67.
17. Gadermaier, F.; Berner, A.; Fließbach, A.; Friedel, J.K.; Mäder, P. Impact of reduced tillage on soil organic carbon and nutrient budgets under organic farming. *Renew. Agric. Food Syst.* **2012**, *27*, 68–80.
18. Peigné, J.; Ball, B.C.; Roger-Estrade, J.; David, C. Is conservation tillage suitable for organic farming? A review. *Soil Use Manag.* **2007**, *23*, 129–144.
19. Casagrande, M.; Peigné, J.; Payet, V.; Mäder, P.; Sans, F.X.; Blanco-Moreno, J.M.; Antichi, D.; Bàrberi, P.; Beeckman, A.; Bigongiali, F.; et al. Organic farmers' motivations and challenges for adopting conservation agriculture in Europe. *Org. Agr.* **2015**.
20. Peigné, J.; Casagrande, M.; Payet, V.; David, C.; Sans, F.X.; Blanco-Moreno, J.M.; Cooper, J.; Gascoyne, K.; Antichi, D.; Bàrberi, P.; et al. How organic farmers practice conservation agriculture in europe. *Renew. Agric. Food Syst.* **2015,** *31,* 72–85.
21. Delate, K.; Cwach, D.; Chase, C. Organic no-tillage system effects on soybean, corn and irrigated tomato production and economic performance in Iowa, USA. *Renew. Agric. Food Syst.* **2012**, *27*, 49–59.
22. Halde, C.; Entz, M.H. Plant species and mulch application rate affected decomposition of cover crop mulches used in organic rotational no-till systems. *Can. J. Plant Sci.* **2016**, *96*, 59–71.
23. Halde, C.; Gulden, R.H.; Entz, M.H. Selecting cover crop mulches for organic rotational no-till systems in Manitoba, Canada. *Agron. J.* **2014**, *106*, 1193.
24. Moyer, J. *Organic No-Till Farming: Advancing No-Till Agriculture: Crops, Soil, Equipment*; Acres U.S.A.: Austin, TX, USA, 2011; p. 204.
25. Parr, M.; Grossman, J.M.; Reberg-Horton, S.C.; Brintin, C.; Crozier, C. Nitrogen delivery from legume cover crops in no-till organic corn production. *Crops Soil Mag. Am. Soc. Agron.* **2012**.
26. Silva, E.M. Screening five fall-sown cover crops for use in organic no-till crop production in the upper midwest. *Agroecol. Sustain. Food Syst.* **2014**, *38*, 748–763.
27. Davis, A.S. Cover-crop roller-crimper contributes to weed management in no-till soybean. *Weed Sci.* **2010**, *58*, 300–309.
28. Kornecki, T.S.; Arriaga, F.J.; Price, A.J. Roller type and operating speed effects on rye termination rates, soil moisture, and yield of sweet corn in a no-till system. *HortScience* **2012**, *47*, 217–223.
29. Kornecki, T.S.; Price, A.J. Effects of different roller/crimper designs and rolling speed on rye cover crop termination and seed cotton yield in a no-till system. *J. Cotton Sci.* **2010**, *14*, 212–220.
30. Mirsky, S.B.; Curran, W.S.; Mortenseny, D.M.; Ryany, M.R.; Shumway, D.L. Timing of cover-crop management effects on weed suppression in no-till planted soybean using a roller-crimper. *Weed Sci.* **2011**, *59*, 380–389.
31. Mischler, R.; Duiker, S.W.; Curran, W.S.; Wilson, D. Hairy vetch management for no-till organic corn production. *Agron. J.* **2010**, *102*, 355–362.
32. Parr, M.; Grossman, J.M.; Reberg-Horton, S.C.; Brinton, C.; Crozier, C. Roller-crimper termination for legume cover crops in north Carolina: Impacts on nutrient availability to a succeeding corn crop. *Commun. Soil Sci. Plant Anal.* **2014**, *45*, 1106–1119.
33. Vaisman, I.; Entz, M.H.; Flaten, D.N.; Gulden, R.H. Blade roller–green manure interactions on nitrogen dynamics, weeds, and organic wheat. *Agron. J.* **2011**, *103*, 879–889.

34. Bernstein, E.R.; Stoltenberg, D.E.; Posner, J.L.; Hedtcke, J.L. Weed community dynamics and suppression in tilled and no-tillage transitional organic winter rye–soybean systems. *Weed Sci.* **2014**, *62*, 125–137.
35. Carr, P.; Gramig, G.; Liebig, M. Impacts of organic zero tillage systems on crops, weeds, and soil quality. *Sustainability* **2013**, *5*, 3172–3201.
36. Carr, P.M.; Anderson, R.L.; Lawley, Y.E.; Miller, P.R.; Zwinger, S.F. Organic zero-till in the northern US great plains region: Opportunities and obstacles. *Renew. Agric. Food Syst.* **2012**, *27*, 12–20.
37. Luna, J.M.; Mitchell, J.P.; Shrestha, A. Conservation tillage for organic agriculture: Evolution toward hybrid systems in the western USA. *Renew. Agric. Food Syst.* **2012**, *27*, 21–30.
38. Mirsky, S.B.; Ryan, M.R.; Curran, W.S.; Teasdale, J.R.; Maul, J.; Spargo, J.T.; Moyer, J.; Grantham, A.M.; Weber, D.; Way, T.R.; et al. Conservation tillage issues: Cover crop-based organic rotational no-till grain production in the mid-atlantic region, USA. *Renew. Agric. Food Syst.* **2012**, *27*, 31–40.
39. Mirsky, S.B.; Ryan, M.R.; Teasdale, J.R.; Curran, W.S.; Reberg-Horton, C.S.; Spargo, J.T.; Wells, M.S.; Keene, C.L.; Moyer, J.W. Overcoming weed management challenges in cover crop–based organic rotational no-till soybean production in the eastern United States. *Weed Technol.* **2013**, *27*, 193–203.
40. Wells, M.S.; Brinton, C.M.; Reberg-Horton, S.C. Weed suppression and soybean yield in a no-till cover-crop mulched system as influenced by six rye cultivars. *Renew. Agric. Food Syst.* **2015**, *31*, 1–12.
41. Wells, M.S.; Reberg-Horton, S.C.; Mirsky, S.B. Cultural strategies for managing weeds and soil moisture in cover crop based no-till soybean production. *Weed Sci.* **2014**, *62*, 501–511.
42. Bernstein, E.R.; Posner, J.L.; Stoltenberg, D.E.; Hedtcke, J.L. Organically managed no-tillage rye–soybean systems: Agronomic, economic, and environmental assessment. *Agron. J.* **2011**, *103*, 1169.
43. Ryan, M.R.; Curran, W.S.; Grantham, A.M.; Hunsberger, L.K.; Mirsky, S.B.; Mortensen, D.A.; Nord, E.A.; Wilson, D.O. Effects of seeding rate and poultry litter on weed suppression from a rolled cereal rye cover crop. *Weed Sci.***2011**, *59*, 438–444.
44. Cooper, J.; Baranski, M.; Stewart, G.; Nobel-de Lange, M.; Bàrberi, P.; Fließbach, A.; Peigné, J.; Berner, A.; Brock, C.; Casagrande, M.; et al. Shallow non-inversion tillage in organic farming maintains crop yields and increases soil C stocks: A meta-analysis. *Agron. Sustain. Dev.* **2016.**
45. Baker, C.J.; Saxton, K.E. *No-Tillage Seeding in Conservation Agriculture*; CABI: Oxfordshire, UK, 2007; p. 340.
46. Derpsch, R.; Friedrich, T.; Kassam, A.; Li, H. Current status of adoption of no-till farming in the world and some of its main benefits. *Int. J. Agric. Biol. Eng.* **2010**, *3*, 1–25.
47. Armengot, L.; Berner, A.; Blanco-Moreno, J.M.; Mäder, P.; Sans, F.X. Long-term feasibility of reduced tillage in organic farming. *Agron. Sustain. Dev.* **2014**, *35*, 339–346.
48. Emmerling, C. Reduced and conservation tillage effects on soil ecological properties in an organic farming system. *Biol. Agric. Hortic.* **2007**, *24*, 363–377.
49. Krauss, M.; Berner, A.; Burger, D.; Wiemken, A.; Niggli, U.; Mäder, P. Reduced tillage in temperate organic farming: Implications for crop management and forage production. *Soil Use Manag.* **2010**, *26*, 12–20.
50. Bilalis, D.; Kakabouki, I.; Karkanis, A.; Travlos, I.; Triantafyllidis, V.; Dimitra, H. Seed and saponin production of organic quinoa (*Chenopodium quinoa* Willd.) for different tillage and fertilization. *Not. Bot. Horti Agrobot. Cluj-Napoca***2012**, *40*, 42.
51. Bilalis, D.; Karkanis, A.; Pantelia, A.; Patsiali, S.; Konstantas, A.; Efthimiadou, A. Weed populations are affected by tillage systems and fertilization practices in organic flax (*Linum usitatissimum* L.) crop. *Aust. J. Crop Sci.* **2012**, *6*, 157.
52. Bilalis, D.; Karkanis, A.; Patsiali, S.; Agriogianni, M.; Konstantas, A.; Triantafyllidis, V. Performance of wheat varieties (*Triticum Aestivum* L.) under conservation tillage practices in organic agriculture. *Not. Bot. Horti Agrobot. Cluj-Napoca* **2011**, *39*, 28.

53. Lefèvre, V.; Capitaine, M.; Peigné, J.; Roger-Estrade, J. Soil Conservation Practices in Organic Farming: Overview of French farmers' Experiences and Contribution to Future Cropping Systems Design. Available online: http://ifsa.boku.ac.at/cms/fileadmin/Proceeding2012/IFSA2012_WS6.3_Lefevre.pdf (accessed on 11 January 2017).
54. Peigné, J.; Cannavaciuolo, M.; Gautronneau, Y.; Aveline, A.; Giteau, J.L.; Cluzeau, D. Earthworm populations under different tillage systems in organic farming. *Soil Tillage Res.* **2009**, *104*, 207–214.
55. Pelosi, C.; Bertrand, M.; Roger-Estrade, J. Earthworm community in conventional, organic and direct seeding with living mulch cropping systems. *Agron. Sustain. Dev.* **2009**, *29*, 287–295.
56. Friedrich, T.; Kassam, A.; Corsi, S.; Jat, R.A.; Sahrawat, K.L.; Kassam, A.H. Conservation agriculture in europe. In *Conservation Agriculture: Global Prospects and Challenges*; CABI: Oxfordshire, UK, 2014; pp. 127–170.
57. Bàrberi, P. Weed management in organic agriculture: Are we addressing the right issues? *Weed Res.* **2002**, *42*, 177–193.
58. Emmerling, C. Response of earthworm communities to different types of soil tillage. *Appl. Soil Ecol.* **2001**, *17*, 91–96.
59. Sans, F.X.; Berner, A.; Armengot, L.; Mäder, P. Tillage effects on weed communities in an organic winter wheat–sunflower–spelt cropping sequence. *Weed Res.* **2011**, *51*, 413–421.
60. Morris, N.L.; Miller, P.C.H.; Orson, J.H.; Froud-Williams, R.J. The adoption of non-inversion tillage systems in the United Kingdom and the agronomic impact on soil, crops and the environment—A review. *Soil Tillage Res.* **2010**, *108*, 1–15.
61. Ball, B.C.; Cheshire, M.V.; Robertson, E.A.G.; Hunter, E.A. Carbohydrate composition in relation to structural stability, compactibility and plasticity of two soils in a long-term experiment. *Soil Tillage Res.* **1996**, *39*, 143–160.
62. Munkholm, L.J.; Schjønning, P.; Rasmussen, K.J.; Tanderup, K. Spatial and temporal effects of direct drilling on soil structure in the seedling environment. *Soil Tillage Res.* **2003**, *71*, 163–173.
63. Alakukku, L.; Uusitalo, R.; Särkelä, A.; Lahti, K.; Valkama, P.; Valpasvuo-Jaatinen, P.; Ventelä, A.-M. Phosphorus Stratification in the Ap Horizon of Ploughed and No-Till Soils and Its Effect on P Forms in Surface Runoff. In Proceedings of the ISTRO 18th Triennial Conference Sustainable Agriculture, Izmir, Turkey, 15–19 June 2009; International Soil Tillage Research Organisation: Izmir, Turkey, 2009.
64. Canali, S.; Diacono, M.; Campanelli, G.; Montemurro, F. Organic no-till with roller crimpers: Agro-ecosystem services and applications in organic mediterranean vegetable productions. *Sustain. Agric. Res.* **2015**, *4*, 70.
65. Berner, A.; Hildermann, I.; Fliesbach, A.; Pfiffner, L.; Niggli, U.; Mader, P. Crop yield and soil fertility response to reduced tillage under organic management. *Soil Tillage Res.* **2008**, *101*, 89–96.
66. Lehocká, Z.; Klimeková, M.; Bieliková, M.; Mendel, L. The effect of different tillage systems under organic management on soil quality indicators. *Agron. Res.* **2009**, *7*, 369–373.
67. Hiltbrunner, J.; Liedgens, M.; Stamp, P.; Streit, B. Effects of row spacing and liquid manure on directly drilled winter wheat in organic farming. *Eur. J. Agron.* **2005**, *22*, 441–447.
68. Hiltbrunner, J.; Liedgens, M.; Bloch, L.; Stamp, P.; Streit, B. Legume cover crops as living mulches for winter wheat: Components of biomass and the control of weeds. *Eur. J. Agron.* **2007**, *26*, 21–29.
69. Hiltbrunner, J.; Streit, B.; Liedgens, M. Are seeding densities an opportunity to increase grain yield of winter wheat in a living mulch of white clover? *Field Crops Res.* **2007**, *102*, 163–171.
70. Bilalis, D.J.; Karamanos, A.J. Organic maize growth and mycorrhizal root colonization response to tillage and organic fertilization. *J. Sustain. Agric.* **2010**, *34*, 836–849.
71. Den Hollander, N.G.; Bastiaans, L.; Kropff, M.J. Clover as a cover crop for weed suppression in an intercropping design: I. Characteristics of several clover species. *Eur. J. Agrono.* **2007**, *26*, 92–103.

72. Hollander, N.G.D. 2012. Growth characteristics of several clover species and their suitability for weed suppression in a mixed cropping design. Ph.D. Thesis, Wageningen University, Wageningen, The Netherlands, 2012.
73. Vian, J.F.; Peigne, J.; Chaussod, R.; Roger-Estrade, J. Effects of four tillage systems on soil structure and soil microbial biomass in organic farming. *Soil Use Manag.* **2009**, *25*, 1–10.
74. Delate, K.; Cwach, D.; Fiscus, M. *Evaluation of an Organic No–Till System for Organic Corn and Soybean Production–Agronomy Farm Trial, 2011*; Organic Ag Program Webpage, Iowa State University: Ames, IA, USA, 2012.
75. Parr, M.; Grossman, J.M.; Reberg-Horton, S.C.; Brinton, C.; Crozier, C. Nitrogen delivery from legume cover crops in no-till organic corn production. *Agron. J.* **2011**, *103*, 1578.
76. Mirsky, S.B.; Curran, W.S.; Mortensen, D.A.; Ryan, M.R.; Shumway, D.L. Control of cereal rye with a roller/crimper as influenced by cover crop phenology. *Agron. J.* **2009**, *101*, 1589.
77. Smith, A.N.; Reberg-Horton, S.C.; Place, G.T.; Meijer, A.D.; Arellano, C.; Mueller, J.P. Rolled rye mulch for weed suppression in organic no-tillage soybeans. *Weed Sci.* **2011**, *59*, 224–231.
78. Lefebvre, M.; Leblanc, M.; Gilbert, P.-A.; Estevez, B.; Grenier, M.; Belzile, L. *Semis Direct Sur Paillis De Seigle Roulé En Régie Biologique;* Institut De Recherche Et De Développement En Agroenvironnement: Québec, QC, Canada, 2011; p. 36.

CHAPTER 12

Cover Crop-Based, Organic Rotational No-Till Corn and Soybean Production Systems in the Mid-Atlantic United States

John M. Wallace[1], Alwyn Williams[2], Jeffrey A. Liebert[3], Victoria J. Ackroyd[2,4], Rachel A. Vann[5], William S. Curran[1], Clair L. Keene[6], Mark J. VanGessel[7], Matthew R. Ryan[3] and Steven B. Mirsky[4,*]

[1] *Department of Plant Science, The Pennsylvania State University, University Park, PA 16802, USA*

[2] *Department of Plant Science and Landscape Architecture, University of Maryland, College Park, MD 20742, USA*

[3] *Section of Soil and Crop Sciences, Cornell University, Ithaca, NY 14853, USA*

[4] *Sustainable Agricultural Systems Laboratory, USDA-ARS, Beltsville, MD 20705, USA*

[5] *Department of Crop and Soil Sciences, North Carolina State University, NC 27695, USA*

[6] *Williston Research Extension Center, North Dakota State University, Williston, ND 58801, USA*

[7] *Department of Plant and Soil Sciences, University of Delaware, Georgetown, DE 19947, USA*

AUTHOR CONTRIBUTIONS

W.S.C., S.B.M., M.R.R. and M.J.V.G. conceived of and designed the experiments. C.L.K., R.A.V., J.A.L., S.B.M., M.R.R., W.S.C. and M.J.V.G.

performed the experiments. J.M.W., A.W. and J.A.L. analyzed the data. J.M.W., A.W., J.A.L., S.B.M. and V.J.A. wrote the paper.

ABSTRACT

Cover crop-based, organic rotational no-till (CCORNT) corn and soybean production is becoming a viable strategy for reducing tillage in organic annual grain systems in the mid-Atlantic, United States. This strategy relies on mechanical termination of cover crops with a roller-crimper and no-till planting corn and soybean into cover crop mulches. Here, we report on recent research that focuses on integrated approaches for crop, nutrient and pest management in CCORNT systems that consider system and regional constraints for adoption in the mid-Atlantic. Our research suggests that no-till planting soybean into roller-crimped cereal rye can produce consistent yields. However, constraints to fertility management have produced less consistent no-till corn yields. Our research shows that grass-legume mixtures can improve N-release synchrony with corn demand and also improve weed suppression. Integration of high-residue inter-row cultivation improves weed control consistency and may reduce reliance on optimizing cover crop biomass accumulation for weed suppression. System-specific strategies are needed to address volunteer cover crops in later rotational phases, which result from incomplete cover crop termination with the roller crimper. The paucity of adequate machinery for optimizing establishment of cash crops into thick residue mulch remains a major constraint on CCORNT adoption. Similarly, breeding efforts are needed to improve cover crop germplasm and develop regionally-adapted varieties.

Keywords: cereal rye; conservation tillage; cover crop; hairy vetch; high-residue cultivation; manure; nitrogen; poultry litter; sustainable agriculture; weed suppression

1. INTRODUCTION

Demand for organic grains continues to increase in the United States in response to growing consumer preference for organic feed (required for meat and milk production) and food-grade agricultural products (e.g., soymilk and tofu) [1]. Increasing societal demand for sustainable agricultural production practices likely contributes to observed increases in consumer preference for organic food [2]. Recent analysis suggests that with current premium prices, organic crop production can be highly profitable [3] even though crop yields are typically lower than in conventional systems [4]. However, production of organic grain has not kept up with demand. This has resulted in a national shortage of organic grain for direct human consumption and for livestock feed, thus requiring high levels (>60%) of imported organic grain [5]. While there are infrastructure limitations to expanding organic grain production, the inability of organic grain producers to scale up their operations due to time committed to weed and soil fertility management, as well as aversion to tillage by many conventional farmers has also contributed to the limited adoption of organic grain production. There is potential for increased domestic production of organic grains in regions that support integrated crop and livestock systems within or among farms, as organic grain production relies heavily on inexpensive supplies of organic nutrient amendments for crop production [6]. Organic production in the mid-Atlantic U.S. (encompassing the region between New York and North Carolina) has potential for expansion due to a prevalence of agricultural production systems that include small integrated crop and livestock (dairy, poultry, swine) farms, large cash-crop farms without animals, consolidated livestock operations that purchase feed grain off-farm and ready access to major metropolitan markets [7]. Annual grain production primarily occurs in the Atlantic Coastal Plain, Piedmont, and Ridge and Valley regions of the mid-Atlantic, which range from sandy to high clay content soils. Annual precipitation is relatively high (760–1012 mm) and evenly distributed, allowing for the use of cover crops in the spring or fall growing seasons.

Development of sustainable cropping systems is critical for improving soil and water conservation. In the mid-Atlantic, agriculture is currently the largest contributor of nutrient and sediment pollutants in the Chesapeake Bay watershed [8]. In response, conservation tillage (e.g., no-till) is widely promoted and incentivized in conventional grain production systems to reduce nutrient and sediment loading into the Chesapeake Bay from agricultural lands. Adoption of no-till corn and soybean production varies across the mid-Atlantic, but is estimated to range from 28%–64% of the total production area [9]. In comparison, organic corn and soybean production in the mid-Atlantic typically relies on frequent use of full-inversion tillage to incorporate nutrient amendments and crop residues, as well as to control weeds. Additional soil disturbance occurs with the use of secondary tillage for seedbed preparation and blind- or inter-row cultivation to control weeds in the cash crop. Frequent tillage and cultivation degrades soil health, increases the potential for soil erosion and is both labor and energy intensive [10,11]. Consequently, successful organic cropping systems in the mid-Atlantic rely on soil building practices, such as the integration of cover crops, the return of crop residues, use of carbon (C)-rich amendments and diversified crop rotations to mitigate the effects of frequent tillage [12]. Organic grain producers are interested in integrating conservation tillage practices with these soil building practices to increase soil quality in their production systems.

Cover crop-based, organic rotational no-till (CCORNT) corn and soybean production is one approach to integrating the soil conservation practices of no-till with the soil building practices of organic systems. CCORNT systems offer a viable alternative to tillage-based production and are being adopted by an increasing number of organic farmers [13,14]. In these systems, winter annual cover crops are mechanically terminated in the spring using a roller-crimper, and the cash crop is then no-till planted into the cover crop mulch. Labor and fuel savings are maximized when a front-mounted roller-crimper is used to terminate the cover crops, and cash crops are no-till planted in the same pass. Soybean is typically no-till planted into fall-planted cereal rye (*Secale cereale* L.) or a similar winter cereal, and corn is no-till planted into hairy vetch (*Vicia villosa* Roth) or a grass-hairy vetch mixture. Similar rolled cover crop strategies have also been developed for grain and vegetable crop systems in the southeastern [15] and the Midwestern U.S. [16]. Central features common to rolled cover crop strategies include reliance on roll-crimped cover crop mulches (grasses and legumes) as the primary mechanism of within-season weed suppression and the contribution of nitrogen (N)-fixing legume cover crops to meet the high N-demand of the subsequent cash crop (e.g., corn). Optimizing cover crop biomass accumulation to enhance physical suppression of weeds is critical to the success of no-till cash crop production. Consequently, practices that improve cover crop performance (e.g., tillage for cover crop seedbed preparation, early seeding and high seeding rates) are typically employed. Simulation models indicate that a rotational no-till approach has the potential to increase C sequestration, reduce greenhouse gas emissions and increase energy efficiency via reduced diesel fuel and labor needs in comparison to current tillage-based organic annual grain systems that rely on primary tillage before both cover and cash crops [13].

Improved crop, nutrient and pest management strategies are needed to develop CCORNT grain production systems in the mid-Atlantic that successfully meet production and conservation goals. Soil and climate influence the feasibility and performance of various integrated, multi-tactic approaches. In the mid-Atlantic, CCORNT corn and soybean research has primarily focused on developing general principles for optimizing cover crop performance and management, devising multi-tactic weed management strategies and improving cash crop planter configurations [13,14]. The aim of this paper is to update past reviews [13,14] by collating recent and ongoing research that addresses the constraints that limit producer adoption and the practices that improve the performance of both cover and cash crops. First, we present an overview of the production viability of CCORNT in the mid-Atlantic, by comparing the yield output and yield stability of these systems against county-level averages and tillage-based organic production systems. We then evaluate cover crop practices and integrated, multi-tactic pest management strategies, focusing our discussion on the two dominant regional grain crops: corn and soybean. Lastly, we identify major constraints and research priorities needed to further improve the potential of CCORNT in the mid-Atlantic U.S.

2. CORN AND SOYBEAN YIELD PERFORMANCE IN CCORNT

The viability of CCORNT production systems hinges on the development of integrated cover crop, nutrient and pest management strategies that result in consistent high yielding cash crops across a range of soil and environmental conditions. Recent research has evaluated a suite of cover crop and weed management approaches during the organic transition period within a winter wheat (*Triticum aestivum*)-hairy vetch/triticale (*x Triticosecale* Wittm.)-no-till corn-cereal rye-no-till soybean rotation. The research project, known as the Reduced-tillage Organic Systems Experiment (ROSE), was an inter-disciplinary, multi-institutional effort between 2010 and 2013 at Pennsylvania State University, the USDA Agricultural Research Service (ARS), Sustainable Agricultural Systems Laboratory at Beltsville, Maryland, and the University of Delaware (Supplemental Table S1). The ROSE tested the role of cover crop termination date (early, intermediate and late) and high-residue cultivation (+/−) in the no-till corn and soybean phases of the rotation. These treatments were imposed in a full-entry cropping system experimental design. This enabled the evaluation of how cover crop and pest management approaches influenced cash crop performance and pest management objectives within-season and across the three-year rotation. Delayed cover crop termination has the potential to improve cash crop management via higher levels of weed suppression and N provisioning services that result from higher levels of cover crop biomass accumulation. However, this strategy can constrain cash crop growing season length. In the ROSE, shorter season crop varieties were used when cover crop termination was delayed for both corn and soybean systems (Table 1).

Table 1. Cereal rye biomass and soybean yields in an organic cover crop-based rotational no-till experiment (2011–2013) conducted by Penn State University, the USDA Agricultural Research Service (ARS) Sustainable Agricultural Systems Laboratory in Beltsville, MD and University of Delaware. Cereal rye biomass and soybean yields are averaged across high-residue cultivation treatments (*n* = 2), blocks (*n* = 4) and years (*n* = 3; 2011–2013) for the three planting date treatments (early, middle, late) evaluated in the experiment. The coefficient of variation (CV) is reported to describe inter-annual yield variability (2011–2013) among alternative planting date strategies and county-level average conventional yields.

Site (Soybean Production)	Planting Date (Maturity Group) †	Planting Date Range (2011–2013)	Cereal Rye Biomass Mean (CV%)		Soybean Yield Mean (CV%)	
			Mg ha^{-1}			
Penn State University Rock Springs, PA	Early (2.9)	26–31 May	5.3	(21)	3.2	(5)
	Middle (2.7)	3–6 June	5.8	(16)	3.3	(5)
	Late (1.1)	11–17 June	7.0	(20)	2.7	(11)
	County Average Yield		—		3.2	(6)
USDA-Beltsville Beltsville, MD	Early (3.8/4.3)	3–28 May	6.4	(29)	3.3	(27)
	Middle (3.4)	11 May–5 June	7.4	(31)	3.8	(22)
	Late (2.7)	18 May–17 June	8.2	(33)	3.4	(39)
	County Average Yield		—		2.7	(10)
University of Delaware Georgetown, DE	Early (3.8/4.3)	3–21 May	7.7	(15)	1.7	(16)
	Middle (3.4)	8–28 May	8.6	(6)	2.6	(1)
	Late (2.7)	15 May–5 June	9.9	(11)	1.9	(16)
	County Average Yield		—		2.7	(7)

† Maturity groups are based on ~150 km latitudinal gradients from south (4 s) to north (1 s) in the mid-Atlantic region.

Delayed cover crop termination in CCORNT soybean systems resulted in greater cereal rye biomass across mid-Atlantic locations, with step-wise increases in biomass levels observed across planting dates at more southern (Maryland and Delaware) locations (Table 1). However, cereal rye biomass is only one of several interacting factors that can influence soybean yields in CCORNT systems. For example, lower soybean yields were observed when planting dates were delayed until mid-June at the Pennsylvania location, which can be attributed in part to the shorter growing season in the northern mid-Atlantic. In comparison, later planting dates at the southern locations produced soybean yields comparable to early planting dates. At each mid-Atlantic location, the highest yielding planting dates were comparable to county-level averages for conventional soybean production (Table 1). Moreover, inter-annual yield variability was also comparable to county averages.

Delayed cover crop termination in no-till corn systems produced similar regional trends to soybean (Table 2). In general, hairy vetch/triticale biomass increased as cover crop termination was delayed, but observed trends suggest that corn yields were not positively correlated with cover crop performance. In more northern areas of the mid-Atlantic

(Pennsylvania), the relatively short growing season required corn to be harvested for silage to enable primary tillage, seedbed preparation and cereal rye establishment prior to no-till soybean production. Earlier planting dates resulted in higher corn silage yields at the Pennsylvania location, but yields were consistently below (16%–34%) county-level averages for conventional corn silage production over the same period. Longer fall growing seasons in southern areas of the mid-Atlantic (Maryland and Delaware) enabled cereal rye establishment following corn grain production. In the ROSE, corn grain yields were well below county averages in Delaware, but exceeded county averages in Maryland. Despite variations in yield, organic rotational no-till production of both corn and soybean was profitable without the need to factor in organic price premiums at the Pennsylvania and Maryland locations each year and in one year at Delaware [17].

Table 2. Cover crop (hairy vetch/triticale) biomass and corn yields in an organic cover crop-based rotational no-till experiment (2011–2013) conducted by Pennsylvania State University (Rock Springs, PA), USDA Agricultural Research Service (ARS; Beltsville, MD, USA) and University of Delaware (Georgetown, DE, USA). Cover crop biomass and corn yields are averaged across treatments (*n* = 2; +/– high-residue cultivation), blocks (*n* = 4) and years (*n* = 3; 2011–2013) for the three planting date treatments (early, middle, late) evaluated in the experiment. The coefficient of variation (CV) is reported to describe inter-annual yield variability (2011–2013) among alternative planting date strategies and county-level average conventional yields.

Site (Corn Grain/Silage)	Planting Date (Hybrid Maturity)	Planting Date Range (2011–2013)	Cover Crop Biomass Mean (CV%)		Corn Yield Mean (CV%)	
			Mg ha^{-1}			
Penn State University	Early (99 day)	31 May–1 June	5.9	(8)	35.8	(10)
Rock Springs, PA	Middle (95 day)	7–12 June	6.4	(2)	31.2	(2)
(silage)	Late (85 day)	15–18 June	6.4	(1)	28.0	(14)
	County Average Yield		—		42.2	(26)
USDA-ARS	Early (104 day)	13 May–4 June	5.6	(17)	7.5	(8)
Beltsville, MD	Middle (99 day)	24 May–12 June	5.8	(3)	7.0	(31)
(grain)	Late (88 day)	31 May–18 June	6.0	(15)	7.6	(13)
	County Average Yield		—		6.0	(48)
University of Delaware	Early (104 day)	4–30 May	5.5	(27)	3.8	(35)
Georgetown, DE	Middle (99 day)	11 May–6 June	6.4	(12)	5.2	(9)
(grain)	Late (88 day)	16 May–17 June	7.2	(6)	3.6	(14)
	County Average Yield		—		9.3	(11)

Compared with mid-Atlantic county-level averages, soybean yields in the ROSE were generally more consistent than corn yields. This difference in yield stability between corn and soybean has been observed in previous studies of CCORNT systems [13]. Inconsistent organic no-till corn performance in the mid-Atlantic could be due, in part, to regional constraints on fertility management. Southern areas of the mid-Atlantic have access to poultry litter for corn crop production. Pelletized poultry litter was broadcast at planting in Delaware and subsurface-banded at side-dress using custom-fabricated banding equipment at the Maryland location. In comparison, the Pennsylvania location utilized liquid dairy manure and applied it during the late-summer hairy vetch establishment phase the previous summer, approximately nine months prior to corn establishment. The effect of these different fertility practices likely influences N availability to corn during corn peak N demand. Further research is needed to determine if asynchrony between corn peak N demand and N supply from fall-applied liquid dairy manure is a significant constraint that may limit adoption of CCORNT corn production in dairy-based regions of the mid-Atlantic.

Beyond comparisons with county-level averages, which are predominantly based on yield data for conventionally-managed farms, it is critical to compare the yield performance of organic rotational no-till against tillage-based organic systems, in order for traditional tillage-based organic grain farmers to determine the value of CCORNT. Experiments comparing CCORNT against tillage-based organic systems have been conducted at the Agricultural Research Center in Beltsville, Maryland since 2008 (Supplemental Table S1). Data collection has primarily focused on corn (2008–2016 growing seasons), but new data are now emerging for soybean (2015–2016 growing seasons). These data show no difference in soybean grain yields between CCORNT and tillage-based systems, while corn grain yields in CCORNT increased compared with tillage-based systems (Figure 1). These results suggest that, despite the greater operational challenges associated with CCORNT, there is an opportunity to increase yield output in these organic systems. However, maintaining consistent corn yields across years in CCORNT systems remains a major challenge. Planting through thick mulch can interfere with achieving satisfactory seed-soil contact, which can result in inconsistent establishment of targeted corn populations and low yields. Soybean can recover from inconsistent germination and establishment due to having a comparatively higher seeding rate [18]. In addition, the later planting dates required in more northern parts of the mid-Atlantic (due to climate conditions) can further negatively impact CCORNT corn yields.

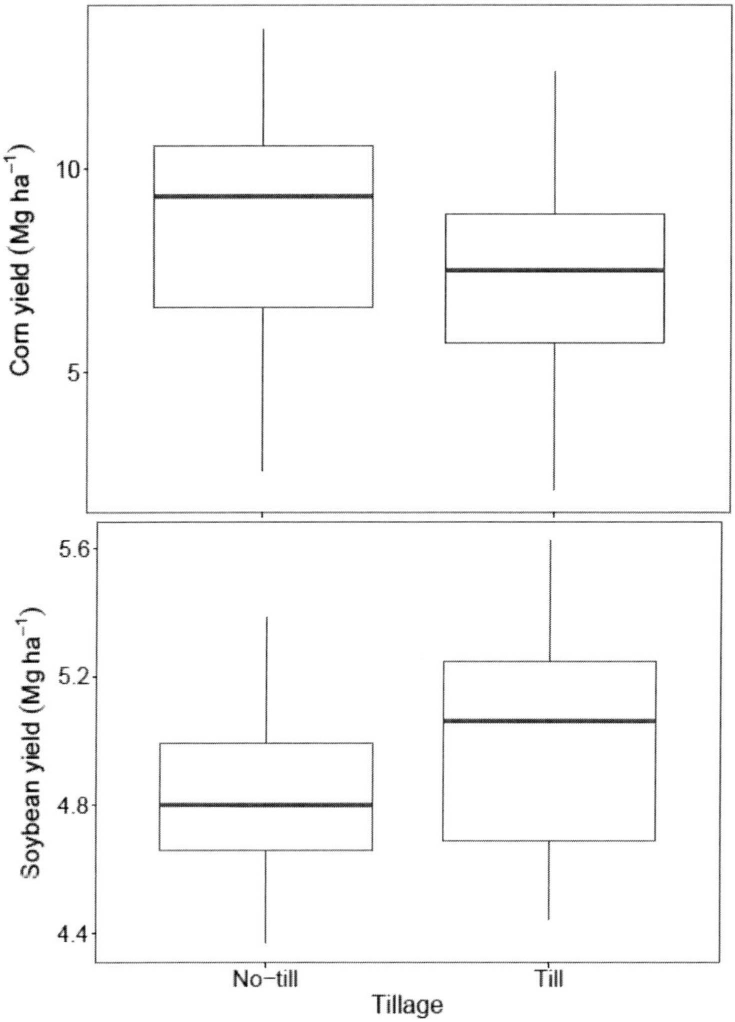

Figure 1. Corn and soybean grain yields in organic rotational no-till systems (No-till) and tillage-based organic systems (Till) at the Agricultural Research Center, Beltsville, Maryland. Corn data span the 2008–2016 growing seasons (2008–2010 with monoculture hairy vetch cover crop; 2011–2016 with cereal rye/hairy vetch cover crop mixtures); soybean data are from the 2015–2016 growing seasons. Corn yields were significantly greater in no-till compared with till ($F_{1,50}$ = 6.45, p = 0.01); soybean yields did not differ between systems. Box plots represent the median (horizontal line), interquartile range (box) and the maximum and minimum values (end of each vertical line).

3. COVER AND CASH CROP MANAGEMENT IN CCORNT

Management strategies that produce high cover crop biomass and optimal cash crop yields are critical to the success of CCORNT corn and soybean production. Critical challenges of CCORNT systems include: (1) poor cover crop establishment, which can negatively impact biomass accumulation and subsequent weed suppression; (2) poor establishment of legume cover crops, which reduces N-fixation; (3) failure to adequately terminate the cover crop, which negatively affects cash crops via light, water and nutrient competition; and (4) planter configurations that cannot sufficiently cut through or remove cover crop biomass, which can lower cash crop populations and yields. Consequently, understanding the potential agronomic tradeoffs related to cover and cash crop management strategies is critical for improving the consistency of CCORNT soybean and corn systems.

3.1. Soybean

Agronomic factors influencing the success of no-till planting soybean into roller-crimped cereal rye in the mid-Atlantic have been well documented [19,20]. In our experience, CCORNT soybean systems have produced competitive yields more consistently than corn. This can be attributed in part to the fact that soybeans have less management complexity than corn. The cereal rye monocultures used prior to soybean are typically easier to manage than legume-grass mixtures utilized prior to corn. Here, we discuss best management practices for optimizing cereal rye biomass accumulation and subsequent termination with a roller crimper, alternative winter grains for no-till soybean systems and multi-tactic strategies that improve weed management.

3.1.1. Cover Crop Establishment Prior to Soybean

Previous research has identified agronomic practices that improve cereal rye establishment and performance in no-till soybean systems [14]. These practices include manipulating establishment timing, seeding rates and soil fertility to maximize cereal rye biomass production and ground cover [21,22]. Soybean typically follows corn in rotation in the mid-Atlantic. In the northern mid-Atlantic (e.g., Pennsylvania and New York), late-fall corn grain harvest limits the agronomic window for cereal rye establishment, but a longer fall growing season in southern mid-Atlantic states (e.g., Maryland and North Carolina) allows for consistent cereal rye establishment. In our experience, adequate seedbed preparation is critical to maximizing cereal rye performance. Prioritizing earlier corn harvest to ensure sufficient time for corn stover management and primary and secondary tillage is often necessary. Based on the window of time afforded by the previous crop, producers can either prepare a seedbed with tillage prior to drilling or no-till drill cover crops. Processing previous corn stover residues with a turbo-till often suffices to ensure optimal cover crop establishment. Historically, we have advocated tillage prior seeding fall cover crops, as perennial weeds and pest cycles are most effectively controlled with primary tillage using a chisel- or moldboard-plow. However, during wet falls, it may be best to avoid tillage and no-till plant cover crops, as seeding into tilled ground can be greatly stalled by a wet fall. However, research and experience have repeatedly demonstrated that cover crop establishment and growth are greater with a prepared seedbed.

Based on our body of research, best management practices for cereal rye establishment include seeding at high rates (approximately 134 kg ha^{-1}) with a grain drill. Increasing cereal rye seeding rates has been shown to increase weed suppression due to greater ground cover in early spring [22]. The use of a grain drill improves establishment and provides a more uniform plant population in comparison to broadcast methods. Alternative practices that have been employed include a dual seeding method in which seed is both drilled and broadcast to encourage increased ground cover in inter-row spaces. Finally, while cereal rye is an efficient scavenger of residual inorganic soil N, it may also be necessary to fertilize cereal rye to maximize biomass and ground cover when soil mineralizable N levels are low.

3.1.2. Cover Crop Termination Prior to Soybean

Optimizations of cover crop termination efficacy and biomass accumulation are the primary factors that influence cover crop termination timing decision-making. Mechanical control of cereal rye using a roller-crimper is most effective when terminating after anthesis, with >85% control of cereal rye occurring once the Zadoks growth Stages 60–65 are reached [21]. Cover crop mulches suppress weed germination and growth of summer annual weeds via physical obstruction and by lowering light and temperature at the soil surface [14]. The level of cover crop biomass typically needed to provide high levels of weed suppression (>8000 kg ha^{-1}) requires delayed cash crop planting and good base soil fertility [23,24]. Delaying cover crop termination by only 7–14 days can result in significant gains to cover crop biomass accumulation. In the mid-Atlantic region, cereal rye biomass accumulation can increase 200 kg ha^{-1} day^{-1} after stem elongation [21]. Consequently, delayed cereal rye termination is a viable strategy for increasing the weed suppression potential of surface mulch residue.

Other cultural strategies that can be integrated with delayed cereal rye termination to maximize weed suppression include earlier fall planting [21] and increased seeding rates to enhance cereal rye ground cover in early spring [22].

Cover crop termination timing can also shift the composition of weed species in CCORNT systems by altering both cover crop biomass accumulation (increasing with delayed termination) and the timing of mechanical control in relationship to weed emergence. For example, in a Pennsylvania experiment, insufficient canopy cover from cereal rye resulted in the emergence of common ragweed (*Ambrosia artemisiifolia* L.) prior to cover crop termination with the roller crimper, which did not control emerged common ragweed seedlings [19]. In comparison, later emerging summer annual species such as giant foxtail (*Setaria faberi* Herrm.) and smooth pigweed (*Amaranthus hybridus* L.) were less abundant. The selection of dominant weed flora was associated with germination prior to rolling and survival following the rolling operation.

The ROSE demonstrated that volunteer cover crops have the potential to limit the viability of CCORNT practices in annual grain rotations [25]. Although previous studies have identified the optimal termination timing for limiting regrowth of roller-crimped cover crops, most research has been conducted using annual experiments that did not track management legacy. Consequently, the agronomic impacts of volunteer cover crop recruitment that may result from cover crop seed production following even low levels of regrowth are not well understood. In the ROSE, within-season control of cereal rye with the roller-crimper improved as termination was delayed, but cereal rye continued to mature on the ground when termination occurred after the onset of the milk stage (>Zadoks 70). Regardless of cover crop termination timing, volunteer cereal rye was detectable across locations in winter wheat (Figure 2); in some cases, grain contamination levels approached the 5% maximum threshold allowed for foreign materials in food-grade wheat. Consequently, there is likely a narrow agronomic window from 50% anthesis (Zadoks 65) to the milk stage (Zadoks 70–79) for optimizing cereal rye control and minimizing the effect of volunteer cover crops in subsequent years. System-specific strategies are needed to manage volunteer cover crops in CCORNT systems. For example, the ROSE revealed that a comparatively longer fall growing season in Delaware allowed for false seed-bedding before winter wheat, which minimized volunteer cereal rye issues. Alternatively, following the soybean phase of the rotation with a spring- or summer-sown crop provides an opportunity to manage volunteer winter cover crops with tillage prior to planting the next crop in rotation. Other strategies may include multiple roller passes or more aggressive roller-crimper technology.

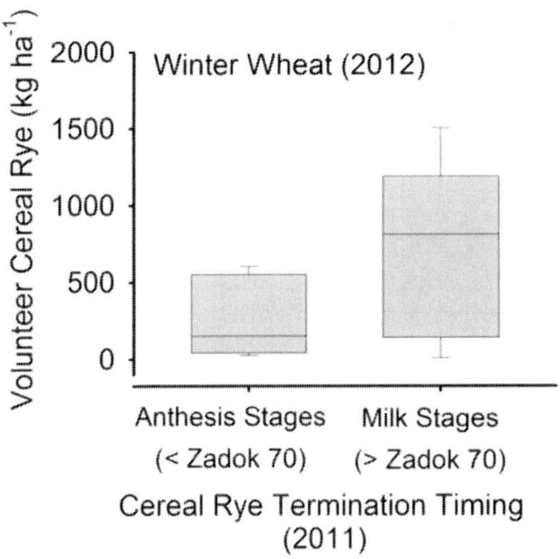

Figure 2. Effect of cereal rye termination timing in 2011 on volunteer cereal rye abundance in the following winter wheat cash crop (2012) at the Pennsylvania and Maryland locations (data from the Reduced-tillage Organic Systems Experiment (ROSE)). Cereal rye termination ranged from the onset of anthesis (Zadoks 59) to the soft dough stage (Zadoks 85).

Although cereal rye is the most commonly-used cover crop species in CCORNT soybean production, other winter cereal species might provide advantages over cereal rye. As the timing of cover crop termination is based on cover crop phenology, using an earlier-maturing species such as barley (*Hordeum vulgare* L.) might enable earlier termination and soybean planting and, thus, a longer soybean growing season and greater yield potential. In an experiment conducted in Maryland and New York from 2013–2014 (Supplemental Table S1), delaying termination by one week resulted in greater biomass for both cereal rye and six-row barley (Figure 3A). For each species at each site, terminating the cover crop later reduced both weed biomass (Figure 3B) and the percentage of uncontrolled cover crop plants, or "bounce-back" (i.e., plants that survived rolling-crimping and returned to their upright growth habit). In Maryland, cover crop bounce-back decreased from 5% down to 0% for barley and 3% down to 0% for cereal rye as termination timing was delayed by one week. In New York, the later termination time reduced cover crop bounce-back from 17% down to 11% for barley and 9% down to 6% for cereal rye. Although termination efficacy tends to be proportional to cover crop growth stage, we did not observe less bounce-back with barley compared to cereal rye, despite barley maturing earlier in the spring. Interestingly, we also did not observe increased weed suppression in cereal rye despite greater cover crop biomass production than barley. Soybean yield was similar across all combinations of species and termination times at each site (Figure 3C), suggesting that competition between uncontrolled cover crops and soybean was negligible. Results from this experiment indicate that lower levels of cover crop biomass (5000–6000 kg ha^{-1}) can provide adequate weed suppression in the northern mid-Atlantic, which might help alleviate some of the challenges associated with soybean seed placement through thick layers of mulch. Our results also suggest that barley might be able to compensate for lower biomass production and achieve relatively high levels of weed suppression due to a more rapid canopy closure and wider leaf blades compared to cereal rye [26].

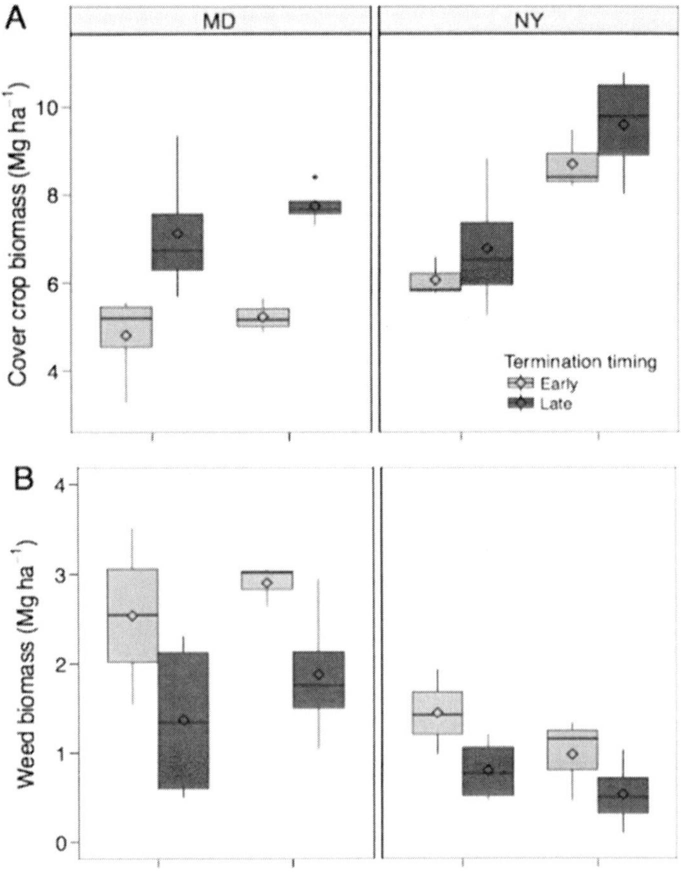

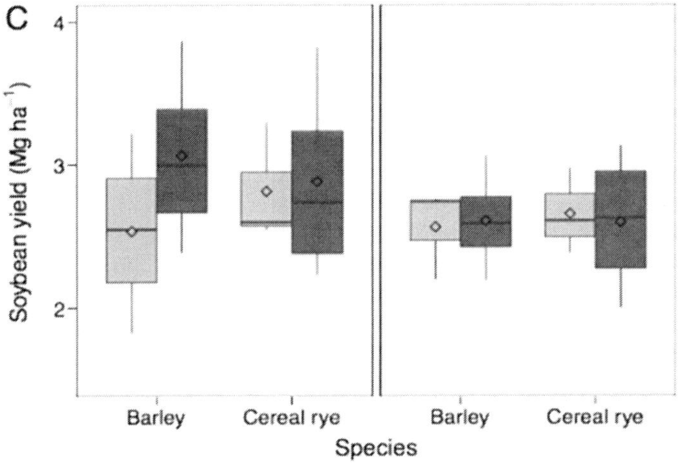

Figure 3. The impact of cover crop termination timing on cover crop biomass (**A**), weed biomass (**B**) and soybean grain yield (**C**) using six-row winter barley and cereal rye cover crops in Maryland (MD) and New York (NY) in 2014. Late termination timing occurred one week after the early date. Box plots represent the mean (open diamond symbol), median (horizontal line), interquartile range (box) and the maximum and minimum values (end of each vertical line). For each response, species and termination timing were included as fixed effects: cover crop biomass ($F_{1,25}$ = 12.68, p = 0.0015 for species and $F_{1,25}$ = 12.33, p = 0.0017 for termination timing); weed biomass ($F_{1,25}$ = 0.032, p = 0.86 for species and $F_{1,25}$ = 11.65, p = 0.0022 for termination timing); and soybean yield ($F_{1,20}$ = 0.0041, p = 0.95 for species and $F_{1,20}$ = 0.90, p = 0.35 for termination timing).

3.1.3. Soybean Planting and Management

In contrast to conventional and tillage-based organic soybean production, farmers that manage soybean in CCORNT systems must overcome the unique challenges associated with planting through thick cover crop mulch. In particular, achieving adequate soybean seed-to-soil contact can be difficult [27]. Soil moisture can be depleted through cover crop transpiration, thereby inhibiting soybean germination [27,28,29], and soybean lodging can be problematic as hypocotyl elongation may weaken stem strength [30].

It is recommended to plant soybean in the same direction as roller-crimping (ideally in a single pass) so that the coulters on a no-till planter can part the residue rather than be forced to cut through cover crops rolled perpendicularly to the direction of planting. However, cover crop residue can still be pushed into the seed furrow (i.e., hair-pinning), which can impede seed-to-soil contact [31,32]. In such instances, it is common to observe desiccated, non-germinated soybean seed on top of the hair-pinned residue, especially when the soil is dry and the closing wheels are unable to cover the furrow. Row cleaners (trash wheels) can reduce hair-pinning by moving cover crop mulch away from the soybean row [29]. There are several commercially-available trash wheels, but units must have a narrow blade angle and ideally overlap in order to adequately both cut and move trash. We have customized planters to include an extra toolbar in front of the double-disk opener and unit mounted coulter. This extra toolbar has residue slicers put in line with the unit mounted coulters. The residue slicers have depth wheels on each side of the coulter, which help press residue down and improve the cutting action. The frame mounted residue slicers provide additional cutting action. Trash wheels have also been mounted to the frame and thereby provide two sets of coulter and trash wheels to adequately move residue and cut the remaining material prior to planting. This can improve soybean seed placement and germination, but in some cases, the residue can become wrapped around the row cleaners, impairing their function [13]. Moreover, mulch covering the crop row is important for in-row weed suppression. If row cleaners move too much residue, emerging crop seedlings can suffer from increased weed competition. Many of these machinery considerations also apply to CCORNT corn.

Using higher soybean planting rates is an effective cultural weed management practice [20,33,34,35]. Higher soybean planting rates enhance weed-crop competition primarily through earlier crop canopy closure and greater shading [33,36,37]. High planting rates can also help maintain maximum yields [34,38]. The price premium for organic soybean (five-year average for feed-grade soybean = \$0.92 kg^{-1} [39]) helps ensure that high planting rates are profitable for organic farmers despite the increase in seed cost. Narrower row spacing (<76 cm) can also be used to hasten canopy closure, but this precludes the option of high-residue, inter-row cultivation to manage inter-row weeds.

Multi-tactic approaches that combine cultural strategies for maximizing cover crop biomass with supplemental high-residue, inter-row cultivation have been investigated [26,40,41]. High-residue cultivators for no-till soybean and corn systems typically have two depth-gauge wheels with a residue-slicing coulter between the wheels, which are followed by a single wide flat sweep. The sweep cuts through the soil at a 2.5–5-cm depth, slicing the root systems of weeds while leaving the surface residue mostly intact. Unlike most inter-row cultivators that are most effective on small weed seedlings before their root systems become established, high-residue cultivators are more effective on weeds after they have become established. Standard cultivators rely on dislodging these small seedlings, while the high-residue cultivator's sweeps separate the roots from the above-ground portion of the plant. Nord et al. [41] found that cereal rye biomass alone provided high levels of weed suppression when weed infestations were low, but high-residue cultivation was needed to achieve acceptable levels of weed suppression when weed infestations were high (i.e., supplemented weed seedbank), regardless of cereal rye biomass accumulation. High-residue cultivation can effectively control summer annual weeds that emerge prior to cover crop termination, such as common ragweed [26,41]. The Rodale Institute (PA) tested the effects of high-residue cultivator timing on weeds in rotational no-till soybeans at 3–4, 5–6 and 7–8 weeks after soybean planting [42]. Averaged across timings, high-residue cultivation resulted in a 67% reduction in weed biomass compared to the no-cultivation control, and soybean yields were 12%–22% higher in high-residue cultivation treatments compared to the control.

The ROSE investigated the use of high-residue inter-row cultivation and delayed cover crop termination as a multi-tactic weed management approach. Delaying cereal rye termination by 7–14 days to increase surface residue mulch produced variable results as a weed management strategy. Total cover crop biomass accumulation and biomass gain with delayed termination varied by location and year (Table 1). Regional differences in weed species response to delays in cover crop termination were also observed, similar to that found by Nord et al. [19] and Teasdale and Mirsky [43]. In comparison, high-residue, inter-row cultivation consistently lowered summer annual weed biomass across various termination timings and study locations in no-till corn and soybean systems (Figure 4). Summer annual weed biomass was also consistently low in the winter wheat phase of the rotation (<100 kg ha^{-1}), regardless of weed biomass levels in no-till soybean across different planting dates. Our experience in the mid-Atlantic suggests that high-residue cultivation will be a necessary tool to prevent exponential increases in weed seedbanks during years when poor cover crop performance or other agronomic factors lead to higher weed infestation levels. Furthermore, mulches have a minimal effect on perennial weeds, which become more challenging in CCORNT systems. High-residue cultivation can serve as an important tool to manage perennial weeds in CCORNT soybean and corn systems [13,14].

3.2. Corn

Tillage-based organic corn production has typically out performed tillage-based organic soybean production, likely because corn rapidly canopies and is therefore more competitive against weeds [44]. In organic no-till systems, however, corn production involves greater management complexity compared to soybean production due to higher N demand. Legume-based cover crops do not typically provide sufficient N in no-till corn, and supplemental organic sources are required. Organic nutrient sources are more difficult to apply than inorganic sources and are also more tightly regulated in the Chesapeake Bay watershed (e.g., Maryland mandates incorporation or subsurface banding of poultry litter). While cover crops in the soybean phase must be managed primarily to control weeds, cover crops in the corn phase must be managed to control weeds and provide a source of N. This necessitates the use of cover crop mixtures prior to corn, for which selection and management is more complex. Below, we review corn production management intricacies and constraints.

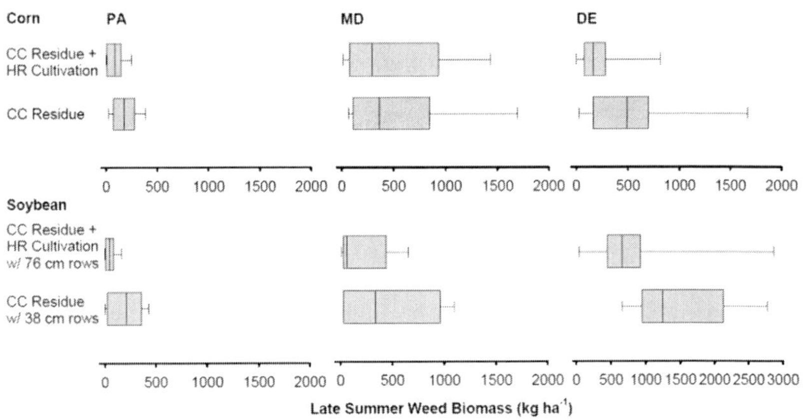

Figure 4. Effect of rolled cover crop (CC) surface residue on weed biomass with or without the use of high-residue (HR) inter-row cultivation; data from the Reduced-tillage Organic Systems Experiment (ROSE). In Pennsylvania (PA), HR cultivation in no-till corn decreased weed abundance ($F_{1,55} = 8.1$, $p = 0.006$). In Delaware (DE, $F_{1,55} = 13.7$, $p < 0.001$) and Maryland (MD, $F_{1,55} = 3.7$, $p = 0.05$), HR cultivation in no-till soybean decreased weed abundance compared to when soybean was planted on 38-cm rows in the absence of HR cultivation, which was evaluated as an alternative weed control strategy.

3.2.1. Cover Crop Establishment Prior to Corn

In the mid-Atlantic region, organic corn is typically planted after a wheat or soybean grain crop. The timing of summer grain harvest dictates the time allowed for legume cover crop establishment. Legume cover crops require earlier planting dates than grasses to produce acceptable levels of biomass and to survive the winter, particularly in the northern mid-Atlantic [24,45]. Initial organic no-till research focused on preceding corn with a hairy vetch monoculture [45]. Hairy vetch is a winter-hardy legume with rapid growth that produces large amounts of biomass, has an optimal suite of characteristics for adding considerable N to the system, and provides some weed control [45]. However, more recent research has switched focus to mixtures of hairy vetch and winter grains, as hairy vetch alone was found to not provide sufficient weed control.

Hairy vetch monocultures or winter grain/hairy vetch mixtures are best established using a grain drill. These species mix well in seed hoppers with minimal separation, resulting in optimal cover crop stands. The proportion of each species in the mixture should be chosen carefully to balance tradeoffs in managing both nitrogen and weed suppression. For example, a recent mid-Atlantic study demonstrated that a wide range of hairy vetch/cereal rye mixture proportions produced >8 Mg ha^{-1}, which is considered to be the biomass threshold for acceptable levels of weed suppression in the southern mid-Atlantic, but a hairy vetch/cereal rye seeding rate of 27/34 kg ha^{-1} was needed to achieve maximum N content and low C:N ratios [46]. Selection of mixture seeding rates should also consider initial field conditions. Sites with low residual soil fertility will favor the growth of the legume(s) in a mix due to the ability of legumes to fix atmospheric N, whereas grasses in a mix are likely to dominate on high fertility sites due to their rapid growth and ability to scavenge N [47]. Seeding rates should be adjusted according to the primary needs of the site (e.g., N-fixation, weed control, etc.) to take advantage of site conditions, minimize seed costs, and maximize the benefits provided by the cover crops.

3.2.2. Cover Crop Termination and Corn Planting

Cereal rye and hairy vetch are terminated most effectively when roller-crimped at anthesis. However, asynchrony between cereal rye and hairy vetch anthesis presents a termination challenge. Hairy vetch is best controlled through the combined action of a roller-crimper (which initially damages the cover crop) and a cash crop planter (which aids termination by adding additional wheel traffic and cutting the cover crop). Research has thus focused on a range of termination strategies to optimize the control of the cover crop mixture using a roller alone and in combination with a planter. We have rolled immediately before planting (roll/planted), rolled twice before planting (roll-roll/planted) and rolled, planted and rolled again before crop emergence (roll/planted roll). We have found that hairy vetch is not completely killed by rollers alone and that a second rolling three to ten days later (roll/plant-roll), depending on site conditions, provides an opportunity to kill any cover crops not terminated with the first pass. This approach is also better than the roll-roll/plant approach, as a rain event between the first rolling and second can greatly delay planting since a terminated cover crop mulch does not transpire and lowers evaporation.

Corn planting date is strongly affected by timing of cover crop termination, and performance has been noted to decline in more northern regions when the later flowering of hairy vetch delays planting [13,48]. In this case, lower corn yields are due to decreased growing degree days (GDD, i.e., a shorter growing season). Recent research in Maryland and North Carolina has found that planting corn when hairy vetch reaches 50% flowering had little to no negative effect on corn yields as compared with earlier corn planting dates [49,50]. These results may reflect the greater flexibility of CCORNT corn production in more southern regions of the mid-Atlantic compared with more northern regions, due to their longer growing season. Matching corn cultivar maturity groups with local climate (i.e., typical patterns of GDD accumulation) can better accommodate shorter growing seasons in the north.

3.2.3. Cover Crops and Weed Control in Corn

Cover crop termination method and timing also have important implications for weed management in no-till corn systems [43]. In a Pennsylvania experiment, perennial weed species were found to be less suppressed by hairy vetch cover crop residues than annual weed species [51], and full-season weed suppression via cover crop residues can be inconsistent [19,21,41]. Consequently, inter-row high-residue cultivation is often needed to provide greater weed suppression. For example, high-residue cultivation in corn planted after a hairy vetch monoculture improved weed control, relative to no cultivation, by 23%–78% over a three-year study [49]. Optimization of the timing and frequency of high-residue cultivation will be critical for designing multi-tactic strategies that result in consistent levels of weed control. A recent Pennsylvania study suggests that two high-residue, inter-row cultivation passes at approximately four and six weeks after planting are needed to provide consistent levels of weed control in reduced-tillage corn with or without cover crop surface mulch residue [40]. This strategy allows for killing both early emerging or fast growing species and later emerging species that were too small to be susceptible in the first pass. However, in addition to increased fuel and labor costs, additional cultivation passes result in less persistent cover crop residue, particularly in the case of more readily-decomposable hairy vetch cover crops. Though not yet documented for high-residue cultivation, disturbance to the legume mulch and soil, and the resulting increase in soil-cover crop contact, has also been shown to increase rates of N availability in the soil [52].

In comparison to legume monocultures, legume/grass cover crop mixtures may also improve weed control and are therefore recommended as the preferred cover crop for corn production. Hairy vetch alone can stimulate weed germination and growth and is an unreliable tool for weed control [51]. In a two-year experiment conducted by USDA-ARS at the Beltsville Agricultural Research Center in Beltsville, Maryland (Supplemental Table S1), hairy vetch monocultures significantly increased weed biomass as compared to cereal rye/hairy vetch mixtures (Figure 5). Strikingly, cover crop mixtures containing only 20% cereal rye and 80% hairy vetch produced the same reductions in weed biomass relative to a hairy vetch monoculture (approximately 70%) as a cereal rye monoculture (Figure 5). We speculate that cereal rye provides structural support for hairy vetch, minimizing plant tissue contact with the soil surface. The soil surface is noticeably wetter and warmer under hairy vetch monocultures, and its prostrate growth habit results in rapid decomposition even while the plant is still growing [53]. Since N is known to stimulate both weed germination and growth [54], grass/legume mixtures that delay N release allow producers to address both weed and N management needs. However, as we discuss in greater detail below, grass/legume mixtures decrease the overall N available for a corn crop when compared to a pure legume cover crop and, thus, may require supplemental sources of N.

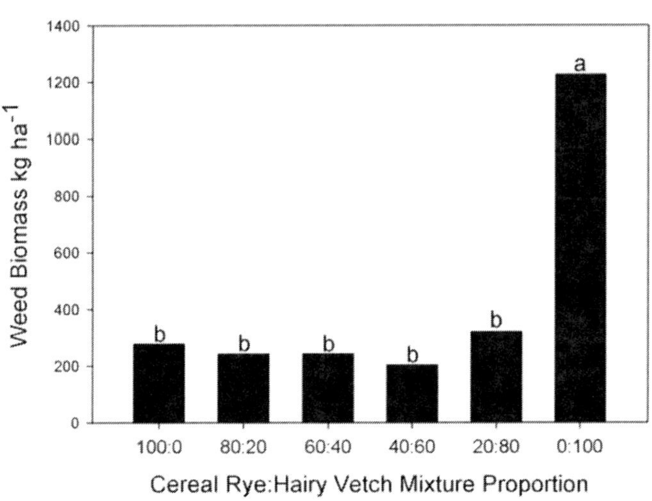

Figure 5. Weed biomass at corn silking over the 2013–2014 growing seasons under different cereal rye/hairy vetch mixture proportions (data from the Agricultural Research Center, Beltsville, MD, USA). Different letters above bars show significant differences ($F_{5,18} = 4.57$, $p < 0.01$).

Failure to effectively terminate cereal rye and hairy vetch can lead to non-terminated cover crops reaching maturity, which may result in competition with the corn crop for resources during early growth and development and in seedbanks that result in volunteer cover crops in future seasons [51]. Hairy vetch produces hard seed that can easily become a volunteer weed problem [25,55]. In our experience, hairy vetch termination needs to be delayed until flowering to achieve high levels of control, but viable seed production occurs very soon after flowering. In the ROSE, the control of hairy vetch with a roller-crimper improved as termination was delayed, but mature seeds were produced for both cover crops (cereal rye and hairy vetch) and emerged as volunteers within the cereal rye-no-till soybean and winter wheat phases of the rotation regardless of termination timing (Figure 6). Volunteer hairy vetch contamination in winter wheat was negligible (<1%) following the three termination timings at the Pennsylvania location, but impacted wheat yields (10%–29%) at the Maryland location. Furthermore, volunteer hairy vetch plants increased in the cereal rye-no-till soybean phase because optimal timing for cereal rye termination with the roller-crimper occurs when hairy vetch is still vegetative and is thus poorly controlled by the roller-crimper pass. Observational evidence suggests that two roller-crimper passes spaced approximately 5–8 days apart can improve the control of hairy vetch/triticale mixtures and thus reduce volunteer hairy vetch populations. Additional labor and fuel costs associated with this extra pass may be warranted to minimize volunteer cover crops in subsequent crop phases of the rotation.

The timing of cover crop termination also has important implications for the management of seed and seedling invertebrate pests in CCORNT systems. At the Pennsylvania ROSE location, corn seedling herbivory increased as cover crop termination was delayed in some years, but was highly variable across termination timings in no-till soybean [56]. The primary herbivores contributing to corn seedling damage were black cutworm (*Agrotis ipsilon* Hufnagel) and slugs (Mollusca). In this case, greater herbivore activity in late-planted corn may be attributed to herbivore response to higher temperatures. However, cover crop management also influenced the activity of ground-dwelling natural enemies of arthropod pests [57]. In particular, greater activity-density of large ground beetles (Coleoptera: Carabidae) occurred as termination of hairy vetch/triticale and cereal rye cover crops were delayed. Moreover, carabid activity-density and species richness increased each year during the three-year transition to organic production. The design of integrated, multi-tactic pest management approaches likely requires a greater understanding of early-season insect pest dynamics in CCORNT systems for the primary pests in the mid-Atlantic region, including wireworms (Coleoptera: Elateridae), seed corn maggot (*Delia platura*) and cutworms

(Lepidoptera: Noctuidae). However, limited evidence suggests that CCORNT systems may provide suitable habitat for predatory arthropods, thereby providing measurable levels of conservation biological control.

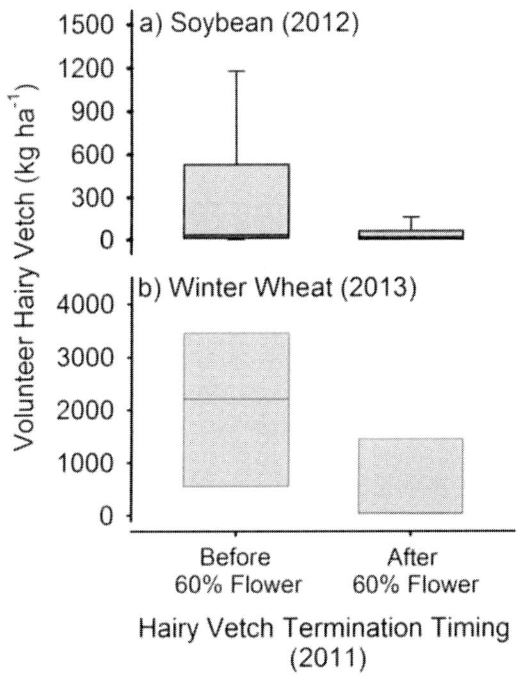

Figure 6. Effect of hairy vetch termination timing in 2011 on volunteer hairy vetch abundance in the following soybean (2012) and winter wheat (2013) cash crops averaged across Pennsylvania and Maryland locations (data from the ROSE). Hairy vetch termination ranged from the vegetative stage to 100% flowering on the last five nodes below the apical meristem.

3.2.4. Corn Fertility Management

Similar to tillage-based corn production, the N needed by the corn crop in CCORNT systems comes from cover crops, soil mineralized N and other sources of supplemental N (e.g., animal byproducts or Chilean nitrate). The proportion of these N sources used by producers depends on the cover crop, its management, past management history and soil type. Hairy vetch can supply over 180 kg ha^{-1} of plant N from the shoots alone, with over 90 kg ha^{-1} of that being plant available [48,52,58,59]. Previous studies have suggested that a monoculture of hairy vetch can supply the full N requirements of a corn crop on heavier soils that have a long history of manure and legume use [60], which can be attributed to the high N supplying capacity of the soil supplementing the legume-based fertility. Local recommendations are that producers not rely on a cover crop alone to meet corn N demand. This is particularly necessary when working with cereal rye/hairy vetch mixtures. Mixtures at 50% legume biomass and higher will produce comparable total shoot N levels to a legume monoculture [58], but with each incremental increase of grass in the mixture, there is a shift in the C:N ratio of the residue. Lower C:N ratio residues decompose rapidly with subsequent early release of N, while higher C:N ratio residues take longer to decompose and delay N release [52,61]. The timing of N release is critical. If N is released too early, prior to when corn enters its peak N demand phase (at approximately the corn six-leaf stage, V6 [62]), the N is susceptible to environmental loss via leaching and gaseous emissions. If N is released too late, it will not provide appropriate nutrition for the corn crop, negatively affecting corn production. Thus, for legume-based cover crops to be an effective N source and to reduce N-loss from agricultural lands, the synchrony of N release with crop physiological demand is vital [63]. The use of a grass in combination with a legume cover crop can improve N-release synchrony. Prior to termination, grass cover crops may take up N produced by legumes, while after termination, the higher C:N ratio of grasses moderates the leaching potential of the low C:N-ratio legumes [64,65]. Recent research in Maryland demonstrated that mixtures of hairy vetch and cereal rye, residues of which have a substantively higher C:N ratio than hairy vetch alone, can slow the decay rate of cover crop residues and reduce the quantity of early season plant-available N in the soil while still providing an important source of N for corn later in the season [52,58]. However, while maximum N content is maintained up to a 50:50 grass/legume mixture (based on biomass), the quality of the material results in lower overall plant-available N during the growing season and necessitates the addition of poultry litter or dairy slurry as used in the ROSE.

The addition of poultry litter or dairy slurry helps to ensure adequate N for a following corn crop, but adds to management complexity. In tillage-based systems, cover crops and animal byproducts are incorporated prior to cash crop planting. Applying animal byproducts poses numerous logistical and environmental difficulties in no-till production. Applying animal waste onto surface mulches increases their decomposition rates [52], speeding potential N leaching from the cover early in the spring when corn N uptake is small. However, decreasing the amount of surface biomass levels via tillage (i.e., to incorporate animal byproducts) will also reduce the amount of cover crop surface biomass present to suppress weeds. Furthermore, the mid-Atlantic region encompasses multiple sensitive watersheds in which regulations dictate manure application management and timing. For example, in the state of Maryland, animal byproducts like poultry litter are not allowed to be surface-applied unless they are incorporated within 48 h. Therefore, N contributions from cover crops remain important to reduce the amount of supplemental N required from animal byproduct sources. The N:P ratio of legumes or grass/legume mixtures offsets the N:P ratio of animal byproducts, which tend to increase soil P when applied at N replacement rates. Not only is it expensive to apply animal byproducts such as liquid dairy manure and poultry litter at N replacement levels, but soil P levels are highly regulated in the mid-Atlantic region. Organic farmers who have historically applied large quantities of organic amendments can build up their N mineralization potential [66]. In cropping systems with a history of cover crop use and organic amendment application, organic grain farmers can produce medium corn yields of 8000 kg ha^{-1} on soil mineralized N alone [67].

4. ONGOING CHALLENGES AND CONCLUSIONS

CCORNT systems can provide high-value and in-demand agricultural products while potentially improving soil health and protecting regional watersheds via improved nutrient management. In the last five years, significant progress has been made in improving the viability of these cropping systems in the mid-Atlantic region of the U.S. For soybeans, CCORNT production holds particular promise now that improved planter technology has increased crop performance and competitiveness against weeds and high-residue cultivators are commercially available to manage perennial and annual weeds not controlled by the mulch. However, major constraints to CCORNT corn production remain, given the complexity of balancing soil fertility and weed management demands. Large amounts of cover crop residue (especially grass residue) are needed to suppress weeds, but such residue can interfere with cash crop establishment and immobilize soil N, decreasing corn yields. In the case of corn, legume-grass cover crop mixtures better balance soil fertility and weed management goals, but present management complexities of their own, including: (1) current crop planter designs struggle to plant into cover crop mixtures; (2) there are few legume options that are compatible with roller-crimping, and viable species like hairy vetch present the risk of regrowth and weediness in future cash crops; and (3) limited economically viable, production-ready equipment that can apply animal by-products in high-residue no-till production systems. Moreover, CCORNT systems suffer from reduced flexibility relative to tillage-based organic systems, due to the limited windows of opportunity for controlling cover crops.

The lack of availability of effective machinery and machinery configurations to plant cash crops into heavy cover crop residue continues to be a major constraint [14]. Unlike conventional producers, organic farmers must maximize the above-ground cover crop biomass for weed suppression. Planter technology must effectively deal with in-row plant material. A recent survey of organic farmers in upstate New York found that many are having to custom develop their own cultivation equipment in response to the lack of research and development on new machinery [68]. As outlined above, large quantities of cover crop biomass are generally required for effective weed control. This constitutes substantial surface mulch that must be cut through to ensure effective seed-soil contact to maximize cash crop germination and establishment. Planter configurations require flexibility to be optimized for effective seed placement depending on local environmental conditions and cover crop performance in a given year. Moreover, more research is required to elucidate region-specific practices related to the timing and combination of roll/plant and roll-roll/plant operations. This will ensure the efficacy of cover crop termination (i.e., minimizing volunteer cover crops), weed suppression and cash crop establishment [35,50]. As data from the ROSE demonstrate, the development of effective high-residue cultivation machinery can potentially allow for lower quantities of cover crop biomass while still providing comparable levels of weed control. However, a tradeoff exists between achieving sufficient cover crop biomass and conducting more frequent high-residue cultivation. Greater cover crop biomass provides effective weed suppression, soil protection and a valuable N source; more frequent high-residue cultivation reduces residue residence time and soil protective benefits and can increase asynchrony between residue N release and cash crop N demand.

Beyond machinery development and its effective implementation, there is also a need for improved cover crop germplasm and management practices [69,70]. Cover crops have not benefited from the same intensive breeding scrutiny as cash crops and consequently show wide variation in agronomic traits and the ability to provide desirable ecosystem services. A recent survey of farmers across the U.S. identified numerous desirable cover crop traits upon which breeding efforts should focus, including N-fixation (for legumes), winter hardiness, early vigor and establishment, biomass production and weed suppression [71]. Improved cover crop germplasm, including the development of highly adapted regional varieties [72], would allow for greatly improved integration and management of cover crops within existing cropping systems, particularly as they relate to cover crop phenology and termination timing [21,29]. Greater standardization of leguminous cover crops would aid the development of more reliable N management recommendations and advance our ability to predict how fixed N contributes to subsequent corn yield [73], particularly when legumes are grown in mixtures with grasses. More research is also required to understand the allelopathic properties of cover crops and how they can be effectively managed to maximize weed suppression while minimizing negative impacts on cash crops. For example, zonal management of cover crop residues, where residues are moved from crop rows to inter-rows prior to cash crop planting, may help minimize negative allelopathic effects on the cash crop while enhancing weed suppressive effects in inter-rows [74]. However, this would have to be balanced with the need to ensure adequate mulch for weed suppression in crop rows.

Lastly, more research is required to elucidate the potential of CCORNT production systems to improve soil health. Tillage-based organic agriculture has often been found to build soil health, e.g., have increased soil organic matter (SOM), soil C and N and microbial biomass and activity relative to conventional systems [75,76]. However, organic systems typically rely on frequent and intensive tillage operations to control weeds, which can undermine the ability of organic systems to build soil health [12,68,77]. No-till agriculture, which eschews tillage and typically relies on herbicides for weed control, has been found to conserve soil health compared with tillage-based agriculture, leading to enhanced SOM in surface soil, microbial biomass and activity [78,79]. However, over-reliance on herbicides has resulted in a growing list of herbicide-resistant weed species that threatens crop production across the U.S. [80]. CCORNT may offer a bridge between tillage-based organic and no-till systems. By attempting to integrate the soil building properties of tillage-based organic agriculture with the soil conservation properties of no-till, CCORNT may enhance soil health while minimizing the drawbacks of tillage-based organic systems. In addition, by employing a multi-tactic weed management approach, CCORNT may represent a long-term sustainable crop production system and offer valuable insights for contemporary conventional and no-till production systems on methods to counter the increasing prevalence of herbicide-resistant weeds.

ACKNOWLEDGMENTS

We would like to thank Ruth Mangum, Chris Pelzer, Brian Caldwell and all other members of the Sustainable Cropping Systems Lab at Cornell University and the Sustainable Agricultural Systems Lab at the USDA Beltsville Agricultural Research Center for their assistance with data collection in New York and Maryland. We additionally acknowledge Mary Barbercheck, Ariel Rivers, Christina Mullen, Mark Dempsey and all other members of the Reduced-tillage Organic Systems Experiment for their assistance in the completion of the project. This material is partially based on work that was supported by the National Institute of Food and Agriculture, USDA, Hatch Project 2016-17-252. Funding was also received from Natural Resource Conservation Service (NRCS)-Conservation Innovation Grant program and Agricultural and Food Research Initiative Grant # 2016-67012-24711 from the USDA National Institute of Food and Agriculture.

REFERENCES

1. McBride, W.D.; Greene, C.; Foreman, L.; Ali, M. *The Profit Potential of Certified Organic Field Crop Production*; ERR-188, United States Department of Agriculture, Economic Research Service: Washington, DC, USA, 2015.
2. Foley, J.A.; Ramankutty, N.; Brauman, K.A.; Cassidy, E.S.; Gerber, J.S.; Johnston, M.; Mueller, N.D.; O'Connell, C.; Ray, D.K.; West, P.C.; et al. Solutions for a cultivated planet. *Nature* **2011**, *478*, 337–342.
3. Crowder, D.W.; Reganold, J.P. Financial competitiveness of organic agriculture on a global scale. *Proc. Natl. Acad. Sci. USA* **2015**, *112*, 7611–7616.
4. Seufert, V.; Ramankutty, N.; Foley, J.A. Comparing the yields of organic and conventional agriculture. *Nature* **2012**, *485*, 229–232.
5. Oberholtzer, L.; Dimitri, C.; Jaenicke, E.C. International trade of organic food: Evidence of US imports. *Renew. Agric. Food Syst.* **2013**, *28*, 255–262.
6. Sulc, R.M.; Franzluebbers, A.J. Exploring integrated crop–livestock systems in different ecoregions of the United States. *Eur. J. Agron.* **2014**, *57*, 21–30.
7. Winsten, J.R.; Kerchner, C.D.; Richardson, A.; Lichau, A.; Hyman, J.M. Trends in the Northeast dairy industry: Large-scale modern confinement feeding and management-intensive grazing. *J. Dairy Sci.* **2010**, *93*, 1759–1769.
8. National Research Council. *Achieving Nutrient and Sediment Reduction Goals in the Chesapeake Bay: An Evaluation of Program Strategies and Implementation*; The National Academies Press: Washington, DC, USA, 2011.
9. Wade, T.; Claassen, R.; Wallander, S. *Conservation-Practice Adoption Rates Vary Widely by Crop and Region*; EIB-147, United States Department of Agricultre, Economic Research Service: Washington, DC, USA, 2015.
10. Camargo, G.G.T.; Ryan, M.R.; Richard, T.L. Energy use and greenhouse gas emissions from crop production using the farm energy analysis tool. *Bioscience* **2013**, *63*, 263–273.
11. Reicosky, D.C.; Kemper, W.D.; Langdale, G.W.; Douglas, C.L.; Rasmussen, P.E. Soil organic matter changes resulting from tillage and biomass production. *J. Soil Water Conserv.* **1995**, *50*, 253–261.
12. Teasdale, J.R.; Coffman, C.B.; Mangum, R.W. Potential long-term benefits of no-tillage and organic cropping systems for grain production and soil improvement. *Agron. J.* **2007**, *99*, 1297–1305.
13. Mirsky, S.B.; Ryan, M.R.; Curran, W.S.; Teasdale, J.R.; Maul, J.; Spargo, J.T.; Moyer, J.; Grantham, A.M.; Weber, D.; Way, T.R.; et al. Conservation tillage issues: Cover crop-based organic rotational no-till grain production in the mid-Atlantic region, USA. *Renew. Agric. Food Syst.* **2012**, *27*, 31–40.
14. Mirsky, S.B.; Ryan, M.R.; Teasdale, J.R.; Curran, W.S.; Reberg-Horton, C.S.; Spargo, J.T.; Wells, M.S.; Keene, C.L.; Moyer, J.W. Overcoming weed management challenges in cover

crop-based organic rotational no-till soybean production in the eastern United States. *Weed Technol.* **2013**, *27*, 193–203.
15. Reberg-Horton, S.C.; Grossman, J.M.; Kornecki, T.S.; Meijer, A.D.; Price, A.J.; Place, G.T.; Webster, T.M. Utilizing cover crop mulches to reduce tillage in organic systems in the southeastern USA. *Renew. Agric. Food Syst.* **2012**, *27*, 41–48.
16. Delate, K.; Cwach, D.; Chase, C. Organic no-tillage system effects on soybean, corn and irrigated tomato production and economic performance in Iowa, USA. *Renew. Agric. Food Syst.* **2012**, *27*, 49–59.
17. Wallace, J.M. Personal communication, 2017.
18. Curran, W.S. Personal communication, 2017.
19. Nord, E.A.; Ryan, M.R.; Curran, W.S.; Mortensen, D.A.; Mirsky, S.B. Effects of management type and timing on weed suppression in soybean no-till planted into rolled-crimped cereal rye. *Weed Sci.* **2012**, *60*, 624–633.
20. Ryan, M.R.; Mirsky, S.B.; Mortensen, D.A.; Teasdale, J.R.; Curran, W.S. Potential synergistic effects of cereal rye biomass and soybean planting density on weed suppression. *Weed Sci.* **2011**, *59*, 238–246.
21. Mirsky, S.B.; Curran, W.S.; Mortensen, D.M.; Ryan, M.R.; Shumway, D.L. Timing of cover-crop management effects on weed suppression in no-till planted soybean using a roller-crimper. *Weed Sci.* **2011**, *59*, 380–389.
22. Ryan, M.R.; Curran, W.S.; Grantham, A.M.; Hunsberger, L.K.; Mirsky, S.B.; Mortensen, D.A.; Nord, E.A.; Wilson, D.O. Effects of seeding rate and poultry litter on weed suppression from a rolled cereal rye cover crop. *Weed Sci.***2011**, *59*, 438–444.
23. Mirsky, S.B.; Spargo, J.T.; Curran, W.S.; Reberg-Horton, S.C.; Ryan, M.R.; Schomberg, H.H.; Ackroyd, V.J. Characterizing cereal rye biomass and allometric relationships across a range of fall available N rates in the eastern US. *Agron. J.* 2017.
24. Mirsky, S.B.; Ackroyd, V.J.; Cordeau, S.; Curran, W.S.; Hashemi, M.; Reberg-Horton, S.C.; Ryan, M.R. Hairy vetch biomass across the eastern US: Effects of latitude, seeding rate and date, and termination timing. *Agron. J.* **2017**.
25. Keene, C.L.; Curran, W.S.; Wallace, J.M.; Ryan, M.R.; Mirsky, S.B.; VanGessel, M.J.; Barbercheck, M.E. Cover crop termination timing is critical in organic rotational no-till systems. *Agron. J.* **2017**, *109*, 272–282.
26. Liebert, J.A.; DiTommaso, A.; Ryan, M.R. Rolled mixtures of barley and cereal rye for weed suppression in cover crop-based organic no-till planted soybean. *Weed Sci.* **2017**.
27. Liebl, R.; Simmons, F.W.; Wax, L.M.; Stoller, E.W. Effect of rye (*Secale cereale*) mulch on weed control and soil moisture in soybean (*Glycine max*). *Weed Technol.* **1992**, *6*, 838–846.
28. De Bruin, J.L.; Porter, P.M.; Jordan, N.R. Use of a rye cover crop following corn in rotation with soybean in the Upper Midwest. *Agron. J.* **2005**, *97*, 587–598.
29. Wells, M.S.; Brinton, C.M.; Reberg-Horton, S.C. Weed suppression and soybean yield in a no-till cover-crop mulched system as influenced by six rye cultivars. *Renew. Agric. Food Syst.* **2016**, *31*, 429–440.
30. Smith, A.N.; Reberg-Horton, S.C.; Place, G.T.; Meijer, A.D.; Arellano, C.; Mueller, J.P. Rolled rye mulch for weed suppression in organic no-tillage soybeans. *Weed Sci.* **2011**, *59*, 224–231.
31. Hovermale, C.H.; Camper, H.M.; Alexander, M.W. Effects of small grain stubble height and mulch on no-tillage soybean production. *Agron. J.* **1979**, *71*, 644–647.
32. Williams, M.M.; Mortensen, D.A.; Doran, J.W. No-tillage soybean performance in cover crops for weed management in the western Corn Belt. *J. Soil Water Conserv.* **2000**, *55*, 79–84.
33. Place, G.T.; Reberg-Horton, S.C.; Dunphy, J.E.; Smith, A.N. Seeding rate effects on weed control and yield for organic soybean poduction. *Weed Technol.* **2009**, *23*, 497–502.
34. Liebert, J.A. Developing the System: Cultural Practices to Enhance Weed Suppression and Profitability in Organic No-Till Planted Soybean. Master's Thesis, Cornell University, NY, USA, 2016.

35. Wells, M.S.; Reberg-Horton, S.C.; Mirsky, S.B. Cultural strategies for managing weeds and soil moisture in cover crop based no-till soybean production. *Weed Sci.* **2014**, *62*, 501–511.
36. Ball, R.A.; Purcell, L.C.; Vories, E.D. Optimizing Soybean Plant Population for a Short-Season Production System in the Southern USA. *Crop Sci.* **2000**, *40*, 757–764.
37. Ball, R.A.; Purcell, L.C.; Vories, E.D. Short-Season Soybean Yield Compensation in Response to Population and Water Regime. *Crop Sci.* **2000**, *40*, 1070–1078.
38. Ball, R.A.; McNew, R.W.; Vories, E.D.; Keisling, T.C.; Purcell, L.C. Path Analyses of Population Density Effects on Short-Season Soybean Yield. *Agron. J.* **2001**, *93*, 187–195.
39. USDA National Organic Grain and Feedstuffs: Bi-Weekly Reports. Available online: https://www.ams.usda.gov/market-news/search-market-news (accessed on 23 January 2017).
40. Keene, C.L.; Curran, W.S. Optimizing high-residue cultivation timing and frequency in reduced-tillage soybean and corn. *Agron. J.* **2016**, *108*, 1897–1906.
41. Nord, E.A.; Curran, W.S.; Mortensen, D.A.; Mirsky, S.B.; Jones, B.P. Integrating multiple tactics for managing weeds in high residue no-till soybean. *Agron. J.* **2011**, *103*, 1542–1551.
42. Zinati, G.; Mirsky, S.B.; Seidel, R.; Grantham, A.; Moyer, J.; Ackroyed, V.J. High-residue cultivation timing impact on organic no-till soybean weed management. *Weed Technol.* **2017**.
43. Teasdale, J.R.; Mirsky, S.B. Tillage and Planting Date Effects on Weed Dormancy, Emergence, and Early Growth in Organic Corn. *Weed Sci.* **2015**, *63*, 477–490.
44. Ryan, M.R.; Smith, R.G.; Mortensen, D.A.; Teasdale, J.R.; Curran, W.S.; Seidel, R.; Shumway, D.L. Weed-crop competition relationships differ between organic and conventional cropping systems. *Weed Res.* **2009**, *49*, 572–580.
45. Teasdale, J.R.; Devine, T.E.; Mosjidis, J.A.; Bellinder, R.R.; Beste, C.E. Growth and Development of Hairy Vetch Cultivars in the Northeastern United States as Influenced by Planting and Harvesting Date. *Agron. J.* **2004**, *96*, 1266–1271.
46. Poffenbarger, H.J.; Mirsky, S.B.; Weil, R.R.; Maul, J.E.; Kramer, M.; Spargo, J.T.; Cavigelli, M.A. Biomass and Nitrogen Content of Hairy Vetch–Cereal Rye Cover Crop Mixtures as Influenced by Species Proportions. *Agron. J.* **2015**, *107*, 2069–2082.
47. Finney, D.M.; White, C.M.; Kaye, J.P. Biomass Production and Carbon/Nitrogen Ratio Influence Ecosystem Services from Cover Crop Mixtures. *Agron. J.* **2016**, *108*, 39–52.
48. Cook, J.C.; Gallagher, R.S.; Kaye, J.P.; Lynch, J.; Bradley, B. Optimizing vetch nitrogen production and corn nitrogen accumulation under no-till management. *Agron. J.* **2010**, *102*, 1491–1499.
49. Teasdale, J.R.; Mirsky, S.B.; Spargo, J.T.; Cavigelli, M.A.; Maul, J.E. Reduced-tillage organic corn production in a hairy vetch cover crop. *Agron. J.* **2012**, *104*, 621–628.
50. Wells, M.S.; Reberg-Horton, S.C.; Mirsky, S.B. Planting date impacts on soil water management, plant growth, and weeds in cover-crop-based no-till corn production. *Agron. J.* **2016**, *108*, 162–170.
51. Mischler, R.A.; Duiker, S.W.; Curran, W.S.; Wilson, D. Hairy vetch management for no-till organic corn production. *Agron. J.* **2010**, *102*, 355–362.
52. Poffenbarger, H.J.; Mirsky, S.B.; Weil, R.R.; Kramer, M.; Spargo, J.T.; Cavigelli, M.A. Legume proportion, poultry litter, and tillage effects on cover crop decomposition. *Agron. J.* **2015**, *107*, 2083–2096.
53. Mirsky, S.B. Personal communication, 2017.
54. Dyck, E.; Liebman, M. Soil fertility management as a factor in weed control: The effect of crimson clover residue, synthetic nitrogen fertilizer, and their interaction on emergence and early growth of lambsquarters and sweet corn. *Plant Soil* **1994**, *167*, 227–237.
55. Mirsky, S.B.; Wallace, J.M.; Curran, W.S.; Crockett, B.C. Hairy Vetch Seedbank Persistence and Implications for Cover Crop Management. *Agron. J.* **2015**, *107*, 2391–2400.
56. Rivers, A.R. Effects of Management on Arthropod Communities in Organic and Conservation Agricultural Systems in Pennsylvania and Mexico. Ph.D. Thesis, The Pennsylvania State University, University Park, PA, USA, 2016.

57. Rivers, A.; Mullen, C.; Wallace, J.; Barbercheck, M. Cover crop-based reduced tillage system influences Carabidae (Coleoptera) activity, diversity and trophic group during transition to organic production. *Renew. Agric. Food Syst.* **2017**.
58. Poffenbarger, H.J.; Mirsky, S.B.; Kramer, M.; Weil, R.R.; Meisinger, J.J.; Cavigelli, M.A.; Spargo, J.T. Cover crop and poultry litter management influence spatiotemporal availability of topsoil nitrogen. *Soil Sci. Soc. Am. J.* **2015**, *79*, 1660–1673.
59. Clark, A.J.; Decker, A.M.; Meisinger, J.J.; McIntosh, M.S. Kill Date of Vetch, Rye, and a Vetch-Rye Mixture: I. Cover Crop and Corn Nitrogen. *Agron. J.* **1997**, *89*, 427–434.
60. Spargo, J.T.; Cavigelli, M.A.; Mirsky, S.B.; Meisinger, J.J.; Ackroyd, V.J. Organic Supplemental Nitrogen Sources for Field Corn Production after a Hairy Vetch Cover Crop. *Agron. J.* **2016**, *108*, 1992–2002.
61. Ranells, N.N.; Wagger, M.G. Nitrogen release from grass and legume cover crop monocultures and bicultures. *Agron. J.* **1996**, *88*, 777–782.
62. Karlen, D.L.; Sadler, E.J.; Camp, C.R. Dry matter, nitrogen, phosphorus, and potassium accumulation rates by corn on Norfolk loamy sand. *Agron. J.* **1987**, *79*, 649–656.
63. Robertson, G.P.; Vitousek, P.M. Nitrogen in agriculture: Balancing the cost of an essential resource. *Annu. Rev. Environ. Resour.* **2009**, *34*, 97–125.
64. Hayden, Z.D.; Ngouajio, M.; Brainard, D.C. Rye-vetch mixture proportion tradeoffs: Cover crop productivity, nitrogen accumulation, and weed suppression. *Agron. J.* **2014**, *106*, 904–914.
65. Pirhofer-Walzl, K.; Rasmussen, J.; Høgh-Jensen, H.; Eriksen, J.; Søegaard, K.; Rasmussen, J. Nitrogen transfer from forage legumes to nine neighbouring plants in a multi-species grassland. *Plant Soil* **2012**, *350*, 71–84.
66. Spargo, J.T.; Cavigelli, M.A.; Mirsky, S.B.; Maul, J.E.; Meisinger, J.J. Mineralizable soil nitrogen and labile soil organic matter in diverse long-term cropping systems. *Nutr. Cycl. Agroecosyst.* **2011**, *90*, 253–266.
67. Mirsky, S.B. Unpublished data. 2017.
68. Baker, B.P.; Mohler, C.L. Weed management by upstate New York organic farmers: Strategies, techniques and research priorities. *Renew. Agric. Food Syst.* **2015**, *30*, 418–427.
69. Brummer, E.C.; Barber, W.T.; Collier, S.M.; Cox, T.S.; Johnson, R.; Murray, S.C.; Olsen, R.T.; Pratt, R.C.; Thro, A.M. Plant breeding for harmony between agriculture and the environment. *Front. Ecol. Environ.* **2011**, *9*, 561–568.
70. Carlson, S.; Stockwell, R. Research priorities for advancing adoption of cover crops in agriculture-intensive regions. *J. Agric. Food Syst. Community Dev.* **2013**, *3*, 125–129.
71. Wayman, S.; Kissing Kucek, L.; Mirsky, S.B.; Ackroyd, V.; Cordeau, S.; Ryan, M.R. Organic and conventional farmers differ in their perspectives on cover crop use and breeding. *Renew. Agric. Food Syst.* **2016**, in press.
72. Wilke, B.J.; Snapp, S.S. Winter cover crops for local ecosystems: Linking plant traits and ecosystem function. *J. Sci. Food Agric.* **2008**, *88*, 551–557.
73. Parr, M.; Grossman, J.M.; Reberg-Horton, S.C.; Brinton, C.; Crozier, C. Nitrogen delivery from legume cover crops in no-till organic corn production. *Agron. J.* **2011**, *103*, 1578–1590.
74. Williams, A.; Kane, D.A.; Ewing, P.M.; Atwood, L.W.; Jilling, A.; Li, M.; Lou, Y.; Davis, A.S.; Grandy, A.S.; Huerd, S.C.; et al. Soil functional zone management: A vehicle for enhancing production and soil ecosystem services in row-crop agroecosystems. *Front. Plant Sci.* **2016**, *7*, 65.
75. Gattinger, A.; Muller, A.; Haeni, M.; Skinner, C.; Fliessbach, A.; Buchmann, N.; Mäder, P.; Stolze, M.; Smith, P.; Scialabba, N.E.-H.; et al. Enhanced top soil carbon stocks under organic farming. *Proc. Natl. Acad. Sci. USA* **2012**, *109*, 18226–18231.
76. Mäder, P.; Fließbach, A.; Dubois, D.; Gunst, L.; Fried, P.; Niggli, U. Soil fertility and biodiversity in organic farming. *Science* **2002**, *296*, 1694–1697.
77. Williams, A.; Hedlund, K. Indicators of soil ecosystem services in conventional and organic arable fields along a gradient of landscape heterogeneity in southern Sweden. *Appl. Soil Ecol.* **2013**, *65*, 1–7.

78. Balota, E.L.; Calegari, A.; Nakatani, A.S.; Coyne, M.S. Benefits of winter cover crops and no-tillage for microbial parameters in a Brazilian Oxisol: A long-term study. *Agric. Ecosyst. Environ.* **2014**, *197*, 31–40.
79. Varvel, G.E.; Wilhelm, W.W. No-tillage increases soil profile carbon and nitrogen under long-term rainfed cropping systems. *Soil Tillage Res.* **2011**, *114*, 28–36.
80. Mortensen, D.A.; Egan, J.F.; Maxwell, B.D.; Ryan, M.R.; Smith, R.G. Navigating a critical juncture for sustainable weed management. *Bioscience* **2012**, *62*, 75–84.

CHAPTER 13

Improvements in crop water productivity increase water sustainability and food security—a global analysis

Kate A Brauman[1], Stefan Siebert[2] and Jonathan A Foley[1]

[1] *Institute on the Environment (IonE), University of Minnesota, 325 Learning & Environmental*

Sciences, 1954 Buford Avenue, St. Paul, MN 55108, USA

[2] *Institute of Crop Science and Resource Conservation (INRES), University of Bonn, Katzenburgweg 5, D-53115 Bonn, Germany*

ABSTRACT

Irrigation consumes more water than any other human activity, and thus the challenges of water sustainability and food security are closely linked. To evaluate how water resources are used for food production, we examined global patterns of water productivity—food produced (kcal) per unit of water (l) consumed. We document considerable variability in crop water productivity globally, not only across different climatic zones but also within climatic zones. The least water productive systems are disproportionate freshwater consumers. On precipitation-limited croplands, we found that ~40% of water consumption goes to production of just 20% of food calories. Because in many cases crop water productivity is well below optimal levels, in many cases farmers have substantial opportunities to improve water productivity. To demonstrate the potential impact of management interventions, we calculated that raising crop water productivity in precipitation-limited regions to the 20th percentile of productivity would increase annual production on rainfed cropland by enough to provide food for an estimated 110 million people, and water consumption on irrigated cropland would be reduced enough to meet the annual domestic water demands of nearly 1.4 billion people.

1. INTRODUCTION

The world is facing twin challenges of water stress and food insecurity—challenges that are already pressing and are projected to grow [1–3]. Because crop production is the largest global consumer of freshwater, and because water is a key resource in food production, neither of these challenges can be addressed in isolation [4, 5]. Producing more food for each drop of water will be a crucial strategy to address both challenges [6].

Issues of global water availability and scarcity have been considered in a variety of ways, including identifying areas of water availability [7], water stress, [8–10], impacts of water use [11], and projections of future water scarcity [2, 12]. Many recent studies have tied water scarcity to agricultural water consumption [13–15]. For the most part, however, these studies have considered water consumption in relation to generic or country-level measures of crop production. However, a number of recent papers have highlighted the wide variability in crop yields worldwide, and located systems where yield improvements are possible [3, 16, 17]. Studies that link agricultural water consumption with detailed yield data tend to be strongly place-based [14, 18–20]. These studies provide key information about specific interventions to improve yields and reduce water use. Global-scale analysis can augment these studies by facilitating learning among systems that are geographically and culturally dispersed. Recognizing that, several reviews have consolidated and analysed field studies of crop water productivity[21,22]. This study draws on both modelled and empirical data to analyse the global range of crop water productivity, providing insight into the magnitude and location of potential levers of change.

When water is scarce, understanding the magnitude of water consumption is important. In most cases, however, evaluation for decision-making requires information about efficiency—when water is being used, is it being used wisely? With new water consumption[23] and crop production data[17], we can quantitatively assess the range of crop water productivity worldwide. Locating the low end of this range is additionally useful because efficiency improvements tend to be easiest where efficiency is low[24,25]. Finally, global data allow inference making about drivers of variation in crop water productivity.

Here we assess global-scale patterns of crop water productivity for 16 major food crops using spatially distributed, crop-specific water consumption and yield data. We quantify the tradeoffs between water consumption and food production, evaluate the extent to which factors other than climate limit crop water productivity, and illustrate potential benefits of improved water productivity worldwide.

2. METHODS

We analysed crop production, water use, and crop water productivity for 16 staple food crops: wheat, maize, rice, barley, rye, millet, sorghum, soybean, sunflower, potato, cassava, sugarcane, sugar beet, oil palm, rapeseed (canola), and groundnut (peanut). These crops constitute 56% of global crop production by tonnage[26], account for 65% of crop water consumption [23], and cover 68% of all cropland[17]. Crop water productivity is defined here as food (edible) kilocalories produced per litre of evapotranspiration, evaluated separately for rainfed and irrigated crops. We evaluate kilocalories as a proxy for crop production based on FAO weight conversions[27] to facilitate comparison among the range of crop water productivity without regard for weighing conventions and to provide insight into food production. Crop choice represents one important strategy farmers use to adapt to changing climate, resource availability, and socioeconomic drivers, and direct comparison of crop water productivity among crops potentially facilitates this [28]. Our analyses are focused on potentially water-limited regions ($P < PET$) because inefficiencies are more pressing where water is scarce. In precipitation-limited regions, the majority of the 16 crops we analyse goes directly to food production (92%, compared to 79% in more humid regions)[3].

2.1. Variation in crop water productivity

Crop production and water consumption data for irrigated and rainfed crops were provided by Siebert and Döll[23], who computed daily water balances and corresponding crop water requirements using the FAO Penman–Monteith method of reference evapotranspiration [29] modified by crop coefficients and considering crop areas and crop calendars [30]. For irrigated crops, actual evapotranspiration (AET) was assumed to be equal to potential evapotranspiration (PET), with irrigation water compensating for any rainwater deficit; supply-side limitations to irrigation were thus not incorporated into the model. In addition, our analysis of irrigated crops evaluates beneficial irrigation only, as data on use for other irrigation-related activities (salt flushing, soil preparation in rice paddies) are not globally available [23]. The ratio between AET and PET was used to separate and downscale yields between irrigated and rainfed crops. Yield data were originally developed by Monfreda *et al*[17] by combining census data for agricultural production with estimates of cropland extent derived from satellite imagery. The data have a spatial resolution of five arc minutes and report 7-year average yield and 5-year average water consumption centred on the year 2000.

To evaluate climate and non-climate effects on crop water productivity, we divided cropland into 20 climatic zones (SI figure 1 available at stacks.iop.org/ERL/8/024030/mmedia) defined by the ratio of mean annual precipitation (P) to mean annual potential evapotranspiration (PET) [31]. We attributed the difference in maximum water productivity across climatic zones to the effects of climate. In contrast, variation in water productivity within each climatic zone was attributed to non-climate effects. For statistical power, we considered only crops with an extent of >50 000 ha (irrigated) or >100 000 ha (rainfed) in a given climatic zone. Further detail on methods is provided in the supplementary information (SI, available at stacks.iop.org/ERL/8/024030/mmedia).

We calculated the Generalized inequality index (Gini) [32] to quantitatively compare inequalities in water productivity among crops and across regions. Cumulative water consumption and cumulative kilocalorie production for each crop at 5 arc minute resolution were ordered by water productivity; the 'curves' in figure 3(b) are compilations of these points.

2.2. Water productivity gap

For each crop, we calculated the production-weighted 20th percentile value for each climatic zone where P < PET. In each climatic zone, we identified all cropland where water productivity for each crop was less than its 20th percentile value. We then calculated the calories that could be added or water saved were the water productivity of any crop below its 20th percentile value increased to that value (SI table 1 available at stacks.iop.org/ERL/8/024030/mmedia). Because improvements for each crop are measured against its own 20th percentile value in the same climatic zone, improvements are scaled to the range of variability for each crop.

3. RESULTS

3.1. Variation in crop water consumption and productivity

Water consumption by crops varies substantially across the globe (SI figure 2 available at stacks.iop.org/ERL/8/024030/mmedia), reflecting differences in cropping density, crop choice, soil characteristics, irrigation availability, and agricultural management as well as climatic drivers of evapotranspiration. The distribution of crop water consumption among climatic zones (figure 1) illustrates the crops likely to contribute substantially where water resources are scarce. In arid climates (P/PET < 0.2), rainwater consumption is dominated by millet and sorghum while irrigation water is mainly used for wheat and rice.

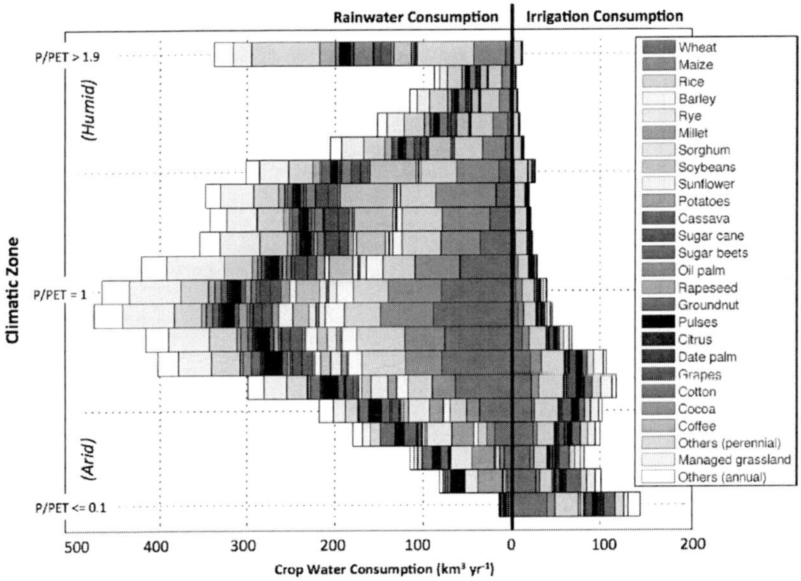

Figure 1. Annual water consumption across climatic zones. Rainwater (left of centre line) and irrigation water (right of centre line) consumption by all crops, including those (in greys) that are not included in this analysis. For the 16 food crops analysed, the majority (83%) of water originates as rainfall on cropland. Climatic zones, most arid at the bottom and most humid at the top, are based on P/PET and are evenly distributed in climate space. Cropland in climatic zones in which P/PET > 1.9 are compressed into a single zone.

In semi-arid and dry-sub-humid climates ($0.2 < P/PET < 0.7$), rainwater consumption is dominated by wheat, rice, and maize while irrigation water use is dominated by wheat, rice, sugarcane, and maize. For the 16 crops studied, 52% of rainwater consumption (by irrigated and rainfed crops) and 82% of irrigation water consumption occur in regions where precipitation is potentially limiting ($P < PET$).

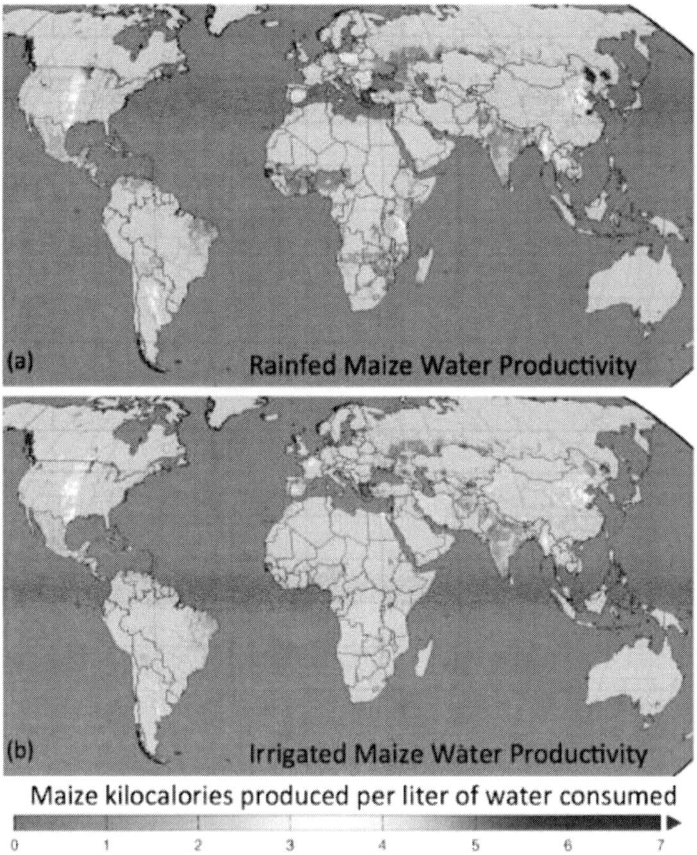

Figure 2. Maize water productivity in precipitation-limited regions. Data for both (a) rainfed and (b) irrigated systems are centred at 3.5 kcal l^{-1}/1.0 kg m^{-3}, close to the median production-weighted water productivity for rainfed (3.2 kcal l^{-1}/0.9 kg m^{-3}) and irrigated (3.5 kcal l^{-1}/1.0 kg m^{-3}) maize where $P < PET$, so that areas of relatively low water productivity are shown in orange and areas of relatively high water productivity are shown in blue.

Crop water productivity is not evenly distributed (figure 3). For example, for irrigated wheat in precipitation-limited areas, the USA has a national-level water productivity (2.9 kcal l^{-1}/0.9 kg m^{-3}) that is close to the world average (2.8 kcal l^{-1}/0.9 kg m^{-3}), but water productivity is unevenly distributed (Gini of 0.32) compared to, for example, China, where irrigated wheat water productivity is uniformly high (4.3 kcal l^{-1}/1.3 kg m^{-3}) and evenly distributed (Gini of 0.09).

We found a large range in crop water productivity. For example, in large parts of Sub-Saharan Africa maize water productivity was below 1 kcal l^{-1} (0.3 kg m^{-3}) while the highest water productivities of maize (>6 kcal l^{-1}/1.7 kg m^{-3}) were computed for the US corn belt and Northern China (figure 2). Field-based studies (such as those surveyed in [21, 22, 33]) document a similar range of water productivities; our analysis extends their insights to a broader set of climates, locations, and crops.

Variation in water productivity causes water consumption in certain areas to be disproportionately large. On the least water productive rainfed cropland in precipitation-limited regions, ~46% of water was consumed in the production of just 20% of maize calories. Irrigated crops consume both rain and irrigation water—for those low productivity irrigated croplands in precipitation-limited regions where irrigation water provides at least 50% of total water, ~44% of irrigation water consumption by maize went to 20% of food production.

Variation in crop water productivity is not solely a function of climate. While each crop we analysed has its highest water productivity in specific climatic zones, water productivity varied within each climatic zone (SI figures 3 and 4 available at stacks.iop.org/ERL/8/024030/mmedia). For most crops, we see as much variability in water productivity resulting from non-climate drivers as from climate (figure 4).

For irrigated crops, we evaluated water productivity based on combined evapotranspiration of rain and irrigation water. Water productivity of only irrigation (blue) water is a function of both fraction of overall water originating as irrigation water and total crop water productivity. The fraction of irrigation water is strongly correlated with aridity (Pearson's r^2 is 0.96 for wheat, 0.85 for maize, and 0.99 for rice), while water productivity of irrigated crops is not (SI figure 4 available at stacks.iop.org/ERL/8/024030/mmedia). We therefore focus on overall water productivity of irrigated crops. Though the full impact of increasing water productivity of irrigated crops would not be gained in irrigation savings, it would result in a lower irrigation water requirement, particularly in more arid climates.

3.2. A water productivity gap

If water productivity on only the least productive, precipitation-limited croplands were increased, the impact would be substantial. For illustrative purposes, we assume all rainfed improvements go towards production and all irrigated improvements go to water savings. On rainfed cropland in precipitation-limited regions, increasing water productivity to the 20th percentile could increase production by ~120 trillion kilocalories per year while holding water consumption constant (figure 5(a)).

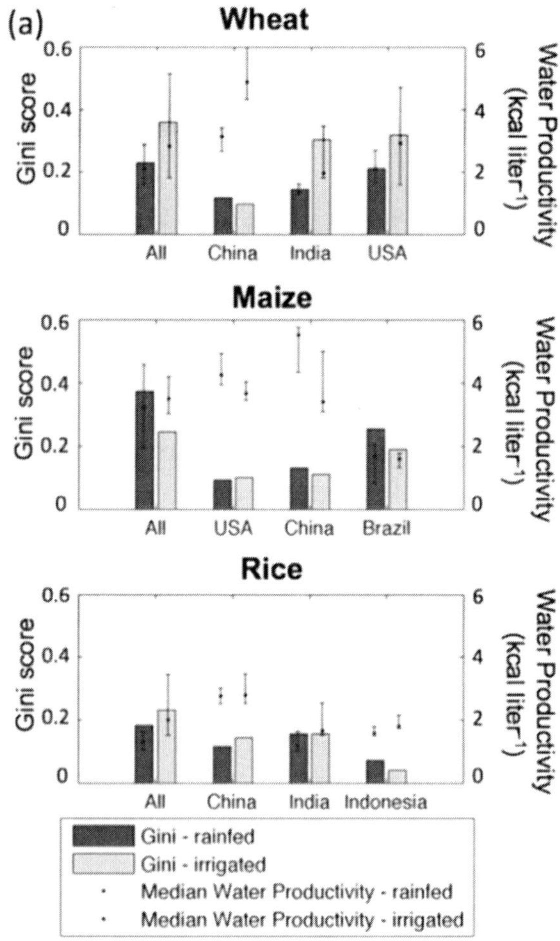

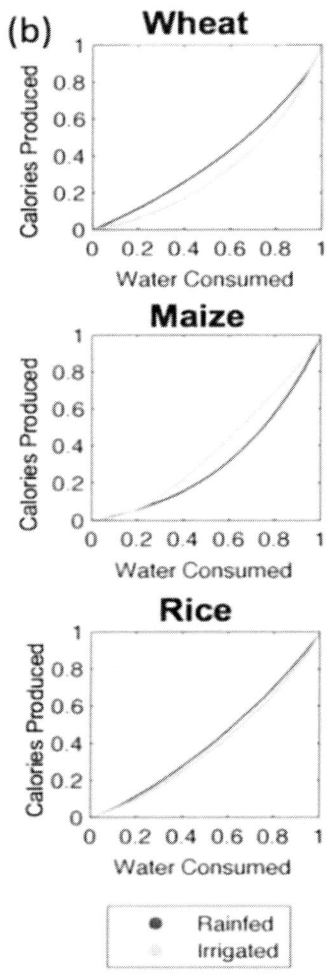

Figure 3. Disproportionate water consumption by major producers in precipitation-limited areas. (a) Bars show the Generalized inequality index (Gini) for crop water consumption for each crop in all precipitation-limited areas (P < PET) and for precipitation-limited areas in countries with the greatest production of each crop. Gini values range from 0 to 1; higher values reflect more disproportionate water consumption. The calorie-weighted median value of water productivity for each group is also shown (black dot, error bars are 25th and 75th percentile values). (b) Fraction of cumulative annual water consumption compared to cumulative annual kilocalorie production for rainfed and irrigated systems in precipitation-limited areas. Curves with more convexity reflect more disproportionate water consumption.

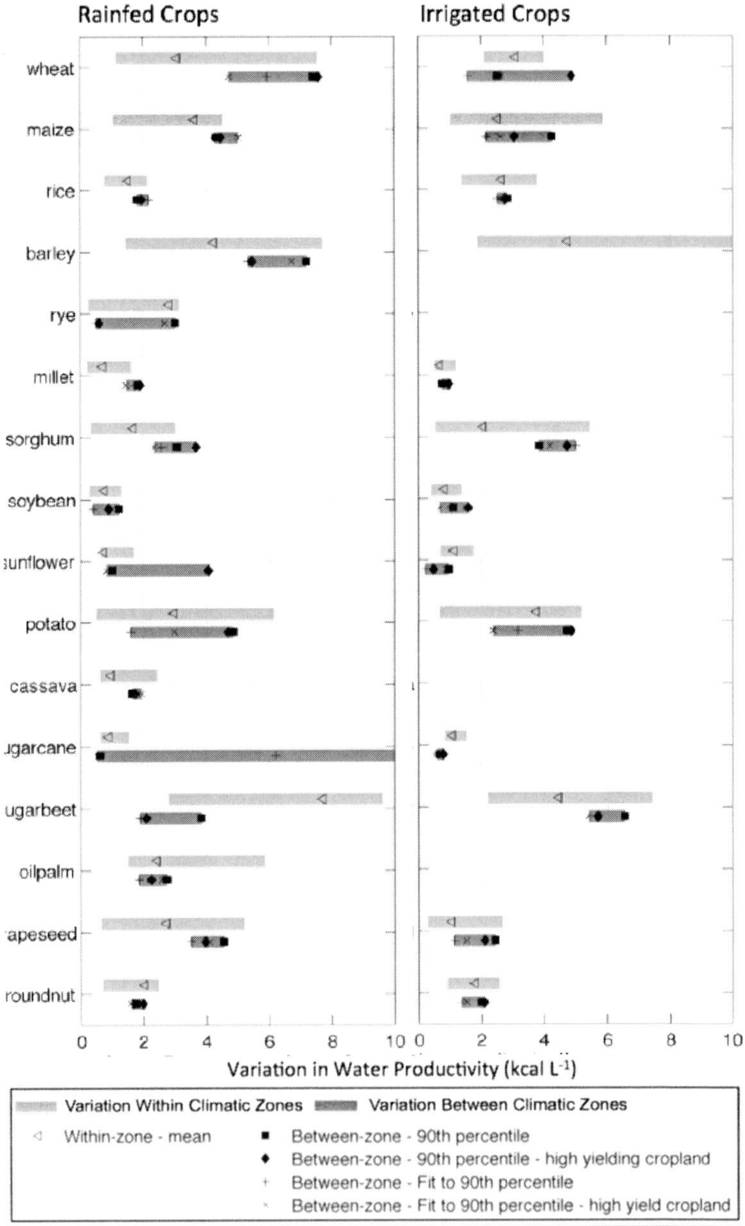

Figure 4. Variation of water productivity within and between climatic zones. Variation in water productivity within climatic zones (light grey bars) is similar to variation between climatic zones (dark grey bars) for both rainfed (left) and irrigated (right) crops. For each crop, we report a range of variability.

This amounts to a ~30% increase over current production in these areas and would provide food for ~110 million people annually (assuming an average food requirement of 3000 kcal d^{-1} cap^{-1} [34]). On irrigated croplands, increasing the lowest water productivity in precipitation-limited regions could save ~77 km^3 yr^{-1} of water without reducing food production (figure 5(b)). This water savings amounts to more than one quarter of current water consumption on low water productivity irrigated cropland and would be enough to meet the annual domestic water demand of ~1.4 billion people (assuming a demand of 150 l d^{-1} cap^{-1} [35]).

To identify countries where water savings on irrigated cropland would be most beneficial, we analysed irrigation water consumption on cropland with the lowest 20% of water productivity in precipitation-limited climates (figure 6). India accounts for more than 40% of total irrigation water consumption in these areas, 67 km^3 yr^{-1}, primarily in wheat and rice cultivation. Because India has so much cropland in these areas, however, irrigation intensity (consumption per hectare) is ~520 mm yr^{-1}, close to the median value of irrigation intensity on all precipitation-limited, low water productivity cropland (~500 mm yr^{-1}). By contrast, in similar climates in Uzbekistan, food crops consume only ~3 km^3 yr^{-1}. However, irrigation intensity (~1400 mm yr^{-1}) is nearly three times greater. Because of low total water consumption, improving irrigation efficiency of food crops in Uzbekistan is unlikely to alleviate basin-wide water shortages, though on-farm impacts may be dramatic. In India, by contrast, small but widespread improvements in irrigation intensity could have a substantial impact. Assessments should not assume that high irrigation consumption is linked to high irrigation intensity.

4. DISCUSSION

4.1. Implications of findings

We sought to quantify the range of tradeoffs between crop water use and production to explore, first, the feasibility of interventions to increase food production and reduce water consumption by improving crop water productivity, and, second, to quantify the potential magnitude of this type of intervention. The extensive variability in crop water productivity within climatic zones indicates that there are many non-climate related drivers of low crop water productivity, and these can potentially be managed. Even when limitations are due to soil or slopes, management techniques such as nutrient addition and terracing or furrowing can improve yields [36]. Water productivity is a ratio, and in many cases low water productivity is likely a function of low yield, so interventions to increase water productivity may focus more on improving yield than on decreasing water consumption.

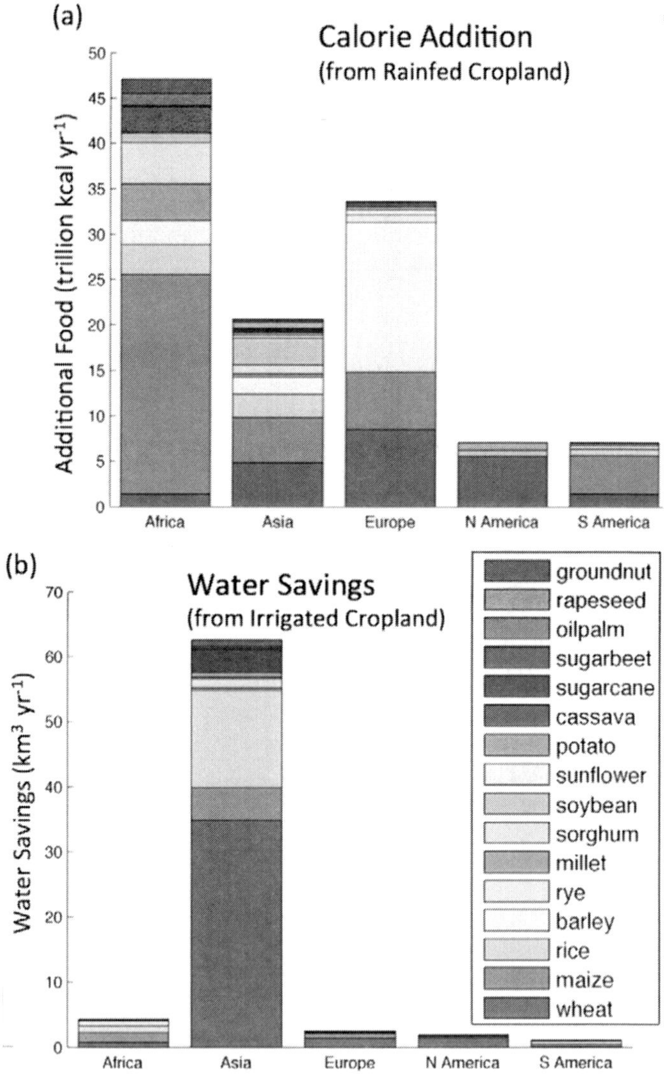

Figure 5. Impact of increasing water productivity on the least water productive cropland in precipitation-limited areas. Regional and crop-specific contribution to (a) increased food production and (b) water savings from increasing water productivity to the calorie-weighted 20th percentile.

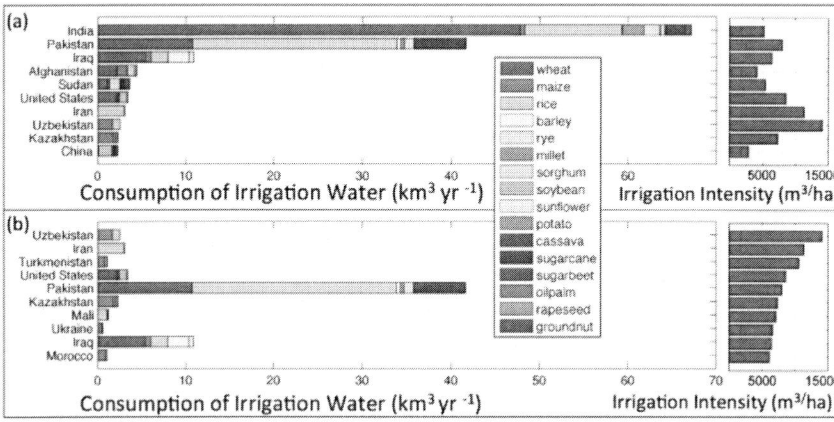

Figure 6. Annual water consumption by country for 16 food crops in precipitation-limited areas with low water productivity. Countries are ranked by (a) irrigation water consumption as well as by (b) irrigation intensity (consumption per hectare). We limited analysis of high-irrigation intensity countries to those where irrigation consumption on precipitation-limited, low water productivity cropland was at least 0.5 $km^3\ yr^{-1}$.

Specific management strategies to increase crop water productivity must be tailored to local contexts; studies in Nigeria, Sudan, and Niger, for example, have shown that nitrogen and phosphorus limitation restrict millet and sorghum yields [37]. Water productivity could be increased by increasing yields via improved soil nutrient conditions and reduced wind-driven erosion. Examples of other interventions that have demonstrated improvements in water productivity include: rainwater harvesting and local water storage, applying drip or deficit irrigation, adjusting planting dates, and modifying tillage practices to reduce evaporation [21, 25, 33, 38–40]. In many cases, techniques for increasing water productivity, especially when it is low, require little or no additional water [41].

Any improvement in crop water productivity will have a positive effect on either food production or water savings. We demonstrate the magnitude of increasing low water productivity to the 20th percentile because this is a plausible level of increase. Allocative efficiency demonstrates that it is easier to increase resource use efficiency when resources are used inefficiently; improving the very lowest performing systems therefore should take less effort than improvements in high-performing systems [24, 25]. Implications of crop water productivity improvements to other levels are discussed in the SI (available at stacks.iop.org/ERL/8/ 024030/ mmedia).

4.2. Limitations

By limiting our analysis to 16 crops, our study does not provide a complete picture of all global crop water use or food production. Further, due to data limitations, we are unable to assess water requirements to prevent salinization or for non-beneficial irrigation. Our analysis thus underestimates the volume of water required for irrigated agriculture, particularly in arid regions where irrigation water is often the primary source of water for irrigated crops. In addition, we assume all irrigation demand is met and that irrigation is not limited by water supply. These issues loom large in local and regional management, and we hope to address them at the global scale in the future. We also do not assess production losses due to spoilage, animal feed, or post-consumer waste. The end points of food production provide important context in decision-making about water allocation but are outside the scope of the current research.

We address limitations of methodology in the SI (available at stacks.iop.org/ERL/8/024030/mmedia) in detail.

5. CONCLUSIONS

By reducing the amount of water required for crop production, increasing crop water productivity is a key strategy for achieving food security and water sustainability in a world with growing demands for both. If, for example, low water productivity in precipitation-limited regions were raised to the 20th percentile, the total rainfed food production in Africa could be increased by more than 10% without exploiting additional cropland. Similar improvements in water productivity on irrigated cropland could reduce total water consumption by ~8–15% in precipitation-limited regions of Africa, Asia, Europe, and South America.

The estimated effects of changing management on water productivity are close to the mean water productivity observed for each crop, demonstrating that in many low-productivity systems on-farm interventions can be a leverage point for increasing water productivity. Global analyses like the one presented here can help generalize findings from field-level or regional agronomic studies. However, appropriate management activities to increase crop water productivity will vary regionally and even locally.

ACKNOWLEDGMENTS

Research support was provided by a grant from the Institute on the Environment and by the Gordon and Betty Moore Foundation. This work also benefited from contributions by General Mills, Mosaic, Cargill, Google, PepsiCo, and Kellogg to support stakeholder outreach and public engagement. We also wish to thank two anonymous reviewers for thoughtful comments on an earlier draft.

REFERENCES

1. Rockstrom J et al 2009 A safe operating space for humanity Nature 461 472–5
2. Alcamo J, Florke M and Marker M 2007 Future long-term changes in global water resources driven by socio-economic and climatic changes Hydrol. Sci. J. 52 247–75
3. Foley J A et al 2011 Solutions for a cultivated planet Nature 478 337–42
4. Postel S L 2003 Securing water for people, crops, and ecosystems: new mindset and new priorities Nat. Resour. Forum 27 89–98
5. Rosegrant M W, Ringler C and Zhu T 2009 Water for agriculture: maintaining foods under growing scarcity Annu. Rev. Environ. Resources 34 205–22
6. Molden D 2007 Water for Food, Water for Life: A Comprehensive Assessment of Water Management in Agriculture (London: Earthscan/James & James)
7. Oki T and Kanae S 2006 Global hydrological cycles and world water resources Science 313 1068–72
8. Vorosmarty C J et al 2010 Global threats to human water security and river biodiversity Nature 467 555–61
9. Falkenmark M and Widstrand C 1992 Population and Water Resources: A Delicate Balance(Population Bulletin vol 46) (Washington, DC: Population Reference Bureau)
10. Wada Y, van Beek L P H, Viviroli D, Durr H H, Weingartner R and Bierkens M F P 2011 Global monthly water stress: 2. Water demand and severity of water stress Water Resources Res. 47 W07518
11. Pfister S, Koehler A and Hellweg S 2009 Assessing the environmental impacts of freshwater consumption in LCA Environ. Sci. Technol. 43 4098–104
12. Murray S J, Foster P N and Prentice I C 2012 Future global water resources with respect to climate change and water withdrawals as estimated by a dynamic global vegetation model J. Hydrol. 448/449 14–29
13. Falkenmark M and Lannerstad M 2005 Consumptive water use to feed humanity-curing a blind spot Hydrol. Earth Syst. Sci. 9 15–28
14. Rockstrom J, Lannerstad M and Falkenmark M 2007 Assessing the water challenge of a new green revolution in developing countries Proc. Natl Acad. Sci. 104 6253–60
15. Hoff H 2010 The global water challenge: modeling green and blue water J. Hydrol. 384 175–6
16. Licker R et al 2010 Mind the gap: how do climate and agricultural management explain the 'yield gap' of croplands around the world? Glob. Ecol. Biogeogr. 19 769–82
17. Monfreda C, Ramankutty N and Foley J A 2008 Farming the planet: 2. Geographic distribution of crop areas, yields, physiological types, and net primary production in the year 2000 Glob. Biogeochem. Cycles 22 GB1022
18. Speelman S, D'Haese M, Buysse J and D'Haese L 2008 A measure for the efficiency of water use and its determinants, a case study of small-scale irrigation schemes in North-West Province, South Africa Agric. Syst. 98 31–9

19. Wang X-Y 2010 Irrigation water use efficiency of farmers and its determinants: evidence from a survey in Northwestern China Agric. Sci. China 9 1326–37
20. Frija A, Chebil A, Speelman S, Buysse J and Van Huylenbroeck G 2009 Water use and technical efficiencies in horticultural greenhouses in Tunisia Agric. Water Manag. 96 1509–16
21. Zwart S J and Bastiaanssen W G M 2004 Review of measured crop water productivity values for irrigated wheat, rice, cotton and maize Agric. Water Manag. 69 115–33
22. Zwart S J, Bastiaanssen W G M, de Fraiture C and Molden D J 2010 A global benchmark map of water productivity for rainfed and irrigated wheat Agric. Water Manag. 97 1617–27
23. Siebert S and Doll P 2010 Quantifying blue and green virtual water contents in global crop production as well as potential production losses without irrigation J. Hydrol. 384 198–217
24. Rockstrom J and Barron J 2007 Water productivity in rainfed systems: overview of challenges and analysis of opportunities in water scarcity prone savannahs Irrig. Sci. 25 299–311
25. Ali M H and Talukder M S U 2008 Increasing water productivity in crop production—a synthesis Agric. Water Manag. 95 1201–13
26. FAOSTAT 2011 [15 August 2011] available from: http://faostat.fao.org/
27. Tilman D, Balzer C, Hill J and Befort B L 2011 Global food demand and the sustainable intensification of agriculture Proc. Natl Acad. Sci. 108 20260–4
28. Seo S N and Mendelsohn R 2008 An analysis of crop choice: adapting to climate change in South American farms Ecol. Econ. 67 109–16
29. Allen R G, Pereira L S, Raes D and Smith M 1998 Crop evapotranspiration—guidelines for computing crop water requirements Report No. 56 (Rome: Food and Agriculture Organization of the United Nations)
30. Portmann F T, Siebert S and Doll P 2010 MIRCA200—global monthly irrigated and rainfed crop areas around the year 2000: a new high-resolution data set for agricultural and hydrological modeling Glob. Biogeochem. Cycles 24 GB1011
31. Middleton N J and Thomas D S G 1997 World Atlas of Desertification (London: Arnold)
32. Weymark J 1981 Generalized gini inequality indices Math. Social Sci. 1 409–30
33. Rockstrom J et al 2009 Conservation farming strategies in East and Southern Africa: yields and rain water productivity from on-farm action research Soil Tillage Res. 103 23–32
34. Bruinsma J 2003 World Agriculture: Towards 2015/2030: An FAO Perspective (London: Earthscan/James & James)
35. Voss F, Florke M and Alcamo J 2009 Preliminary spatially explicit estimates of past and present domestic water use WATCH Technical Report No. 17
36. van Ittersum M K, Leffelaar P A, van Keulen H, Kropff M J, Bastiaans L and Goudriaan J 2003 On approaches and applications of the Wageningen crop models Eur. J. Agron. 18 201–34
37. Gandah M, Stein A, Brouwer J and Bouma J 2000 Dynamics of spatial variability of millet growth and yields at three sites in Niger, West Africa and implications for precision agriculture research Agric. Syst. 63 123–40
38. Hatfield J L, Sauer T J and Prueger J H 2001 Managing soils to achieve greater water use efficiency: a review Agron. J. 93 271–80
39. Hoff H, Falkenmark M, Gerten D, Gordon L, Karlberg L and Rockstrom J 2010 Greening the global water system J. Hydrol. 384 177–86
40. Passioura J 2006 Increasing crop productivity when water is scarce—from breeding to field management Agric. Water Manag. 80 176–96
41. Rockstrom J et al 2010 Managing water in rainfed agriculture: the need for a paradigm shift Agric. Water Manag. 97 543–50

CHAPTER 14

Farmer's Knowledge and Perceptions on Rice Insect Pests and Their Management in Uganda

Simon Alibu *, Michael H. Otim, Stella E. A. Okello, Jimmy Lamo, Moses Ekobu and Godfrey Asea

National Crops Resources Research Institute (NaCRRI), Namulonge, P.O. Box 7084, Kampala, Uganda

AUTHOR CONTRIBUTIONS

All the authors completed the study conception and design. M.H.O., S.E.A.O., J.L., M.E. and G.A. developed the study tools. S.A., M.H.O. and S.E.A.O. analyzed and interpreted the data. S.A., M.H.O., S.E.A.O. and M.E. wrote the draft of the manuscript. S.A., M.H.O., J.L. and G.A. did the critical revision of the manuscript. J.L. and G.A. supervised the EAAPP Project.

ABSTRACT

Rice is a new crop in Uganda, but has quickly grown in importance. Between 2000 and 2010, total area under rice cultivation in the country grew by 94% from 140,000 ha. Changes in the agro ecosystem due to expansion in rice area may have altered the pest status of rice insect pests. However, far too little attention has been paid to assessing the prevalence and importance of rice insect-pests in Uganda. In this study, we interviewed 240 lowland-rice farming households from eight districts within the north, east and central regions of Uganda about their perceived insect-pest problems and control measures employed, if any. A semi-structured questionnaire was used. The farmers ranked rice insect pests as the most important biotic constraint in rice production, with stem borers and the African rice gall midge (AfRGM) perceived to be the 1st and 2nd most detrimental insect pests, respectively. In spite of this, only 36% of the respondents could positively identify symptoms of AfRGM damage on rice plants, while 64% were familiar with stem borer damage. Over 60% of interviewed farmers expressed confidence in the effectiveness of insecticides for controlling rice insect pests. Cultural control measures were not popular among the farmers.

Keywords: Uganda; rice field pests; farmer perceptions; alternative hosts; IPM

1. INTRODUCTION

Rice is a relatively new crop in Uganda, but has quickly grown in importance. The importance of rice in Uganda is highlighted in the national rice development strategy (NRDS), which seeks to make the country self-sufficient in rice by doubling production from 177,800 Mt in 2008 to 680,000 Mt by 2018 [1]. Projections in the NRDS indicate that doubling rice production will respond to the increasing demand of Ugandans for rice. Per capita annual consumption of rice in Uganda went up from 3 kg in 1990 to 8 kg in 2010, representing a 62.5% rise. Correspondingly, total area under rice cultivation in Uganda increased by over 70% from 39,000 ha in 1990 to 140,000 ha in 2010 [2]. The rapid increase in the area under rice cultivation is largely attributed to the introduction of upland NERICA (New Rice for Africa), which found a niche especially among farmers who until recently thought of rice as a crop adapted only to irrigated/rainfed lowland ecosystems such as Kibimba, Doho, Olweny, and Agoro irrigation schemes. Today, many small-scale farmers can be found growing rice across the country alongside a few large scale rice growers. This out-scaling of rice cultivation into areas that were initially thought of as "non-rice producing regions" makes it increasingly difficult to ignore emerging biotic constraints or yield-reducing factors like pests, diseases and weeds, which may jeopardize future productivity. Insect pests constitute one of the major yield reducing factors in rice [3]. Crop losses due to insect pests on rice in the developing countries of Africa have been estimated at about 20% [4]. In 2008, about half of the rice farmers in Sub Saharan Africa (SSA) experienced insect pest attacks, affecting 27% of the rice area and causing an estimated 20% yield loss [4]. However, rice insect pests can cause even much higher yield losses. For instance, yield losses of up to 45% were reported in Kenya [4]. Recent developments in the rice industry in Uganda have heightened the need to understand key insect-pest threats on rice and the yield losses that they cause, in addition to exploring available management options. This is against the background that the increasing area under rice production in Uganda may have altered insect pest complexes and changed the pest status of some insects. Thus, the purpose of this study was to assess farmers' knowledge and perceptions about rice insect pests, and how they manage rice insect pests. Understanding farmers' perceptions is important in guiding the development of sustainable and cost effective integrated pest management (IPM) strategies. Sound pest management practices can stabilize and secure rice production, to the benefit of the poor rice-farming households [5,6]. Moreover, farmer pest management strategies identified in this study (if any) may accelerate technology development and promotion, since farmer pest management practices represent the direct result of decisions they make every season to enable them to remain in production.

2. MATERIALS AND METHODS

2.1. Study Area

The survey was conducted in the north, east and central regions of Uganda. Four districts were selected from the north, i.e., Lira, Dokolo, Otuke and Alebtong; three districts from the east, i.e., Iganga, Bugiri and Kamuli; and one district from the central area, i.e., Kayunga (Figure 1).

The north and east had a larger share of selected districts because they have a long history of lowland rice cultivation, which started in the 1950s with the establishment of Kibimba, Doho and Olweny irrigation schemes. Lowland rice cultivation is only a recent phenomenon in the central region. From each district, one rice producing sub county was selected; Barr for Lira district; Agwata for Dokolo district; Okwang for Otuke district; Apala for Alebtong district; Bukanga for Iganga district, Nabukalu for Bugiri district, Butansi for Kamuli district and Kangulumira for Kayunga district, altogether totaling to eight sub-counties. Within each sub county, three parishes were selected and 10 households drawn from each parish, constituting an overall total of 240 households.

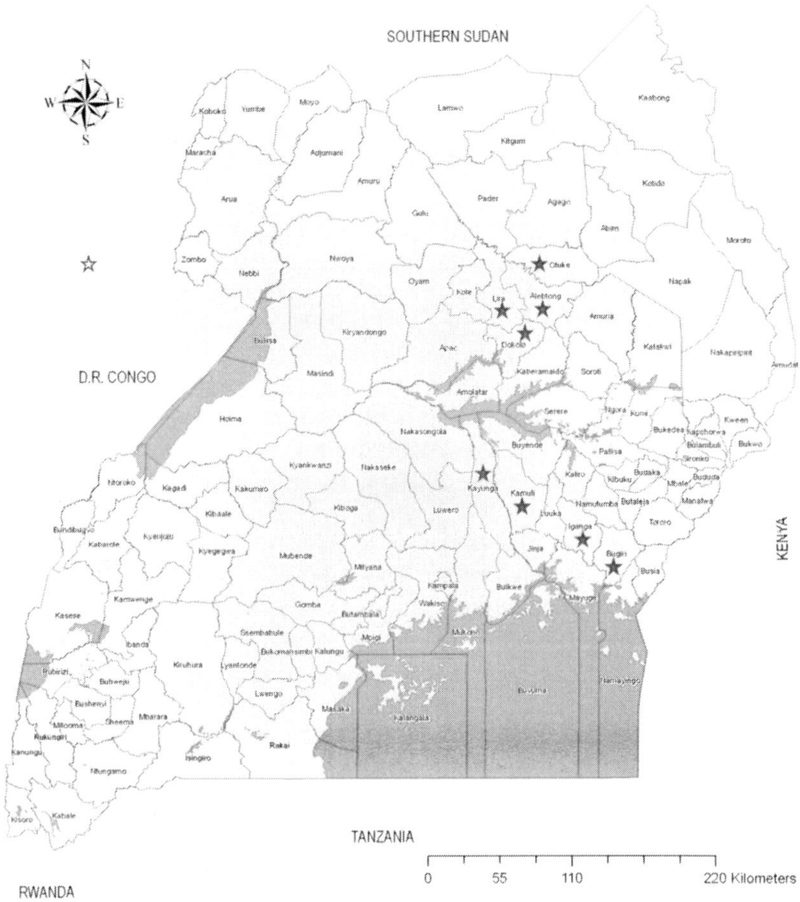

Figure 1. Map showing all districts in Uganda. Districts that were surveyed are marked with red stars.

2.2. Survey Method and Data Analysis

Rice-farming households were interviewed using a semi-structured questionnaire that was developed by a multi-disciplinary team comprising socio-economists, agronomists, entomologists and breeders. Before beginning the survey, the questionnaire was first pre-tested in Luwero District and then reviewed—based on pre-test findings. The review enabled clarification and streamlining of the interview questions for a smooth interviewing process. During the interviews, farmers were asked about their age, level of education, years of farming experience and their land tenure. They were also asked about their experiences with rice farming. Open-ended and closed questions were used to determine their knowledge and perceptions on a range of issues that included general and biotic constraints in rice production as well as insect pest problems. The farmers were showed pictures of rice insect pests and then asked to identify the pests and indicate the damage that the insect pests cause. They were then asked to rank the production constraints and rice insect pests in order of importance (1 being most important and 5 least important), and then indicate how they manage the rice insect pests, their source of information on how to manage rice insect pests, and effectiveness of the management options. A weighting score was used to determine overall importance of the various rice insect pests. The weight score of an insect pest was computed by multiplying its rank weight by the frequency with which it is cited by respondents. For instance, the most important insect pest was assigned the highest rank weight, while the least important insect pest was assigned the lowest rank weight of 1. A weight score was then calculated by multiplying the rank weight of a variable by its response frequency, and the various weight scores summed up to give the total weight score. Farmer's knowledge and perceptions of weeds as alternate hosts of rice insect pests was also sought.

Responses to the questions were coded before analysis. The open-ended questions were critically analyzed to identify recurrent themes which could be quantified to determine farmers' knowledge and perceptions. The statistical analysis was performed using Stata 10 [7], and statistics such as frequency distributions and percentages were used to analyze and report responses from the farmers.

3. RESULTS

3.1. Farmers Socio-Economic Profile

Table 1 summarizes the socio-economic profile of the farmers interviewed. One third of the farmers were female. Approximately, sixty two percent (61.7%) of the respondents were aged between 18 and 40 years while 35.6% fell within the 41–50 years age bracket. Only 2.7% of the respondents were over 60 years of age. Almost two thirds (61.6%) of the respondents went to primary school whereas 24.1% attended secondary school. A minority of the farmers (12.9%) interviewed indicated not getting any formal education and a negligible number (3) had attended college/university.

Table 1. Socio-economic profile of the respondents.

Characteristics	Summary of Responses	
	Response Frequency (*n*)	Response Percentage (%)
Sex		
Female	69	30.8
Male	155	69.2
Age (years)		
18–30	68	30.6
31–40	69	31.1
41–50	50	22.5
51–60	29	13.1
>60	6	2.7
Level of Education		
No Education	29	12.9
Primary	138	61.6
Secondary	54	24.1
College/University	3	1.3

Table 2. Average land holding of the respondents by district and region.

Region	District	Average Land Holding (Acres)	
		By District	By Region
Northern	Lira	5.7	8.2
	Dokolo	6.0	
	Otuke	15.2	
	Alebtong	5.8	
Eastern	Iganga	3.7	7.2
	Bugiri	10.5	
	Kamuli	7.4	
Central	Kayunga	3.7	3.7

Overall, the average land holding of the respondents was 6.4 acres. However, land holding varied from region to region. Farmers in the northern region owned larger pieces of land on average when compared to their counterparts in the east and central regions (Table 2).

Figure 2 compares the rice-growing experience of farmers in the three regions that were surveyed. Generally, most of the farmers had between 0–10 years of experience in rice farming. Clearly, farmers in Eastern Uganda had the longest history of rice cultivation as indicated by their relatively higher response percentage in the 16–20 and >20 years' experience-categories. In contrast, farmers from the central region were the least experienced. None of the farmers in the central region had more than 10 years' experience in rice cultivation.

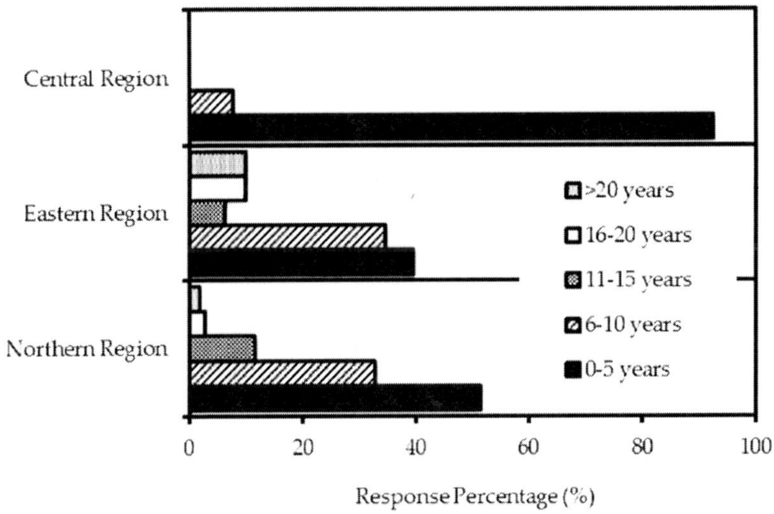

Figure 2. Years of experience of the respondents in rice farming. The total number of respondents for the northern, eastern and central regions was 113, 81 and 26, respectively.

3.2. Rice Production Constraints

The five most important constraints in rice production were insect pests, diseases, weeds, financial constraints and birds, as shown in Figure 3. Financial limitations were related to production costs such as tractor hire and hire of labor for various production operations performed, especially when family labor is inadequate.

3.3. Farmers Knowledge and Perceptions on Rice Insect Pests

When asked to list and rank rice insect pests according to their importance, respondents ranked stem borers and the African rice gall midge (AfRGM) as the 1st and 2nd most detrimental insect pests on rice (Figure 4). These were closely followed in rank by the rice bugs, leaf folders, armyworm, cutworm and termites in decreasing order. Other insect pests included rice hoppers, rice hispa, mole crickets, root weevils and grasshoppers among others.

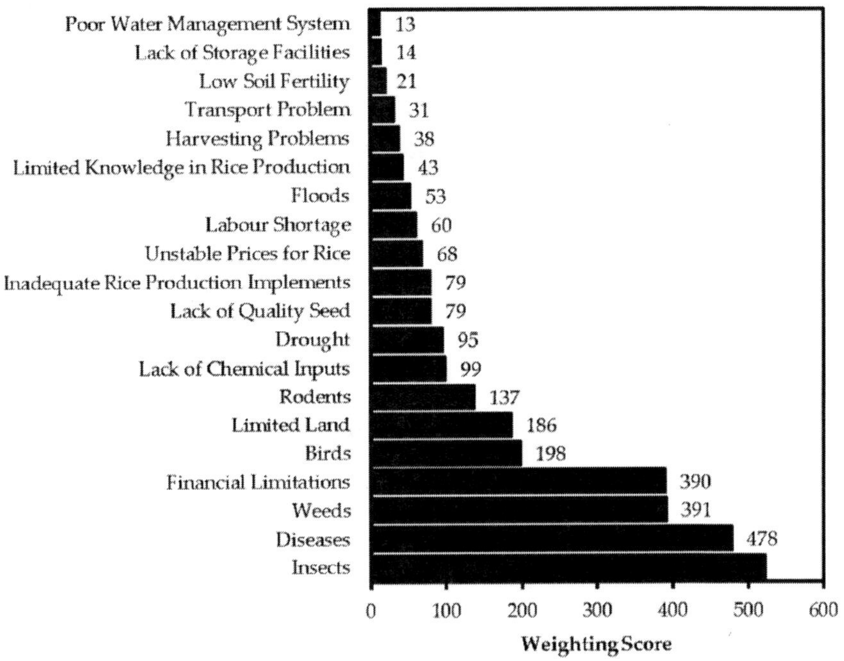

Figure 3. Rice production constraints in the survey areas.

Rice insect pests were broadly categorized into three groups, namely, leaf/stem feeding insects, grain sucking insects and root feeding insects. Generally the leaf/stem feeding insects or defoliators seemed to dominate over other categories—going by the high average weighting score (340). Grain sucking insects came second in importance with an average weighting score of 223, while root sucking insects were considered the least important insect-pest problem scoring only 109. The most important leaf/stem feeding insect-pests according to the farmers were the stem borer, African rice gall midge (AfRGM) and leaf folders. Rice bugs were considered the most significant pest among the grain sucking insects while root feeding insects such as the root weevil and whorl maggot were perceived as minor pests.

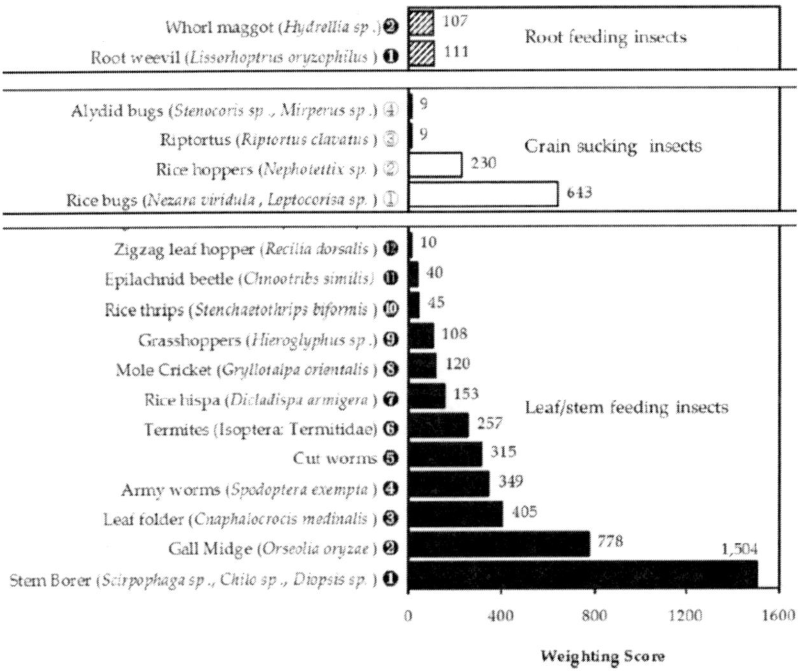

Figure 4. Farmer ranking of rice insect pests by category and according to importance on rice.

3.4. How Farmers Control Common Rice Insect Pests

The five most important rice-insect pests as perceived by the farmers were stem borers, African rice gall midge (AfRGM), rice bugs, leaf folders and the armyworm. One interesting finding that emerged from this survey is that only about 40% of the farmers who were interviewed could correctly identify major insect pests of rice or symptoms that they cause (Figure 5).

Over two thirds (72.3%) of the 224 farmers who were interviewed did not implement any measure to control these pests. Less than one third of the respondents (27.7%, 62 farmers) indicated practicing some kind of control measure against rice insect pests. Less than one third of the respondents (51 farmers, 22.8%) targeted the five major insect pests mentioned above. The major strategies that these farmers adopted for managing the five insect pests were; early sowing, burning rice stubble, use of insecticide and adjusting fertilizer. Table 3 presents a breakdown of the response frequencies and percentages of farmers using control measures against the five major insect pests identified.

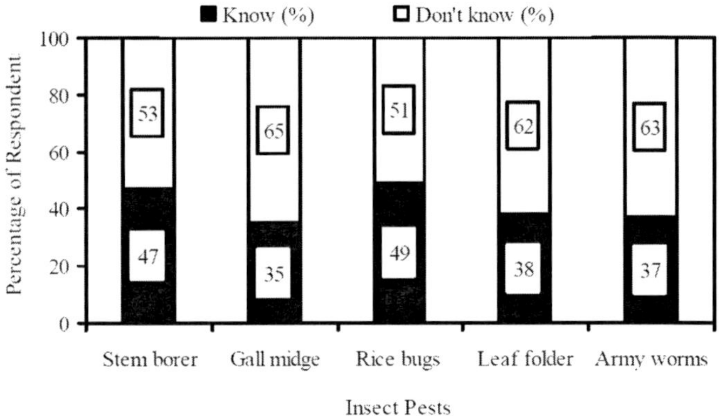

Figure 5. Proportion of farmer respondents who could correctly identify either the major insect pests of rice or their symptoms.

It is apparent from this table that most farmers (62.7% of all the respondents controlling the five major rice insect pests) rely on the use of insecticides to control rice insect pests. The least used pest management practice was early sowing. The burning of rice stubbles and adjusting fertilizer rates were moderately used. On the other hand, the only cultural control method that featured more (second to insecticide use) in the responses of the farmers was burning of rice stubble.

3.5. Effectiveness of Control Measures

The level of effectiveness of the control measure used on these five most common insect pests varied. Of the four respondents who reported controlling by sowing early (refer to Table 3), one said it is effective; while two said it was somewhat effective. One respondent said it was not effective. Seven respondents mentioned adjusting fertilizer rates as a way of managing rice insect pests. Two of these farmers noted that it is moderately effective and five said it was ineffective. Burning stubble as a way of managing insect pests was practiced by 10 farmers, seven of whom reported it to be moderately effective and three said that it is ineffective. Forty one farmers reported using insecticides to control insect pests. Of these, 15 farmers declared that insecticides were effective, while 24 said they were moderately effective. Two farmers reported that pesticides are not effective.

Table 3. Proportion of farmer respondents who use different control measures for the 5 major insect-pests of rice.

Control Measure	Stem Borer	Gall Midge	Rice Bugs	Leaf Folders	Army Worm
	Response Frequency (%)	Response Frequency (%)	Response Frequency (%)	Response Frequency (%)	Response Frequency (%)
Early Sowing	3 (13.0%)	1 (8.3%)	0 (0%)	0 (0%)	0 (0%)
Adjusting fertilizer rate	4 (17.3%)	1 (8.3%)	0 (0%)	2 (33.3%)	0 (0%)
Burning rice stubble	3 (13.0%)	1 (8.3%)	2 (67%)	1 (16.7%)	1 (14.3%)
Using insecticides	13 (56.5%)	9 (75%)	1 (33%)	3 (50.0%)	6 (85.7%)
Total	23 (100%)	12 (100%)	3 (100%)	6 (100%)	7 (100%)

3.6. Source of Information on Insect Pest Control Measures

58% of the farmers who were interviewed received information on insect control measures from fellow farmers, while 40% used their own experience. Only 2% received information from agricultural extension workers (Figure 6).

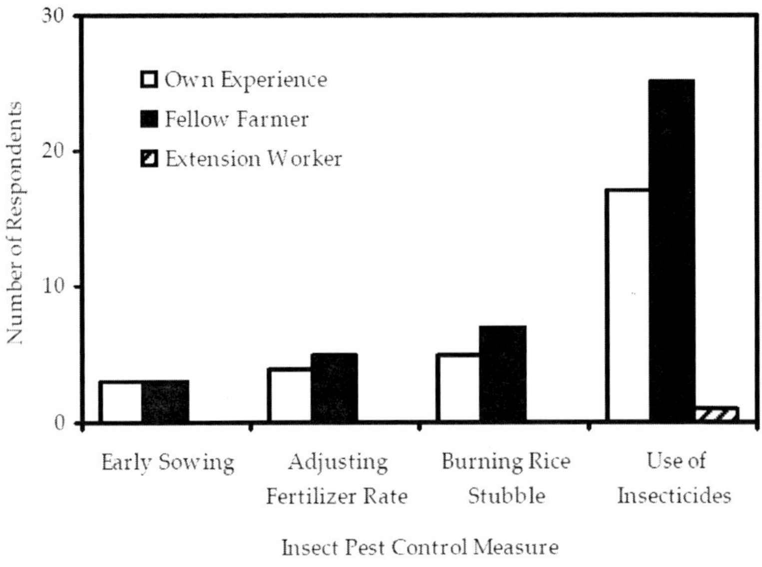

Figure 6. Farmers source of information for rice insect pest management.

3.7. Farmers Knowledge and Perceptions of Weeds as Alternate Hosts of Rice Insect Pests

When the farmers were asked whether they observe any of the rice insects on weeds/vegetation growing in and around their rice farms, 34 reported the presence of weeds, which are alternative hosts of rice insect pests in and around rice fields. Three of these farmers were from Lira, two from Otuke, eight from Alebtong, five from Iganga, seven from Bugiri, three from Kamuli and six from Kayunga. Some of the insects which live on weeds as perceived by farmers include the stem borer, gall midge, termite, leaf rollers and whorl maggot. On the other hand, some of the weeds that served as alternative hosts to rice insect pests included elephant or Napier grass (*Pennisetum purpureum*), sedges (*Cyperus* spp.), black-jack (*Bidens pilosa*), wandering Jew (*Commelina benghalensis*) and star grass (*Cynodon dactylon*). Elephant grass and star grass were infested by all the above insects, while *C. benghalensis* was infested by all except the rice gall midge. Sedges suffered infestation from the rice gall midge, termites and the whorl maggot. However, black jack harbored the gall midge and whorl maggot.

4. DISCUSSION

The purpose of this study was to assess farmers' knowledge and perceptions about rice insect pests and how the farmers cope with them. The farmers interviewed were well aware of the importance of rice insect pests. They regarded insect pests as the number one production constraint, with diseases ranking 2nd and weeds 3rd. This is contrary to the findings of Diagne et al. [4], who found weed infestation to be the most important biotic constraint followed by insect pests and then birds [4]. This inconsistency may be due to the large differences between the 18 major rice-producing countries that were surveyed by Diagne et al. [4]. For instance, they found the average area affected by insect pests ranging from a low 11% in Guinea and a high 49% in Kenya. Hence our findings may be representative of the scenario in Uganda. Something clear from this study however is that insect pests constitute one of the major biotic constraints limiting rice production in Uganda just like any other country in Sub Saharan Africa (SSA), which is in agreement with the observation Lenné [5].

In all the districts surveyed, a majority of the farmers interviewed identified five insect pests as being of major importance. These included stem borers, the African rice gall midge (AfRGM), rice bugs, leaf folders and the army worm. Of these, stem borers were perceived to be the most detrimental of the rice insect pests. This situation appears to be similar across much of SSA [8]. Stem borers belong to the category of insects which feed on the leaves and stems of rice. Second to the stem borers amongst the leaf/stem feeders was the rice gall midge. The rice gall midge also ranks high among the important insect pests in SSA, and is a primary focus for control in rice fields [9]. Farmers rated rice bugs as the most harmful of the grain sucking insect pests. Rice bugs are important insect pests during the milky stage of rice and they feed by sucking the sap of developing spikelets causing serious rice crop losses. Root feeding insect-pests on the other hand were rated rather low in terms of importance.

A disheartening fact that emerged from the study however was that about 60% of the farmers who were interviewed could not correctly identify major insect pests of rice or symptoms they cause, which affects any attempts to control specific insect pest. Perhaps this explains why over two thirds of the farmers interviewed did not manage rice insect pests, presumably because they cannot identify the pests. Among the key pest control measures reported by the farmers were early sowing, burning rice stubble, use of insecticide and adjusting fertilizer rates. The predominant measure, which was also perceived by farmers as most effective, was the use of insecticides, while cultural control methods were under represented. The only cultural control method that was used frequently and also perceived to be effective is the burning of rice stubble. However, the perception that insecticides are the most effective control method for rice insect pests contradicts the principles of IPM. According to the World Bank, the main reason for unnecessary pesticide use is lack of knowledge and information among farmers about other crop protection approaches and the true costs and benefits of pesticide use [10]. This highlights a need to develop and sensitize farmers on aspects of IPM. Insecticides can be effective for controlling rice insect pests if used carefully and according to instructions. Injudicious pesticide applications however may result in insect-pest outbreaks. The best example is the emergence of the rice brown plant hopper (BPH). Prior to 1970, the BPH was a virtually unnoticed member of the fauna of tropical rice fields in Asia. Its rise as one of the most important pests of rice in South-East Asia was strongly associated with the overuse of pesticides [11].

Farmers were aware of the fact that weeds in and around rice fields are potential alternative hosts of rice insect-pests. The farmers identified elephant grass, sedges, black-jack wandering Jew and star grass as some of the alternative hosts of rice insect pests. Such farmer knowledge is valuable in understanding interactions between weeds and insect populations, and thereby aid in developing sustainable and cost effective IPM strategies.

5. CONCLUSIONS

Theoretically drafted strategies to control pest populations are bound to fail unless farmers, who are the major stakeholders, are involved from the beginning. The value of farmers' knowledge is their long and intimate association with the crop, their unlimited opportunities for observation during the whole cropping season and their comprehension of pest-crop dynamics within the agro-ecological landscape. By engaging them, a pest management strategy that is sustainable, safe and cost-effective can be designed and implemented [12]. For the development and implementation of sound insect-pest management practices, the following findings of our study are important: (1) stem borers, AfRGM, rice bugs, leaf folders, and army-worms are perceived by farmers as the major insect pests of rice in Uganda; (2) rice farmers are aware of the damage potential of insect pests; (3) rice farmers believe that insecticides are the most effective insect pest control measure; and (4) vegetation including weeds in and around the rice fields are perceived as alternative hosts of rice insect pests.

ACKNOWLEDGMENTS

This study was supported by a World Bank funded Eastern Africa Agricultural Productivity Project (EAAPP).

REFERENCES

1. Ministry of Agriculture, Animal Industries and Fisheries (MAAIF). *Uganda National Rice Research Strategy (NRDS)*; MAAIF: Entebbe, Uganda, 2012.
2. Ahmed, M. *Analysis of Incentives and Disincentives for Rice in Uganda*; MAFAP, FAO: Rome, Italy, 2012.
3. Pingali, P.L.; Gerpachio, R.V. Living with reduced pesticide use in tropical rice in Asia. *Food Policy* **1997**, *22*, 107–118.
4. Diagne, A.; Alia, D.Y.; Amovin-Assagba, E.; Wopereis, M.C.S.; Saito, K.; Nakelse, T. Farmer perceptions of the biophysical constraints to rice production in sub-Saharan Africa, and potential impact of research. In *Realizing Africa's Rice Promise*; Wopereis, M.C.S., Johnson, D.E., Ahmadi, N., Tollens, E., Jalloh, A., Eds.; CABI: Accra, Ghana, 2013; pp. 46–68.
5. Lenné, J. Pests and poverty: The continuing need for crop protection research. *Outlook Agric.* **2000**, *29*, 236–250.
6. Warburton, H.; Martin, A.M. Pest management and poor people. In *1999 BCPC Symposium Proceedings No. 73: International Crop Protection: Achievements and Ambitions*; BCPC: Farnham, UK, 1999; pp. 35–55.
7. StataCorp. *Stata Statistical Software: Release 10*; StataCorp LP: College Station, TX, USA, 2007.
8. Seck, P.A.; Diagne, A.; Mohanty, S.; Wopereis, M.C.S. Crops that feed the world: Rice. *Food Secur.* **2012**, *4*, 7–24.
9. Oikeh, S.O.; Nwilene, F.E.; Agunbiade, T.A.; Oladimeji, O.; Ajayi, O.; Mande, S.; Tsunematsu, H.; Samejima, H. *Growing Upland Rice: A Production Handbook*; Africa Rice Center: Cotonou, Benin, 2008; p. 40.
10. The World Bank Group. Integrated Pest Management. Available online: http://go.worldbank.org/F98SD7NLZ0 (accessed on 24 June 2014).
11. Joshi, R.C.; Shepard, B.M.; Kenmore, P.E.; Lydia, R. Insecticide-induced resurgence of brown planthopper (BPH) on IR62. *Int. Rice Res. Newsl.* **1992**, *17*, 9–10.
12. Agullar, C.H.; Lasalita-Zapico, F.; Namocatcat, J.; Fortich, A.; Bojadores, R.M. Farmers' Perceptions about Banana Insect Pests and Integrated Pest Management (IPM) Systems in SocSarGen, Mindanao, Philippines. In Proceedings of the International Conference on Intelligent Agriculture, IPCBEE, Hong Kong, China, 13–14 February 2014; IACSIT Press: Singapore, Singapore, 2014; Volume 63.

CHAPTER 15

Can Novel Management Practice Improve Soil and Environmental Quality and Sustain Crop Yield Simultaneously?

Upendra M. Sainju*

USDA-ARS, Northern Plains Agricultural Research Laboratory, Sidney, MT 59270, United States of America

AUTHOR CONTRIBUTIONS

Conceived and designed the experiments: UMS. Performed the experiments: UMS. Analyzed the data: UMS. Contributed reagents/materials/analysis tools: UMS. Wrote the paper: UMS.

ABSTRACT

Little is known about management practices that can simultaneously improve soil and environmental quality and sustain crop yields. The effects of novel and traditional management practices that included a combination of tillage, crop rotation, and N fertilization on soil C and N, global warming potential (GWP), greenhouse gas intensity (GHGI), and malt barley (*Hordeum vulgarie* L.) yield and quality were examined under non-irrigated and irrigated cropping systems from 2008 to 2011 in eastern Montana and western North Dakota, USA. In loamy soil under non-irrigated condition in eastern Montana, novel and traditional management practices were no-till malt barley-pea *(Pisum sativum* L.) with 80 kg N ha^{-1} and conventional till malt barley-fallow with 80 kg N ha^{-1}, respectively. In sandy loam soil under irrigated and non-irrigated conditions in western North Dakota, novel and traditional management practices included no-till malt barley-pea with 67 (non-irrigated) to 134 kg N ha^{-1}(irrigated) and conventional till malt barley with 67 (non-irrigated) to 134 kg N ha^{-1}(irrigated), respectively. Compared with the traditional management practice, soil organic C (SOC) and total N (STN) at 0–120 cm were 5% greater with the novel management practice under non-irrigated condition in eastern Montana and under irrigated condition in western North Dakota, but were not different under non-irrigated condition in western North Dakota. In both places under irrigated and non-irrigated conditions, total applied N rate, residual soil NO_3-N content at 0–120 cm, global warming potential (GWP), and greenhouse gas intensity (GHGI) were 15 to 70% lower with the novel than the traditional management practice. Malt barley yield and quality were not different between the two practices in both places. Novel management practices, such as no-till malt barley-pea with reduced N rate,can simultaneously enhance soil and environmental quality, reduce N input, and sustain crop yield compared with traditional practices in the northern Great Plains, USA.

1. INTRODUCTION

Traditional practices, such as conventional till with crop-fallow or continuous monocropping with N fertilization, have reduced soil C and N storage and crop yields due to residue incorporation into the soil, aggregate disruption, and absence of crops during fallow in the northern Great Plains [1,2]. Residue incorporation into the soil increases its contact with microorganisms, thereby increasing its mineralization and reducing soil C and N storage. Both tillage and fallow can increase soil organic matter mineralization due to increased microbial activity, thereby increasing residual soil NO_3-N content and the potential for N leaching [2, 3]. Continuous monocropping can reduce crop yields compared with crop rotation due to increased infestation of weeds, diseases, and pests [4]. Traditional practices can also increase net greenhouse gas (GHG) emissions compared with improved practices, such as no-till with crop rotation and reduced N rate [5,6].

Improved management practices, such as no-till, diversified crop rotation, increased cropping intensity, cover cropping, and reduced rate of N fertilization, can increase soil C and N storage and reduce GHG emissions compared with traditional practices [1,7,8,9]. No-till can increase soil organic C (SOC) and soil total N (STN) and reduce GHG emissions compared with conventional till by reducing soil disturbance, residue incorporation, and microbial activity, which in turn, lower CO_2 emissions [6,10,11]. Increased quality and quantity of crop residue returned to the soil from increased cropping intensity, diversified crop rotation, cover cropping, and N fertilization can increase SOC and STN compared with fallow, monocropping, and no cover crop or N fertilization [6,12,13]. Nitrogen fertilization stimulates N_2O emissions [6, 14, 15], but has a variable effect on CO_2 and CH_4 emissions [13,16,17,18]. Therefore, practices that can reduce N fertilization rates without influencing crop yields can substantially lower net global warming potential (GWP) and greenhouse gas intensity (GHGI) [6,19].

No-till can reduce crop residue and soil organic N mineralization and residual soil NO_3-N content which is subjected to leaching in the groundwater compared with conventional till [3]. Most crops are unable to uptake NO_3-N from subsurface soils unless they have deep root systems [20]. Soil profile NO_3-N content can be reduced with increased cropping intensity due to greater N immobilization, less summer fallow, and greater amount of N removed by crops [20].Nitrogen loss below the root zone can be enhanced due to increased fallow period as a result of leaching and absence of crops to utilize N [3].

Soil water conservation is also greater in no-till than conventional till [21]. As a result, crop yields can be sustained by utilizing soil water more efficiently in no-till systems [2,21,22,23]. Inclusion of pea in rotation with spring wheat (*Triticum aestivum* L.) and barley can sustain their yields by efficiently utilizing soil water and reduce N fertilization rate due to N supplied from pea residue which is rich in N as a result of biological N fixation from the atmosphere [4,24,25]. In addition, pea uses less soil water than spring wheat and barley, thereby leaving more water available for succeeding crops and increasing their yields [4, 26]. Other benefits of crop rotation compared with monocropping include decreased pressure from weeds, diseases, and pests [4,27], reduced chemical inputs, and improved economic and environmental sustainability [28].

Although information on the effects of tillage, cropping systems, and N rates on soil C and N and crop yields in the northern Great Plains is available [13,23], little is known about novel management practices that can simultaneously enhance soil organic matter, reduce N rate, potential for N leaching, and GHG emissions, and sustain crop yields and quality in the northern Great Plains, USA. This study compared novel and traditional management practices that included combinations of tillage, cropping systems, and N fertilization rates on soil C and N, GWP, GHGI, and malt barley yield and quality under non-irrigated and irrigated conditions from

2008 to 2011 in eastern Montana and western North Dakota, USA. It was hypothesized that the novel management practice (no-till barley-pea with reduced N rate) can increase soil C and N storage, reduce residual soil NO_3-N content, GWP, and GHGI, and sustain malt barley yield and quality compared with the traditional management practice (conventional till malt barley-fallow or continuous malt barley with recommended N rate).

2. MATERIALS AND METHODS

The details of the experiment were described by Sainju et al. [29]. The studies were conducted on government farms owned by USDA-ARS, Sidney, MT and no specific permission was needed to conduct the study and collect data. Also, the field study did not involve protected or endangered species. Briefly, the experiments were conducted under non-irrigated condition from 2006 to 2011 in Sidney, eastern Montana and under irrigated and non-irrigated conditions from 2005 to 2011 in Nesson Valley, western North Dakota, both were located in the northern Great Plains, USA. In Sidney, mean monthly air temperature ranges from -8°C in January to 33°C in August and mean (68-yr average) annual precipitation is 340 mm. The soil was Williams loam (fine-loamy, mixed, frigid, Typic Argiborolls) with 350 g kg^{-1} sand, 325 g kg^{-1}silt, 325 g kg^{-1} clay, 11.5 g kg^{-1} SOC, and 7.2 pH at the 0–15 cm depth. In Nesson Valley, mean monthly air temperature ranges from -5°C in January to 32°C in August and mean (56-yr average) annual precipitation is 405 mm. The soil was Lihen sandy loam (sandy, mixed, frigid, Entic Haplustoll) with 720 g kg^{-1} sand, 120 g kg^{-1} silt, 160 g kg^{-1} clay, 11.2 g kg^{-1} SOC, and 7.7 pH at the 0–15 cm depth.

In Sidney, treatments included a combination of four tillage and cropping sequences (no-till continuous malt barley, no-till malt barley-pea, no-till malt barley-fallow, and conventional till malt barley-fallow), each with a split-plot treatment of four N rates (0, 40, 80, and 120 kg N ha^{-1}). In Nesson Valley, treatments included two irrigation systems (irrigated vs. non-irrigated), each with a split-plot treatment of five management practices (conventional till malt barley with 67 or 134 kg N ha^{-1}, conventional till malt barley with 0 kg N ha^{-1}, no-till malt barley-pea with 67 or 134 kg N ha^{-1}, no-till malt barley with 67 or 134 kg N ha^{-1}, and no-till malt barley with 0 kg N ha^{-1}). The recommended N rates for irrigated and non-irrigated malt barley in Nesson Valley were 134 and 67 kg N ha^{-1}, respectively. Actual amount of N fertilization rates applied to malt barley, however, varied from year to year (Tables 1–3), because N rates were adjusted to soil NO_3-N content to a depth of 60 cm that was measured after crop harvest in the fall of the previous year. No N fertilizer was applied to pea. Both experiments were arranged in randomized complete block design with three replications. For this study, only no-till malt barley-pea with 80 kg N ha^{-1} as the novel management practice and conventional-till malt barley-fallow with 80 kg N ha^{-1} as the traditional management practice were selected in Sidney. Under non-irrigated condition in Nesson Valley, only no-till malt barley with 67 N ha^{-1}as the novel management practice and conventional till malt barley with 67 kg N ha^{-1} as the traditional practice were selected. Similarly, under irrigated condition in Nesson Valley, only no-till malt barley with 134 N ha^{-1} as the novel management practice and conventional till malt barley with 134 kg N ha^{-1} as the traditional practice were selected.

In both sites, six-row malt barley and pea were planted in April every year. Conventional-tilled plots were tilled as needed and no-tilled plots were left undisturbed. Recommended rates of P (from triple superphosphate, 45% P) and K fertilizers (from muriate of potash, 60% K) were broadcast in all plots at planting and herbicides and pesticides were applied as needed to control weeds, pests, and diseases. All N fertilizer (from urea, 46% N) was broadcast at planting under non-irrigated condition in Sidney and Nesson Valley. Under irrigated condition in Nesson Valley, half of N fertilizer was broadcast at planting and other half at six weeks later. Under irrigated condition in Nesson Valley, plots were irrigated with measured amount of water according to soil water content and crop demand [30]. Malt barley and pea grain yields, protein concentration, and barley kernel plumpness were determined in August every year. Grain yields were determined from a swath of 11.0 × 1.5 m using a combine harvester after oven-drying samples at 60°C for 3 d. Samples of malt barley grain were ground to 1.0 mm and total N concentration (g N kg^{-1} plant) was determined by using the high induction furnace C and N analyzer (LECO, St. Joseph, MI). Protein concentration in malt barley grain was determined by multiplying N concentration by 6.25. Grain kernels were separated into plump, normal, and thin fractions by sieving 100 g grains in a nest of sieves containing 2.4 and 2.0 mm sieves in a strand sieve shaker (Seedburo Equipment Co., Des Plains, IL) for 3 min. Plump kernel refers to the proportion of grains retained in the 2.4 mm sieve, normal to those in the 2.0 mm sieve, and thin to those that passed through the 2.0 mm sieve. After grain harvest, crop residue was returned to the soil. For this experiment, only data from 2008 to 2011 were used.

Soil samples from 0–120 cm in Sidney and from 0–85 cm in Nesson Valley were collected after crop harvest in October from 2008 to 2011. Because of the sandy texture of the soil in the subsoil layers, we were able to sample only to a depth of 85 cm in Nesson Valley. Samples were collected from five places within the plot, composited, air-dried, and sieved to 2 mm. An additional undisturbed soil core was collected from each plot to determine bulk density, which was measured by dividing the oven-dried weight of the soil core at 105°C for 24 h by the volume of the core.

The SOC and STN concentrations (g kg^{-1}) in the samples were determined by using a C and N analyzer (LECO Corp., St Joseph, MI) after grinding the soil to <0.05 mm and pretreating with 6 mole L^{-1} HCl to remove inorganic C. The NO_3-N concentration in soil samples was determined by extracting samples with 2 mole L^{-1} KCl for 1 h and analyzing the extract colorimetrically with an autoanalyzer (Lachat Instruments, Loveland, CO). The SOC, STN, and NO_3-N contents (kg or Mg C or N ha^{-1}) were determined by multiplying their concentrations by the bulk density and the thickness of the soil layer. The soil C and N sequestration rates were determined from the slope of the regression line between SOC and STN contents and year.

Static chambers were installed immediately after crop planting in 2008 to measure N_2O and CH_4 fluxes throughout the year until 2011 in both sites [29]. Gas samples were collected at 0, 20, and 40 min intervals in an observation date in the field and their concentrations were determined using a gas chromatograph in the laboratory. Samples were collected at 3–30 d intervals throughout the year where closer sampling (3 d) was done immediately following tillage, fertilization, planting, precipitation, irrigation, and snowmelt when gas fluxes peaked and wider sampling (7 to 30 d) in other periods when gas fluxes slowed. The fluxes of N_2O and CH_4as concentrations gradient over time were calculated by using linear or nonlinear regressions and fluxes for missing dates were calculated by using linear interpolation [31, 32]. Total fluxes in a year were calculated by summing the fluxes from all observation dates. The GWP in each year was calculated by summing CO_2 equivalents of total annual N_2O and CH_4 fluxes, farm operations, and N fertilization minus CO_2 equivalent of soil C sequestration rate [6, 9, 33]. The GHGI was calculated by dividing GWP by malt barley grain yield [6].

A paired "t" test was used to compare the effect of novel vs. traditional management practice on mean SOC, STN, NO_3-N contents, GWP, GHGI, and malt barley yield and quality in each year and across years under non-irrigated condition in Sidney and under irrigated and non-irrigated conditions in Nesson Valley [34]. Because irrigation and its interaction with management practice were significant ($P \leq 0.05$) for all measured parameters as tested by the mixed-model of SAS [34], data were discussed separately for irrigated and non-irrigated conditions in Nesson Valley. Statistical significance was evaluated at $P \leq 0.05$, unless otherwise stated.

Table 1. Comparison of novel (no-till malt barley-pea with 80 kg N ha-1) and traditional (conventional till malt barley-fallow with 80 kg N ha-1) management practices on ecosystemservices under non-irrigated condition in Sidney, eastern Montana.

Parameter	Novel management practice					Traditional management practice					Difference†
	2008	2009	2010	2011	Mean	2008	2009	2010	2011	Mean	
Soil organic C (Mg C ha^{-1})‡	180	182	185	183	183	178	176	180	166	175	8 (5)*
Soil total N (Mg N ha^{-1})‡	16.0	16.8	17.3	17.1	16.8	15.8	16.6	17.0	16.6	16.5	0.3 (2)
Residual soil NO_3-N (kg N ha^{-1})‡	75	90	95	60	80	82	95	95	132	101	-21 (-21)**
Total N rate (kg N ha^{-1})§	60	58	49	43	210	57	49	53	23	182	28 (15)
GWP (kg CO_2 eq. ha^{-1} yr^{-1})§	188	-124	-125	541	120	565	350	124	573	403	-283 (-70)***
GHGI (kg CO_2 eq. kg^{-1} grain yield)¶	0.08	-0.05	-0.05	0.22	0.05	0.23	0.14	0.05	0.18	0.15	-0.10 (-67)***
Barley grain yield (Mg ha^{-1})	2.40	2.55	2.75	2.50	2.55	2.45	2.50	2.72	2.85	2.63	-0.08 (-3)
Barley grain concentration (g kg^{-1})	127	145	140	128	135	129	140	135	132	134	1 (1)
Barley grain plumpness (g kg^{-1})	768	770	750	800	772	710	758	687	793	737	35 (5)*

*Significant at $P = 0.05$

**Significant at $P = 0.01$

***Significant at $P = 0.001$

† Number in parenthesis denotes value in percentage.

‡ Measured at the 0- to 120-cm soil depth.

§ Global warming potential

¶ Greenhouse gas intensity

Table 2. Comparison of novel (no-till malt barley-pea with 67 kg N ha-1) and traditional (conventional till malt barley with 67 kg N ha-1) practices on ecosystemservices under non-irrigated condition in Nesson Valley, western North Dakota.

Parameter	Novel management practice					Traditional management practice					Difference†
	2008	2009	2010	2011	Mean	2008	2009	2010	2011	Mean	
Soil organic C (Mg C ha^{-1})‡	53	59	65	67	61	54	61	67	70	63	-2 (-3)
Soil total N (Mg N ha^{-1})‡	9.0	9.7	10.5	10.0	9.8	9.3	9.9	10.8	10.4	10.1	-0.3 (-3)
Residual soil NO_3-N (kg N ha^{-1})‡	25	40	53	22	35	51	60	75	78	66	-31 (-47)**
Total N rate (kg N ha^{-1})§	30	25	20	30	105	50	40	35	35	160	-55 (-34)
GWP (kg CO_2 eq. ha^{-1} yr^{-1})§	503	795	324	362	496	1501	2005	1121	1145	1443	-947 (-66)***
GHGI (kg CO_2 eq. kg^{-1} grain yield)¶	0.18	0.23	0.09	0.10	0.15	0.55	0.60	0.32	0.29	0.44	-0.29 (-66)***
Barley grain yield (Mg ha^{-1})	2.81	3.52	3.45	3.22	3.25	2.75	3.35	3.45	3.65	3.30	-0.05 (-2)
Barley grain concentration (g kg^{-1})	140	151	155	142	147	145	139	148	140	143	4 (3)
Barley grain plumpness (g kg^{-1})	760	800	823	777	790	775	798	802	789	791	-1 (-1)

**Significant at $P = 0.01$

***Significant at $P = 0.001$

† Number in parenthesis denotes value in percentage.

‡ Measured at the 0- to 85-cm soil depth.

§ Global warming potential

¶ Greenhouse gas intensity

Table 3. Comparison of novel (no-till malt barley-pea with 134 kg N ha-1) and traditional (conventional till malt barley with 134 kg N ha-1) management practices on ecosystem services under irrigated condition in Nesson Valley, western North Dakota.

Parameter	Novel management practice					Traditional management practice					Difference†
	2008	2009	2010	2011	Mean	2008	2009	2010	2011	Mean	
Soil organic C (Mg C ha^{-1})‡	55	61	60	64	60	54	55	59	60	57	3(5)*
Soil total N (Mg N ha^{-1})‡	9.1	9.5	10.2	10.0	9.7	8.9	9.3	9.0	8.4	8.9	0.8 (9)*
Residual soil NO_3-N (kg N ha^{-1})‡	27	34	36	39	34	30	65	87	78	65	-31 (-48)**
Total N rate (kg N ha^{-1})§	75	70	70	50	265	90	85	70	75	320	-55 (-17)
GWP (kg CO_2 eq. ha^{-1} yr^{-1})§	745	1530	957	948	1045	1505	1950	1803	1170	1607	-562 (-35)***
GHGI (kg CO_2 eq. kg^{-1} grain yield)¶	0.21	0.20	0.18	0.37	0.24	0.41	0.36	0.36	0.39	0.38	-0.14 (-37)***
Barley grain yield (Mg ha^{-1})	3.51	5.12	5.34	3.55	4.38	3.65	5.35	5.00	2.80	4.20	0.18 (4)
Barley grain concentration (g kg^{-1})	131	138	135	132	134	133	139	136	132	135	-1 (-1)
Barley grain plumpness (g kg^{-1})	850	875	880	843	862	842	880	828	790	835	27 (3)

*Significant at $P = 0.05$

**Significant at $P = 0.01$

***Significant at $P = 0.001$

† Number in parenthesis denotes value in percentage.

‡ Measured at the 0- to 85-cm soil depth.

§ Global warming potential

¶ Greenhouse gas intensity

3. RESULTS AND DISCUSSION

3.1 Non-irrigated Cropping System, Eastern Montana

At the 0–120 cm depth, SOC and STN increased from 2008 to 2010 and then either remained similar or slightly decreased in 2011 with both novel and traditional management practices for non-irrigated cropping system in Sidney, eastern Montana (Table 1, S1 Table). This resulted in soil C sequestration rates of 1.8 and -3.2 Mg C ha^{-1} yr^{-1} and N sequestration rates of 0.38 and 0.28 Mg N ha^{-1} yr^{-1} at 0–120 cm for novel and traditional management practices, respectively. Mean SOC across years was 5% greater with the novel than the traditional management practice, but mean STN was not affected by management practices. Undisturbed soil, followed by continuous cropping likely increased SOC (resulting in a positive C sequestration rate) with the novel compared with the traditional management practice. In contrast, residue incorporation into the soil and aggregate disruption due to tillage, followed by increased microbial activity as a result of greater soil temperature and water content as well as absence of crops during fallow probably reduced SOC (resulting in a negative C sequestration rate) with the traditional management practice [1, 13]. Because SOC and STN are key components of soil organic matter that influences soil physical, chemical, and biological properties [1, 24, 35], increased SOC and STN with the novel management practice suggest that this practice can improve soil health and quality compared with the traditional management practice under non-irrigated or dryland cropping systems in eastern Montana.

Residual soil NO_3-N content at 0–120 cm after crop harvest varied with years with both novel and traditional management practices (Table 1, S1 Table). This was probably due to variations in organic N mineralization from crop residue and soil as a result of differences in air temperature and precipitation as well as N removal by crops from year to year. As a result, N rate applied from N fertilizers also varied among years (Table 1). Mean residual soil NO_3-N content across years was 21% lower with the novel than the traditional management practice. Because increased residual soil profile NO_3-N content increases the potential for N leaching to the groundwater [3, 20], reduced NO_3-N content suggests that the novel management practice can reduce the potential for N leaching, thereby improving water quality. Total N rate applied from 2008 to 2011, however, was 15% greater with the novel than the traditional management practice. Because applied N rate was based on soil NO_3-N content only to a depth of 60 cm, not 120 cm, it appeared that reduced NO_3-N content to a depth 60 cm increased N rate in the novel management practice. Mineralization of crop residue and soil organic N during tillage and fallow increases soil NO_3-

N content and reduces N fertilization rate in the traditional management practice [20,23,24,26].

The GWP and GHGI also varied among years with both novel and traditional management practices (Table 1, S2 Table). Variations in N_2O and CH_4 fluxes due to differences in soil temperature and water content and N fertilization rates resulted in various GWP and GHGI among years, a case similar to those reported by various researchers [5, 6, 9]. Mean GWP and GHGI across years were 67 to 70% lower with the novel than the traditional management practice (Table 1). Increased C sequestration rate, followed by lower N_2O and CH_4 fluxes and reduced CO_2 emissions associated with farm operations, probably decreased GWP and GHGI with the novel compared with the traditional management practice. Similar results have been reported by various researchers [6, 9, 19, 36]. This indicates that the novel management practice can reduce net GHG emissions compared with the traditional management practice in the loam soil under non-irrigated cropping systems in eastern Montana.

Malt barley grain yield, protein concentration, and grain plumpness also varied among years (Table 1, S3 Table) due to variations in air temperature and precipitation during the growing season. Mean grain yield and protein concentration across years were not significantly different between the novel and traditional management practices, and mean grain plumpness was 5% greater with the novel than the traditional management practice. The results suggest that dryland malt barley grain yield and quality can be similar or superior with the novel compared with the traditional management practice.

3.2 Non-irrigated Cropping System, Western North Dakota

The SOC and STN at 0–85 cm increased from 2008 to 2011 for both novel and traditional management practices under non-irrigated condition in Nesson Valley, western North Dakota, except for a slight decrease in STN in 2011 for the novel management practice (Table 2, S4 Table). This resulted in C sequestration rates of 4.8 and 5.4 Mg C ha^{-1} yr^{-1} and N sequestration rates of 0.38 and 0.42 Mg N ha^{-1} yr^{-1} for novel and traditional management practices, respectively. These values were greater than those reported for SOC and STN at 0–120 cm for novel and traditional management practices under non-irrigated cropping systems in eastern Montana above. Differences in soil and climatic conditions, management practices, and depth of the soil profile may have influenced turnover rates of plant C and N to soil C and N, thereby influencing C and N sequestration rates between locations. Soil in western North Dakota was sandy loam compared to loam in eastern Montana. Western North Dakota also receives 50 mm more precipitation than eastern Montana, although air temperature is similar.

Also, soils were sampled to a depth of 85 cm in western North Dakota compared to a depth of 120 cm in eastern Montana. It is likely that increased turnover rates of plant C and N into soil C and N increased C and N sequestration rates with coarse- compared with fine-textured soil. Both mean SOC and STN across years, however, were not significantly different between novel and traditional management practices.

As in eastern Montana, residual soil NO_3-N content at 0–85 cm varied among years in the novel management practice, but increased with year in the traditional management practice (Table 2, S4 Table). Increased mineralization of crop residue and soil organic N due to tillage likely increased residual NO_3-N from 2008 to 2011 in the traditional management practice. Mean residual soil NO_3-N content across years was 47% lower in the novel than the traditional management practice. Similarly, total N rate applied to malt barley from 2008 to 2011 was 34% lower with the novel than the traditional management practice (Table 2). Inclusion of legumes, such as pea which fixes N from the atmosphere, in rotation with malt barley likely reduced residual soil NO_3-N content and total N rate by supplying N from its residue in the novel management practice.

The GWP and GHGI for both management practices also varied among years (Table 2, S5 Table). The GWP and GHGI values were greater in western North Dakota than eastern Montana due to increased N_2O and CH_4 emissions. Both mean GWP and GHGI across years were 66% lower with the novel than the traditional management practice. Although soil C sequestration rate was slightly lower, reduced N_2O and CH_4 emissions, followed by decreased CO_2 emissions associated with reduced N fertilization rate and farm operations, reduced GWP and GHGI with the novel compared with the traditional management practice.

Malt barley grain yield, protein concentration, and grain plumpness also varied among years due to variations in climatic conditions (Table 2, S6 Table). Grain yield was lower in 2008 than other years due to below-average precipitation (96 mm below the 65-yr average of 272 mm during the growing season from April to August), but rebounded in other years when growing-season precipitation was near or above the normal. Mean malt barley grain yield, protein concentration, and grain plumpness across years were not significantly different between novel and traditional management practices, suggesting that the novel management practice can produce similar levels of crop yield and quality as traditional practice, a case similar to that observed for non-irrigated cropping systems in eastern Montana.

3.3 Irrigated Cropping System, Nesson Valley, Western North Dakota

The SOC and STN at 0–85 cm varied among years for novel and traditional management practices under irrigation condition in Nesson Valley, western North Dakota (Table 3, S4 Table). Regression analysis showed that C sequestration rates were 2.6 and 2.2 Mg C ha^{-1} and N sequestration rates 0.34 and -0.18 Mg N ha^{-1} for novel and traditional management practices, respectively. These values were lower than for the non-irrigated condition above. Leaching of water soluble C and N from the soil profile due to irrigation may have reduced SOC and STN levels, as well as C and N sequestration rates under irrigated compared with non-irrigated condition. Several researchers [37, 38, 39] also reported lower SOC under irrigated than non-irrigated cropping systems due to increased C mineralization and loss of water soluble C through leaching due to irrigation. Mean SOC and STN across years were 5 to 9% greater with the novel than the traditional management practice (Table 3). As in eastern Montana, undisturbed soil condition and reduced mineralization of crop residue and soil organic matter may have increased SOC and STN in the novel compared with the traditional management practice.

Residual soil NO_3-N content at 0–85 cm also varied among years (Table 3, S5 Table). These values were greater than those observed for the non-irrigated condition above, probably the results of increased crop residue and soil organic N mineralization as well as greater N rate applied to irrigated barley. Crops can utilize only 50 to 70% of applied N, and increased N rate can increase residual soil profile NO_3-N content [40, 41]. Mean residual NO_3-N content was 31% lower with the novel than the traditional management practice, a value similar to that obtained for non-irrigated condition. Because of variations in residual soil NO_3-N content, N rate applied to malt barley also varied among years (Table 3). Total N rate from 2008 to 2011 was 17% lower in the novel than the traditional management practice, which is half of that obtained in the non-irrigated condition. Although N supplied from pea residue reduced N rate in the novel compared with the traditional management practice under both irrigated and non-irrigated conditions, lower N rate under the irrigated condition was due to greater residual soil NO_3-N content.

As with other sites, GWP and GHGI varied among years in both management practices (Table 3) due to variations in N_2O and CH_4 emissions and N fertilization rates. These values were greater than those observed under the non-irrigated condition (Table 2), probably due to enhanced microbial activity as a result of increased soil water availability from irrigation and N substrate availability from N fertilization. Mean GWP and GHGI across years were 35 to 37% lower with the novel than the traditional management practice. Increased C sequestration rate, followed by lower N_2O and CH_4 emissions and reduced CO_2 emissions associated with N fertilization and farm operations reduced GWP and GHGI with the novel compared with the traditional management practice.

Malt barley grain yield, protein concentration, and grain plumpness also varied among years due to differences in growing season climatic conditions (Table 3, S6 Table). Grain yield and grain plumpness were higher with the irrigated than the non-irrigated condition, but protein concentration was similar (Tables 2 and 3). Increased soil water availability due to irrigation probably increased grain yield and grain plumpness under the irrigated condition. As with the non-irrigated condition, mean grain yield, protein concentration, and grain plumpness across years were not significantly different between novel and traditional management practices (Table 3). Mean malt barley protein concentration and grain plumpness across years obtained in this study were within the acceptable levels of <135 g kg^{-1} protein concentration and >800 g kg^{-1} grain plumpness for six-row malt barley for malting purpose [42].

4. CONCLUSIONS

The SOC and STN at 0–120 cm increased, but residual soil NO_3-N content, N rate, GWP, GHGI, malt barley grain yield, protein concentration, and grain plumpness varied among years from 2008 to 2011 in both novel and traditional management practices under irrigated and non-irrigated conditions in eastern Montana and western North Dakota. Mean SOC, STN, malt barley grain yield, protein concentration, and grain plumpness across years were greater or similar, but residual soil NO_3-N content, N rate, GWP, and GHGI were lower with the novel than the traditional management practice. These results suggest that the novel management practice, such as no-till malt barley-pea with reduced N rate, can improve soil and environmental quality and sustain malt barley yield simultaneously by increasing soil C and N storage, reducing the potential for N leaching and net GHG emissions, and maintaining malt barley yield and quality compared with the traditional practices, such as conventional till continuous malt barley or malt barley-fallow with recommended N rates under irrigated and non-irrigated cropping systems in the northern Great Plains. The practice may also reduce agricultural inputs, such as the energy requirement for tillage and N fertilizer. Another benefit would be reduced incidences with weeds, diseases, and pests due to crop rotation in the novel management practice compared with monocropping or crop-fallow in the traditional management practice. The results can be applied to other regions with similar soil and climatic conditions where small grains are grown.

ACKNOWLEDGMENTS

I appreciate the excellent support from Joy Barsotti, Michael Johnson, Bill Iversen, Bryan Gephart, and Chris Russell for plot management, irrigation, field data collection, and collection and analysis of gas, soil, and plant samples in the laboratory. I also appreciate the funding for these studies from USDA-ARS, GRACEnet Project.

Mention of trade names or commercial products in this publication is solely for the purpose of providing specific information and does not imply recommendation or endorsement by USDA. The USDA is an equal opportunity employer.

REFERENCES

1. Halvorson AD, Wienhold BJ, Black AL (2002) Tillage, nitrogen, and cropping system effects on soil carbonsequestration. Soil Sci Soc Am J 66:906–912.
2. Peterson GA, Halvorson AD, Havlin JL, Jones OR, Lyon DG, Tanaka DL (1998) Reduced tillage andincreasing cropping intensity in the Great Plains conserve soil carbon. Soil Tillage Res 47:207–218.
3. Eck HV, Jones OR (1992) Soil nitrogen status as affected by tillage, crops, and crop sequences. AgronJ 84:660–668.
4. Miller PR, McConkey B, Clayton GW, Brandt SA, Staricka JA, Johnston AM et al. (2002) Pulse cropadaptation in the northern Great Plains. Agron J 94:261–272.
5. Mosier AR, Halvorson AD, Peterson GA, Robertson GP, Sherrod L (2005) Measurement of net globalwarming potential in three agroecosystems. Nutri Cycl Agroecosys 72:67–76.
6. Mosier AR, Halvorson AD, Reule CA, Liu XJ (2006) Net global warming potential and greenhouse gasintensity in irrigated cropping systems in northeastern Colorado. J Environ Qual 35:1584–1598. PMID:16825479
7. Lal R, Kimble JM, Stewart BA (1995) World soils as a source or sink for radiatively-active gases. In: LalR, editor, Soil management and greenhouse effect. Advances in soil science. CRC Press, BocaRaton, FL, pp. 1–8.
8. Paustian K, Robertson GP, Elliott ET (1995) Management impacts on carbon storage and gas fluxes inmid-latitudes cropland. In: Lal R, editor, Soils and global climate change. Advances in soil science.CRC Press, Boca Raton, FL, pp. 69–83.
9. Robertson GP, Grace PR (2004) Greenhouse gas fluxes in tropical and temperate agriculture: Theneed for a full-cost accounting of global warming potentials. Environ Develop Sustain 6:51–63.
10. Lemke RL, Izaurralde RC, Nyborg M, Solberg ED (1999) Tillage and nitrogen source influence soilemitted nitrous oxide in the Alberta Parkland region. Can J Soil Sci 79:15–24.
11. Drury CF, Reynolds WD, Tan CS, Welacky TW, Calder W, McLaughlin MB (2006) Emissions of nitrous oxide and carbon dioxide: Influence of tillage type and nitrogen placement depth. Soil Sci Soc Am J 70:570–581.
12. Follett RF (2001) Soil management concepts and carbon sequestration in cropland soils. Soil Tillage Res 61:77–92.
13. Sainju UM, Jabro JD, Caesar-TonThat T (2010) Tillage, cropping sequence, and nitrogen fertilization effect on dryland soil carbon dioxide emission and carbon content. J Environ Qual 37:98–106.

14. Dusenbury MP, Engel RE, Miller PR, Lemke RL, Wallander R (2008) Nitrous oxide emissions from a northern Great Plains soil as influenced by nitrogen management and cropping systems. J Environ Qual 37:542–550. PMID: 18389938
15. Robertson GP, Vitousek PM (2009) Nitrogen in agriculture: Balancing the cost of an essential resource.Annu Rev Environ Resour 34: 97–125.
16. Bronson KF, Mosier AR (1994) Suppression of methane oxidation in aerobic soil by nitrogen fertilizers,nitrification inhibitors, and urease inhibitors. Biol Fert Soils 17:263–268.
17. Amos B., Arkebauer TJ, Doran JW (2005) Soil surface fluxes of greenhouse gases in an irrigated maize-based agroecosystem. Soil Sci Soc Am J 69:387–395.
18. Al-Kaisi MM, Kruse ML, Sawyer JE (2008) Effect of nitrogen fertilizer application on growing season carbon dioxide emissions in a corn-soybean rotation. J Environ Qual 37:325–332. doi: 10.2134/ jeq2007.0240 PMID: 18268294
19. Robertson GP, Paul E, Harwood R (2000) Greenhouse gases in intensive agriculture: Contribution ofindividual gases to the radiative forcing of the atmosphere. Science 289:1922–1925. PMID: 10988070
20. Wood CW, Westfall DG, Peterson GA, Burke IC (1990) Impact of cropping intensity on carbon and nitrogen mineralization under no-till agroecosystems. Agron J 82:1115–1120.
21. Farhani HJ, Peterson GA, Westfall DG (1998) Dryland cropping intensification: A fundamental solution to efficient use of precipitation. Adv Agron 64:197–223.
22. Deibert EJ, French E, Hoag B (1986) Water storage and use by spring wheat under conventional tillage and no-tillage in continuous and alternate crop-fallow systems in the northern Great Plains. J Soil Water Conserv 41:53–58.
23. Aase JK, Pikul JL Jr (1995) Crop and soil responses to long-term tillage practices in the northern Great Plains. Agron J 87:652–656.
24. Sainju UM, Caesar-TonThat T, Lenssen AW, Evans RG (2009) Dryland crop yields and soil organic matter as influenced by long-term tillage and cropping sequence. Agron J 101:243–251.
25. Sainju UM, Lenssen AW (2011) Soil nitrogen dynamics under dryland alfalfa and durum-forage cropping sequences. Soil Sci Soc Am J 75:669–677.
26. Lenssen AW, Waddell JT, Johnson GD, Carlson GR (2007) Diversified cropping systems in semiarid Montana: Nitrogen use during drought. Soil Tillage Res 94:362–375.
27. Vigil MF, Anderson RA, Beard WE (1997) Base temperature growing-degree-hour requirements for emergence of canola. Crop Sci 37:844–849.
28. Gregory PJ, Ingram JSI, Anderson R, Betts RA, Brovkin V, Chase TN et al. (2002) Environmental consequences of alternative practices for intensifying crop production. Agric Ecosys Environ 88:279–290.
29. Sainju UM, Caesar-TonThat T, Caesar A (2012) Comparison of soil carbon dioxide flux measurements by static and portable chambers in various management practices. Soil Tillage Res 118:123–131.
30. Stevens WB, Evans RG, Jabro JD, IversenWM(2010) Nitrogen availability for sugarbeet affected by tillage system and sprinkler irrigation method. Agron J 102:1745–1752.
31. Hutchinson GL, Mosier AR (1981) Improved soil cover method for field measurements of nitrous oxide fluxes. Soil Sci Soc Am J 45:311–316.
32. Liebig MA, Tanaka DL, Gross JR (2010) Fallow effects on soil carbon and greenhouse gas flux in central North Dakota. Soil Sci Soc Am J 74:358–365.
33. Lal R (2004) Carbon emission from farm operations. Environ Int 30:981–990. PMID: 15196846
34. Littell RC, Freund RJ, Spector PC (1991) SAS system for linear models. 3rd ed. SAS Inst. Inc., Cary,NC.
35. Kuo S, Sainju UM, Jellum EJ (1997) Winter cover cropping influence on nitrogen in soil. Soil Sci SocAm J 61:1392–1399.
36. Barsotti JL, Sainju UM, Lenssen AW, Montagne C, Hatfield PG (2013) Net greenhouse gas emissions affected by sheep grazing in dryland cropping systems. Soil Sci Soc Am J 77:1012–1025.

37. Denef K, Stewart CE, Brenner J, Paustian K (2008) Does long-term center-pivot irrigation increase soil carbon stocks in semiarid agroecosysyems? Geoderma 145:121–129.
38. Schipper LA, Dodd MB, Pronger J, Mudge PL, Upsdell M, Moss RA (2013) Decadal changes in soil carbon and nitrogen under a range of irrigation and phosphorus fertilizer treatments. Soil Sci Soc Am J 77:246–256.
39. Follett RF, Jantalia CP, Halvorson AD (2012) Soil carbon dynamics for irrigated corn under two till age systems. Soil Sci Soc Am J 77:951–963.
40. Hallberg GR (1989) Nitrate in groundwater in the United States. In: Follett RF, editor, Nitrogen management and groundwater protection. Elsevier, Amsterdam, pp. 35–74.
41. Bergstrom L., Kirchmann H (2004) Leaching and crop uptake of nitrogen from nitrogen-15 labeled green manures and ammonium nitrate. J Environ Qual 33:1786–1792. PMID: 15356239
42. American Malting Barley Association (2005) Malting barley breeding guidelines: Ideal commercial malt criteria. Available at www.ambainc.org/ni/index/htm. Accessed 4 Dec, 2006. Verified 10 Aug. 2007. Miwaukee, WI.

CHAPTER 16

Sustaining Chili Pepper Production in Afghanistan through Better Irrigation Practices and Management

S. Alan Walters [1,*] and Ajay K. Jha [2]

[1] *Department of Plant, Soil, and Agricultural Systems, Southern Illinois University, Carbondale, IL 62901, USA*

[2] *Institute for Global Agriculture & Technology Transfer (IGATT), Fort Collins, CO 80528, USA; igatt2015@gmail.com*

AUTHOR CONTRIBUTIONS

S. Alan Walters collected the mid-season data, analyzed all data, interpreted results, and wrote most of the manuscript. Ajay K. Jha designed the study, was responsible for planting, maintenance, and harvesting activities, and contributed to the writing/editing of the manuscript.

ABSTRACT

Water management and utilization is an ongoing problem in developing countries with semi-arid to arid climates such as Afghanistan. The lack of effective irrigation systems are oftentimes the most limiting factor for maximizing agricultural productivity in these countries. In Afghanistan, the most widely used irrigation methods are basin/border for cereal crops and furrow for vegetables and grapes, although drip irrigation is a technology that could be used to significantly improve water use efficiency (WUE) in horticultural crop production. Therefore, three irrigation methods (basin, furrow, and drip) were evaluated for their influences on chili pepper production and WUE at the Afghanistan Ministry of Agriculture, Irrigation and Livestock (MAIL) Badam Bagh Agricultural Research and Demonstration Farm in Kabul over the 2009 and 2010 growing seasons. Results from this study indicated that both drip and furrow irrigation provided similar high chili pepper plant growth and yield responses compared to the low amounts provided by basin irrigation ($p \leq 0.05$). The drip and furrow irrigation methods provided a similar low incidence of *Phytophthora* blight disease, as 4% and 7% of chili pepper plants were visually afflicted by this disease, respectively, while an astounding 69% of chili peppers grown with basin irrigation had symptoms of this disease. Drip irrigation resulted in the best overall WUE ($p \leq 0.05$), as this water delivery method utilized the least amount of water and provided the highest chili pepper yield. Furrow irrigation provided a lower WUE compared to drip, but was greater than that of basin irrigation. Although this study indicated that drip irrigation had the greatest WUE for chili pepper production, furrow irrigation is still the method of choice by farmers in Afghanistan to provide water to this crop. The associated costs with pressurized drip irrigation systems are too expensive for farmers to purchase and maintain, which has led to the widespread use of surface irrigation. Moreover, the resistance of growers to change to newer and more advanced technologies is commonplace in many developing countries, and without some type of improvement to current water management practices at the farm level, there is a bleak outlook to maximize agricultural productivity in these areas of the world with limited rainfall and minimal water resources. Although it is essential to sustain this important resource through better irrigation management practices, on-farm agricultural economics are often more important than the needs of future generations and the environment.

Keywords: agricultural extension; Badam Bagh; Afghanistan; food security; sustainable agriculture; water-use efficiency; vegetable production

1. INTRODUCTION

Irrigation is crucial to maximize agricultural productivity in arid or semi-arid areas, and is often the most limiting factor for crop production in countries having these climate types. Over the last 40 years, irrigation has been a major contributor to the growth of food and fiber supply for a global population that has more than doubled [1]. Irrigation becomes even more of an issue in developing countries such as Afghanistan given the relatively small amounts and variable rainfall events in areas that have suitable soils for crop production [2]. Although there are significant water resources available for irrigation of crops and water for livestock, Afghanistan suffers from damaged or poorly maintained irrigation infrastructure and poor overall performance of existing irrigation systems [3]. About two-thirds of the cultivated managed land utilizes some type of irrigation practice and irrigated areas produce almost 85% of all agricultural products in Afghanistan, with most waters delivered by various canal networks [2]. Although a complex array of problems influence water availability in this country, most agree that the primary limiting factor to crop production in Afghanistan is the inefficient use of available water [4,5]. Recently, the overall water efficiency was estimated to be only about 25%–30% for crop production in Afghanistan [6]. Due to the low water-use efficiencies and lack of adequate inputs for maintenance of irrigation systems, yields of horticultural crops tend to be very low [2].

Chili pepper is an important horticultural crop in Afghanistan since this crop performs well in hot, dry environments and suffers less postharvest losses compared with many other fleshy fruited crops. This crop is sold either fresh or as a dried product that is often processed in some manner; for the dried product, fruit are either allowed to dry on the plant or are pulled at a mature stage and sun-dried (Figure 1). Water is critically important during the production season for high value crops, such as chili pepper. Moreover, chili pepper is highly sensitive to limited and excessive water in the root zone, with the period from flowering to fruiting being the most critical [7]. Thus, horticultural crops, such as chili pepper, are especially sensitive to moisture stress and maintaining consistent moisture throughout the growing season is essential to maximize fruit quality and yield [8]. Therefore, in arid/semi-arid climates, irrigation is generally used in chili pepper production schemes to maximize fruit quality and yield. The dominant irrigation methods used in Afghanistan are basin/border irrigation for cereal crops and furrow irrigation for vegetables and grapes [2]. Basin/border irrigation is probably the most widely utilized method to supply water to crops in Afghanistan. This irrigation system supplies water onto the surface of a defined land area surrounded by embankments (bunds) with the basin then flooded; water then eventually percolates over time down into the soil profile. Basin irrigation works best in soils with relatively low infiltration rates. Furrow irrigation in the most common irrigation practice used for chili peppers in Afghanistan, with water requirements typically ranging from 7000 to 12,000 m^3/ha [8]. With this type of irrigation system, water runs in small, parallel channels (furrows) between crop rows that are usually elevated on ridges between the furrows. Although furrow irrigation provides reasonable water use efficiency (WUE), the overall proficiency of this system is influenced by several factors including field slope, length of run, soil type, water infiltration rates, length and rate of application, as well as evaporative, seepage, and drainage losses that can occur at multiple locations in the system. Additionally, waterlogging of soils is commonplace with this type of irrigation.

Figure 1. Mature red chili pepper fruit sun-drying in a village near Mazar-Sharif, Afghanistan.

Agricultural water management is one of the most important factors required to achieve sustainable food production throughout the world, especially in more arid regions [9]. Therefore, utilization of the most appropriate irrigation system is crucial to improve agricultural production and irrigation efficiency in both irrigated and rainfed regions of the world [10]. Drip systems are a widely used irrigation method to improve WUE, especially in arid or semi-arid environments, and are valuable production tools in areas where water is limiting. This irrigation method has several advantages over furrow systems including reduced water use, the ability to apply fertilizers through system, more precise water distribution, and reduced soil-borne diseases and weed growth as row middles remain drier than those with furrow irrigation [11]. However, the increased water efficiency in drip irrigation systems is generally related to reduced soil percolation and surface evaporation compared with other irrigation systems. A more precise application of water through drip irrigation systems will provide more efficient water use for certain agricultural activities, especially fruit and vegetable production in Afghanistan. However, numerous considerations must be taken into account when selecting the most suitable irrigation system, as these will vary depending on location, crop grown, production year utilized, and farmer [10]. Moreover, information regarding research on agricultural water management activities in Afghanistan is limiting. Therefore, a field study was conducted in Afghanistan comparing basin, drip, and furrow irrigation for chili pepper production during the 2009 and 2010 growing seasons.

2. MATERIALS AND METHODS

2.1. Field Experiment Location, Setup, and Maintenance

A field study was conducted during 2009 and 2010 at the Afghanistan Ministry of Agriculture, Irrigation and Livestock (MAIL) Badam Bagh Agricultural Research and Demonstration Farm in Kabul, which has a sandy loam soil. This study was initiated to compare the influence of various irrigation methods (basin, drip, and furrow) on the productivity of chili pepper and resulting disease development. The experiment was set up in a completely randomized design (CRD) with 4 replications. The size of each experimental plot area was 20 m in length and 20 m in width, with rows spaced 1.5 m apart. A local chili pepper landrace was utilized, with seed germinated in early April each year and grown in 128 cell plug trays. Transplants at the two- to three-leaf stage were placed into the field during early May each year with 0.3 m in-row spacings. Plants were harvested by hand at the red-ripe stage five to seven times during each growing season, with harvests beginning in July and ending in early October.

Raised beds approximately 0.3 m in height were used for the drip and furrow irrigation methods, while no raised beds were used for basin irrigation. Irrigation was applied about 10–15 times during the growing season (or once or twice per week), which depended on rainfall and chili pepper water requirements. The total amount of water applied over the chili pepper growing season in both years was 130, 184, and 236 m^3 for drip, furrow, and basin, respectively. During the growing season, minimal amounts of rainfall were provided each year (between 70 and 103 mm).

Fertilization practices for chili pepper included the application of urea (46-0-0), diammonium phosphate (DAP; 18-46-0), and potassium (K; 0-0-60). Prior to chili pepper transplanting, urea, DAP and K were incorporated into the soil during tillage operations at 100 kg, 175 kg, and 125 kg/ha, respectively. Urea was also applied by side-dressing to chili pepper plants by hand at 75 kg/ha at one and two months after transplanting.

2.2. Data Collection

Various data were collected for chili pepper regarding specific responses to the irrigation methods evaluated. For each experimental unit, plant height (cm), number of fruit per plant, and *Phytophthora* blight disease incidence (0% to 100% of plants infected) were all collected at mid-season growth (mid-August). The total yield (ton/ha) was determined from all harvests for each experimental unit. The WUE of irrigation systems for chili pepper yields were calculated by the formula: Y/WR, where Y = yield of crop (kg/ha) and WR = total water consumed for crop production (m^3/ha).

2.3. Data Analysis

Data were subjected to analysis of variance procedures using general linear models procedure of SAS (SAS Inst., Cary, NC, USA) appropriate for a completely randomized experimental design to determine the effects of irrigation type on chili pepper growth, yield, *Phytophthora* blight disease incidence, and WUE. Fisher's least significance difference (LSD) test was used to separate differences among irrigation treatments at $p \leq 0.05$.

3. RESULTS

3.1. Statistical Analyses

Both years were similar as there was no interaction ($p > 0.05$) of year by irrigation method for the variables evaluated. Therefore, data were combined and presented over the two years.

3.2. Chili Pepper Growth and Yield

Irrigation type had a tremendous influence on chili pepper growth and yield (Table 1). Both drip and furrow irrigation provided similar higher plant height responses compared to basin irrigation, which was reduced by approximately 40% compared to the other irrigation methods. The lower amount of plant growth observed by basin irrigation had a definite negative influence on the resulting pepper fruit yields. The average number of fruit obtained per plant at peak production (mid-August) was reduced by about 76% in the basin irrigation method compared to the other two irrigation methods evaluated, which were again similar in their responses. Although drip and furrow irrigation provided similar yields (ton/ha), these irrigation methods provided 55% and 50% more fruit weight/ha, respectively, compared with basin irrigation.

Table 1. Chili pepper plant height, fruit produced per plant, yield, and *Phytophthora* blight disease incidence as influenced by irrigation type at Afghanistan Ministry of Agriculture, Irrigation and Livestock (MAIL) Badam Bagh Agricultural Research and Demonstration Farm in Kabul over the 2009 and 2010 growing seasons.

Irrigation Method	Plant Height (cm)	Number Fruit per Plant	Yield (ton/ha)	*Phytophthora* Disease Incidence (%)
Basin	29 b	9 b	5 b	69 b
Drip	49 a	39 a	11 a	4 a
Furrow	48 a	38 a	10 a	7 a

Data are the means of 4 replications for each growing season. Plant height, fruit number per plant, and *Phytopthora* disease incidence (0% to 100% of plants infected) were collected in mid-August each year. Yield was a summation of all harvests over the growing season. Means followed by the same letter are not significantly different according to Fisher's least significant difference (LSD) test at $p < 0.05$.

3.3. Phytophthora Blight Disease Incidence

Irrigation method played a major role regarding the incidence of *Phytophthora* blight that developed in chili peppers (Table 1). The drip and furrow irrigation methods had a similar low incidence of this disease, as 4% and 7% of plants were affected by *Phytophthora* blight, respectively. However, 69% of chili peppers grown with basin irrigation were afflicted by this disease (Figure 2). This indicates that the over-application of water along with growth on flat areas with no raised beds (which restricted water drainage) provided ideal conditions for the development of this disease.

Figure 2. Significant chili pepper plant stand reduction (as noted by missing plants in each row) in basin irrigation due to Phytophthora blight at the Afghanistan Ministry of Agriculture, Irrigation and Livestock (MAIL) Badam Bagh Agricultural Research and Demonstration Farm in Kabul.

3.4. Water Utilization and WUE

The amount of water utilized for chili pepper production differed ($p \leq 0.05$) among all irrigation methods utilized, with basin irrigation using the most (236 m^3/ha) followed by furrow (184 m^3/ha) and drip (130 m^3/ha) (Table 2). These results indicate that furrow irrigation used about 22% less water than basin and 29% more than drip, while drip irrigation used 45% less water than basin. Although the greatest amount of water was utilized with basin irrigation for chili pepper production, this water delivery method provided the lowest growth and yields, as well as the highest *Phytophthora* blight incidence compared with the other irrigation methods.

Table 2. Water utilization and water use efficiency (WUE) by irrigation method for chili pepper grown at Afghanistan Ministry of Agriculture, Irrigation and Livestock (MAIL) Badam Bagh Agricultural Research and Demonstration Farm in Kabul over the 2009 and 2010 growing seasons.

Irrigation Method	Water Utilization (m^3/ha)	WUE
Basin	236 c	21.2 c
Drip	130 a	84.6 a
Furrow	184 b	54.3 b

Data are the means of 4 replications for each growing season. Means followed by the same letter are not significantly different according to Fisher's LSD test at $p < 0.05$. The WUE of irrigation systems were calculated by the formula: Y/WR, where Y = yield of crop kg/ha and WR = total water consumed for crop production (m^3/ha).

The WUE for chili pepper production differed ($p \leq 0.05$) among the irrigation delivery systems evaluated (Table 2). Drip irrigation resulted in the best overall WUE, as this water delivery method utilized the least amount of water and provided the highest chili pepper yield, which resulted in the highest WUE compared with the other irrigation methods. Furrow irrigation provided a lower WUE compared to drip, but was greater than basin, which had the lowest WUE.

4. DISCUSSION

4.1. Water Management and Utilization in Afghanistan

Although Afghanistan faces a complex array of agricultural problems that are detrimental to its economic development, the lack of sufficient water and inefficient use of this limited resource is the primary limiting factor to crop production. Water management and utilization is an ongoing problem in developing countries with semi-arid/arid climates such as Afghanistan, given the relatively small amounts and variable rainfall events in areas that have suitable soils for crop production [2]. This is further exacerbated by the over-exploitation of groundwater resources which has resulted in shrinking water reserves for not only agricultural activities but also for drinking water supplies for both urban and rural populations. Thus, there is definitely a need for better water management and utilization for agricultural activities in this country.

4.2. Crop WUE in Afghanistan and Associated Problems

The most appropriate strategy to accelerate the growth of the rural Afghanistan economy is to increase crop productivity per unit land area rather than expanding into more environmentally unsuitable areas [5]. The most likely way to achieve this goal is through the improvement of irrigation technologies, as well as better field production and management practices, along with the development of human and institutional capacity [2]. During the last 50–60 years, crop WUE's have increased mostly because of greater potential yields through genetic manipulation and improved efficiency of irrigation methods [12]. Drip irrigation is a technology that is well known to significantly improve WUE through the controlled and timely application of water, which leads to a more effective utilization of resources, including available land, fertilizer, and water [13]. Furthermore, this technology is known to provide higher vegetable yields, improved WUE, and higher product quality compared with other irrigation methods [14,15,16,17,18,19]. Results from this study provide further evidence that drip irrigation in semi-arid/arid climates provide the best overall WUE for horticultural crops, such as chili pepper. However, although drip irrigation may provide the best overall yields and WUE, it may not always be the most practical or provide the greatest economical returns. Significant efforts have been made to introduce drip irrigation to communities across Afghanistan over the past decade, but this effort has largely failed [2]. High material costs, insufficient pressures required to effectively operate these systems, drip line and plastic disposal, and clogging of emitters by water sediments or root intrusion are problems commonly associated with attempts to deploy drip irrigation systems in Afghanistan. Similarly, clogging of emitters is the main concern for vegetable growers in semi-arid regions of Botswana, but this problem can be minimized by using appropriate filters and flushing out the main, sub-main, and lateral lines on a regular basis [20].

Most farmers in Afghanistan generally lack knowledge on how to optimize crop water requirements. Although effective water resource management practices are critical for maintaining food security in Afghanistan, under- and over-irrigation of crops is common. While under-irrigation can lead to vegetable yield and quality loss, over-irrigation can lead to devastating disease problems, such as *Phytophthora capsici* in various vegetable crops [5]. Indeed, this study provided further evidence that too much water delivered through basin irrigation is connected to reduced chili pepper growth and yields, as well as the highest *Phytophthora* blight incidence, compared with the other irrigation methods.

4.3. Drip versus Furrow Irrigation Systems

The small increase observed for chili pepper yield in this study using drip systems compared to furrow irrigation is most likely too small to offset the associated establishment, maintenance, and management costs for drip irrigation systems. Therefore, although the WUE is much greater for drip irrigation, furrow irrigation appears to be the most practical irrigation method for growers in Afghanistan with limited availability of inputs. A common assumption often made is that converting from furrow to drip irrigation will significantly improve crop yield, but this is not necessarily true, and this was the case with this study on chili peppers. Most vegetable growers in Afghanistan are reluctant to adopt drip irrigation for multiple reasons, even though it is the best method to obtain the highest WUE for chili peppers. Moreover, drip irrigation using water from an on-farm storage facility is a much better method compared with trying to improve irrigation scheduling for furrow systems based on soil moisture requirements, as scheduling water flow into a field once the soil is depleted of moisture is not a simple process in Afghanistan. Most farmers cannot schedule a furrow irrigation event when it is most needed for their crops (even if they had the equipment required to detect soil moisture levels), as they receive water only at specific, designated times, and this can be less than once per month during the driest portion of the growing season. Furthermore, there is usually a waste of irrigation water in surface systems during the first half of the growing season, through unregulated floodwater entering canals, and a shortage of water during the second half of the season, when river flow decreases to its annual minimum [2].

Currently, furrow irrigation is the most widely utilized method for commercial chili pepper production in Afghanistan (Figure 3); moreover, since growers understand the management requirements for this water application method, they will most likely continue to use this type of irrigation for years to come despite the fact that it has higher water requirements than drip irrigation. Furrow irrigation systems have low associated production costs compared with drip, so it is really the most economical method for farmers in Afghanistan. Based upon chili pepper yields and WUE's obtained from this study, furrow irrigation is an acceptable water delivery system for this vegetable crop in Afghanistan, although careful management is necessary to ensure sufficient water for maximum crop growth, while preventing excessive evaporative, seepage, or drainage losses.

Figure 3. A typical commercial chili pepper field utilizing furrow irrigation near Mazar-Sharif, Afghanistan.

Furrow irrigation is oftentimes not as water-efficient as it is portrayed due to the loss of water that occurs in canals prior to reaching the field. Most times, this is not considered when determining the WUE of this irrigation system, whereas water used for drip irrigation is generally pumped from wells into small reservoirs, making this irrigation method much more efficient compared with the unseen wastes that often occur in transporting furrow-irrigation water to a field. In Afghanistan, rural watercourses and canals (which are used to get water to furrow-irrigated fields) are generally unimproved earthen trenches with water losses occurring via seepage, over-bank spillage, and evaporation [2]. Furthermore, in both basin and furrow irrigation systems in Afghanistan, irrigation waters are often improperly drained from fields, which can lead to reduced crop stands and yields. Inadequate drainage impacts agricultural production by causing soil compaction, soil erosion, salt accumulation, and even crop failure. Due to low water use efficiencies and a lack of adequate inputs for the maintenance of canals and all types of irrigation systems, yields of horticultural crops tend to be very low in Afghanistan [2]. However, WUE and crop yields of furrow irrigation systems can be improved with laser-leveling of production fields, as this facilitates surface-water runoff and eliminates pooling for several years following treatment; for example, chili pepper yields under furrow irrigation in Afghanistan increased by almost 15% in laser-leveled fields compared with those that were not laser-leveled [21].

5. CONCLUSIONS

Water is a crucial resource for food production and ultimately, human survival throughout the world. An inconsistent and inadequate supply of water to farmers is the most important limiting factor that prevents accelerated agricultural expansion in Afghanistan [5]. Moreover, since the majority of the populace resides in rural areas that depend on irrigated agriculture, improving access to this vital resource is essential to improving the quality of life in Afghanistan. Thus, effective water resource management practices are essential for maintaining food security in developing countries like Afghanistan, which is further exacerbated by the arid or semi-arid climate [2]. Agriculture must meet the future food security challenges by increasing production while conserving this important natural resource [1]. However, fresh water is an ever dwindling resource that is oftentimes mismanaged, especially in areas that have limited supplies and low rainfall. Efforts to improve water use management and efficiencies for crop growth must be a high priority for farmers in arid/semi-arid countries. Although current water extraction rates for irrigation is not sustainable in these areas (even without considering future climate warming and precipitation changes), irrigation demand is expected to increase in the future [22]. In this study, drip irrigation resulted in the best overall WUE ($p \leq 0.05$), as this water delivery method utilized the least amount of water and provided the highest chili pepper yield. However, furrow irrigation is still the method of choice for farmers in Afghanistan and many other developing countries with arid/semi-arid climates to provide water to chili pepper and other crops for several reasons. Most importantly, the costs associated with pressurized drip irrigation systems are too expensive to be widely used, which has led to the widespread use of surface irrigation [10]. Moreover, the resistance of growers to change to newer and more advanced technologies is commonplace in many developing countries [20]. Without some type of improvement to current water management practices at the farm level, the outlook of maximizing agricultural productivity in these areas of the world with limited rainfall and minimal water resources is bleak. However, drip irrigation is a relatively simple, low-input technology that can substantially expand the ability of smallholder farmers in developing countries to plant high value horticultural and agronomic crops even with fluctuating water supplies in a changing climate [13]. Growers of horticultural crops in arid/semi-arid climates really need to focus on utilizing more efficient water saving and irrigation technologies to sustain water resources for future generations. Although it is essential to sustain this important resource through better irrigation management practices, on-farm economics are oftentimes more important than the needs of future generations and the environment.

ACKNOWLEDGMENTS

The authors would like to thank the United States Agency for International Development (USAID) for providing financial assistance for the Afghanistan Water, Agriculture and Technology Transfer (AWATT) project. This project provided the funding for the planting, maintenance, harvesting, and data collection required for this research and associated extension activities.

REFERENCES

1. De Wrachien, D.; Goli, M.B. Global warming effects on irrigation development and crop production: A world-wide view. *Agric. Sci.* **2015**, *6*, 734–747.
2. Walters, S.A.; Groninger, J.W. Water distribution systems and on-farm irrigation practices: Limitations and consequences for Afghanistan's agricultural productivity. *Water Int.* **2014**, *39*, 348–359.
3. Torell, G.L.; Ward, F.A. Improved water institutions for food security and rural livelihoods in Afghanistan's Balkh river basin. *Int. J. Water Res. Dev.* **2010**, *26*, 613–637.
4. Thomas, V.; Ramzi, A.M. SRI contributions to rice production dealing with water management constraints in northeastern Afghanistan. *Paddy Water Environ.* **2011**, *9*, 101–109.
5. Walters, S.A.; Groninger, J.W.; Myers, O. Rebuilding Afghanistan's agricultural economy: Vegetable production in Balkh province. *Outlook Agric.* **2012**, *41*, 7–13.
6. Qureshi, A.S. *Water Resources Management in Afghanistan: The Issues and Options*; Working Paper 49, Pakistan Country Series 49; International Water Management Institute (IWMI): Columbo, Sri Lanka, 2002; 25p. Available online: https://cropwatch.unl.edu/documents/Water%20Resource%20Issues%20In%20Afghanistan.pdf (accessed on 20 August 2016).

7. Dagdelen, N.; Yilmaz, E.; Sezgin, F.; Gürbüz, T. Effects of water stress at different growth stages on processing pepper (*Capsicum annum* cv. *Kapija*) yield, water use and quality characteristics. *Pak. J. Biol. Sci.* **2004**, *7*, 2167–2172.
8. Walters, A.; Hughes, C. *Chili Pepper Information Sheet for Afghanistan*; Univ. California Regents: Davis, CA, USA, 2004; Available online: http://afghanag.ucdavis.edu/a_horticulture/row-crops/peppers/IS_Chili_Peppers.pdf (accessed on 20 August 2016).
9. Valipour, M. Evolution of irrigation-equipped areas as share of cultivated areas. *Irrig. Drain. Syst. Eng.* **2013**.
10. Valipour, M. Necessity of irrigated and rainfed agriculture in the world. *Irrig. Drain. Syst. Eng.* **2013**.
11. Locascio, S.J. Management of irrigation for vegetables: Past, present and future. *HortTechnology* **2005**, *15*, 482–485.
12. Passioura, J. Increasing crop productivity when water is scarce: From breeding to field management. *Agric. Water Manag.* **2006**, *80*, 176–186.
13. Jha, A.K.; Malla, R.; Sharma, M.; Panthi, J.; Lakhankar, T.; Krakauer, N.Y.; Pradhanang, S.M.; Dahal, P.; Shrestha, M.L. Impact of irrigation method on water use efficiency and productivity of fodder crops in Nepal. *Climate* **2016**.
14. Aujla, M.S.; Thind, H.S.; Buttar, G.S. Fruit yield and water use efficiency of eggplant (*Solanum melongea* L.) as influenced by different quantities of nitrogen and water applied through drip and furrow irrigation. *Sci. Hortic.* **2006**, *112*, 142–148.
15. Sezen, S.M.; Yazar, A.; Eker, J.L. Effect of drip irrigation regimes on yield and quality of field grown bell pepper. *Agric. Water Manag.* **2006**, *81*, 115–131.
16. Srinivas, K.; Hegde, D.M.; Havanagi, G.V. Irrigation studies of watermelon (*Citrullus lanatus* (Thunb) Matsum et Nakai). *Irrig. Sci.* **1989**, *10*, 293–301.
17. Tiwari, K.N.; Mal, P.K.; Singh, R.M.; Chattopadhyay, A. Response of okra (*Abelmoschus esculentus* L.) to drip irrigation under mulch and non-mulch conditions. *Agric. Water Manag.* **1998**, *38*, 91–102.
18. Tiwari, K.N.; Singh, S.; Mal, P.K. Effect of drip irrigation on yield of cabbage (*Brassica oleracea* L. *var. capitata*) under mulch and non-mulch conditions. *Agric. Water Manag.* **2003**, *58*, 19–28.
19. Yuan, B.Z.; Sun, J.; Kang, Y.; Nishiyama, S. Response of cucumber to drip irrigation water under a rainshelter. *Agric. Water Manag.* **2006**, *81*, 145–158.
20. Imtiyaz, M.; Mgadla, N.P.; Manase, S.K.; Chendo, K.; Mothobi, E.O. Yield and economic return of vegetable crops under variable irrigation. *Irrig. Sci.* **2000**, *19*, 87–93.
21. USAID-AWATT. Component 1—Integrated Water Management: Laser land leveling. In *USAID Afghanistan Water, Agriculture, and Technology Transfer (AWATT) Program Quarterly Report (April-May-June)*; USAID: Washington, DC, USA, 2010; p. 7.
22. Puma, M.J.; Cook, B.I. Effects of irrigation on global climate change during the 20th century. *J. Geophysical. Res.* **2010**.

CHAPTER 17

Combining Limited Multiple Environment Trials Data with Crop Modeling to Identify Widely Adaptable Rice Varieties

Tao Li[1], Jauhar Ali[2], Manuel Marcaida, III[1], Olivyn Angeles[1], Neil Johann Franje[2], Jastin Edrian Revilleza[2], Emmali Manalo[1], Edilberto Redoña[2,3], Jianlong Xu[4], Zhikang Li[4]

ABSTRACT

Multi-Environment Trials (MET) are conventionally used to evaluate varietal performance prior to national yield trials, but the accuracy of MET is constrained by the number of test environments. A modeling approach was innovated to evaluate varietal performance in a large number of environments using the rice model ORYZA (v3). Modeled yields representing genotype by environment interactions were used to classify the target population of environments (TPE) and analyze varietal yield and yield stability. Eight Green Super Rice (GSR) and three check varieties were evaluated across 3796 environments and 14 seasons in Southern Asia. Based on drought stress imposed on rainfed rice, environments were classified into nine TPEs. Relative to the check varieties, all GSR varieties performed well except GSR-IR1-5-S14-S2-Y2, with GSR-IR1-1-Y4-Y1, and GSR-IR1-8-S6-S3-Y2 consistently performing better in all TPEs. Varietal evaluation using ORYZA (v3) significantly corresponded to the evaluation based on actual MET data within specific sites, but not with considerably larger environments. ORYZA-based evaluation demonstrated the advantage of GSR varieties in diverse environments. This study substantiated that the modeling approach could be an effective, reliable, and advanced approach to complement MET in the assessment of varietal performance on spatial and temporal scales whenever quality soil and weather information are accessible. With available local weather and soil information, this approach can also be adopted to other rice producing domains or other crops using appropriate crop models.

1. INTRODUCTION

As the staple food for half of the world population, rice is widely grown in diverse environments worldwide. Rice productivity has doubled since the Green Revolution with more than 50% of the increase attributed to continuous rice breeding efforts primarily in the irrigated system [1]. By practice, any breeding program is aimed at a specific target population of environments (TPE). It is often difficult to define the TPE for breeding efforts inclusive of rainfed areas, because rice growing environments of rainfed ecosystems vary considerably across locations and years. As a predominantly self-pollinated species, yield performances of rice varieties tend to show a significant level of genotype x environment interaction (GEI), particularly in rainfed environments. In any breeding program, it is routine to generate hundreds, or even thousands, of advanced progenies. To identify promising lines for specific TPEs from the huge numbers of advanced progenies, it requires evaluation of yield performances of large number of lines in multi-environment trials (MET), which is normally a time-consuming and very expensive process. An efficient tool that can resolve this limitation could extremely accelerate the process of varietal development.

Green Super Rice (GSR), defined as rice varieties that can produce high and stable yields using less resource input, was conceptualized and developed as the ideal rice varieties for rainfed environments. Considerable progress has been made in developing large number of GSR lines with high yield potential and tolerances/resistances to multiple abiotic and biotic stresses [2, 3, 4, 5, 6]. However, to identify promising GSR varieties suitable for specific tropical rainfed areas in Asian and African countries, there is a need to test and evaluate large number of promising GSR lines across multiple environments and to identify their TPEs. This has been a major challenge in GSR breeding efforts.

To capture GEI using a series of mathematical functions in crop models has been proven as an effective tool for evaluating yield performances of varieties in numerous environments based on key plant eco-physiological principles [7, 8]. For rice, the ORYZA2000 model has been used to predict rice growth and grain yields with high confidence level similar to other rice models in the world [9]. It has also been successfully used for performance evaluation of breeding lines or varieties over large numbers of environments [7] and environment characterization [10].

This study aims to develop and test an efficient strategy for quantifying yield performances and stabilities of GSR varieties with tolerances to multiple stresses by determining their TPEs over temporal and spatial scales using ORYZA version 3.0 (https://sites.google.com/a/irri.org/oryza2000/home).

2. MATERIALS AND METHODS

The modeling approach developed in this study intends to evaluate the performance of rice breeding materials across different environments. The evaluation process has four steps: 1) data collection from field experiments to determine varietal genetic coefficients to feed the crop model and from secondary data source for regional soil and weather information; 2) model calibration and validation to develop varietal genetic coefficients and evaluate the confidence level of model prediction on crop yield using experimental data; 3) conduct simulations on each rice grid cell in a given region using available soil and weather information; and 4) analysis of the simulation results to define the TPEs and evaluate varietal performance among different TPEs.

2.1 Field experiments

The materials used in this study included eight BC_1F_6 GSR introgression lines with improved yields and tolerances to drought, salinity and submergence plus three check varieties (Table 1).

Field experiments were conducted to test the performance of the eight GSR varieties (Table 1) during three dry seasons (2011, 2012, and 2013) and one wet season (2013) in Los Baños, Philippines (LB, 121°15'E and 14°11'N).

The same set of varieties was also tested in Nueva Ecija, Philippines (NE, 120°56'E and 15°42'N) during the 2013 dry and wet seasons (Table 2). The experiments conducted in 2011 and 2012 had two types of water management: 1) continuously flooded throughout the season; and 2) switching from continuously flooded to rainfed condition after panicle initiation.

Table 1. Parentage, breeding methodology and duration of 8 green super rice (GSR) lines and check varieties used in this study.

Variety	Parentage (Cross information)	Breeding method and generation	Maturity(days after sowing) under CF*	Description
Feng Fu Zhan (FFZ)	Feng-Si-Zhan/Fu-Qing-Zhan 4	Pedigree breeding	115	CAAS-bred GSR variety for irrigated conditions
GSR IR1-1-Y4-Y1	Huang-Hua-Zhan/Yue-Xiang-Zhan// Huang-Hua-Zhan	Backcross introgression line BC1F6	120	High-yielding, irrigated cultivar with tolerance of salinity, submergence and drought
GSR IR1-5-S8-D3-SUB1	Huang-Hua-Zhan/OM1723//Huang-Hua-Zhan	Backcross introgression line BC1F6	120	High-yielding, irrigated cultivar with tolerance of salinity, submergence and drought
GSR IR1-5-S10-D1-D1	Huang-Hua-Zhan/OM1723//Huang-Hua-Zhan	Backcross introgression line BC1F6	124	High-yielding, irrigated cultivar with tolerance of salinity and drought
GSR IR1-5-S14-SI2-Y2	Huang-Hua-Zhan/OM1723//Huang-Hua-Zhan	Backcross introgression line BC1F6	115	High-yielding, irrigated cultivar with tolerance of salinity and drought
GSR IR1-8-S6-S3-Y2	Huang-Hua-Zhan/Phalguna//Huang-Hua-Zhan	Backcross introgression line BC1F6	110	High-yielding, irrigated cultivar with tolerance of salinity, submergence and drought
GSR IR1-8-S12-Y2-D1	Huang-Hua-Zhan/Phalguna//Huang-Hua-Zhan	Backcross introgression line BC1F6	121	High-yielding, irrigated cultivar with tolerance of salinity and drought
GSR IR1-12-D10-S1-D1	Huang-Hua-Zhan/Teqing//Huang-Hua-Zhan	Backcross introgression line BC1F6	110	Aromatic, high-yielding, irrigated cultivar with tolerance of salinity and drought
IR74371-70-1-1	IR 55419–4*2/WAY RAREM	Pedigree breeding	110	Drought tolerant [11]
NSICRc158	IR 73885-1-4-3-2-1-6 (MATATAG 9)/ IR70479-45-2-3//IR64680-81-2-2-1-3	Pedigree breeding	120	High Yielding under Irrigated variety [12]
PSB Rc82	IR 47761-27-1-3-6/PSB RC 28 (IR56381-139-2-2)	Pedigree breeding	119	High Yielding under Irrigated variety [13]

* CF = continuous flooding or under irrigated conditions

All experiments in 2013 were under fully irrigated conditions with standard practices of nitrogen fertilizer application and other crop management.

Phenology development was recorded and destructive samples were collected from all field experiments to determine the leaf area index (LAI) and biomass weights for green leaves, dead leaves, stems, and panicles. Grain yields were determined in two 3 m^2 areas at physiological maturity. Soil water potential at soil depth of 15 cm was measured in rainfed fields using MPS 2 sensors (Decagon Devices Inc., Pullman WA) connected to a Decagon EM50 data logger.

The measured data under different crop management strategies and climate-soil conditions were used to parameterize the genetic parameters of the tested GSR lines and checks to evaluate the reliability of ORYZA (v3) in representing the growth, development, and yields of the tested rice varieties across different environments.

2.2 Modeling study

General description of ORYZA version 3.0. Evaluations conducted under various environments and crop management practices have established the reliability of ORYZA2000 in predicting rice growth, development, and yield [14]. This model has been successfully used in evaluating the performance of breeding lines in various environments [7] under different water and nitrogen management systems [5, 10, 15, 16, 17]. The ORYZA (v3) is an improved version of ORYZA2000 with additional functions to quantify the effects of drought, nitrogen deficiency, and irrigation management [18] (https://sites.google.com/a/irri.org/oryza2000/home). Unlike ORYZA2000 [19], most of the variety parameters used by ORYZA (v3) are genetic parameters, and the effects of environment on photosynthesis, assimilate partitioning, growth of crop organs, leaf area development, and water and nitrogen uptake have been integrated into the model with extra parameters representing responses of specific varieties.

Model calibration for varietal parameterization. The observed data from experiments marked by "C" (Table 2) was used for variety parameterization in the process of model calibration. The experiment-specific crop management practice and weather information were used correspondingly in each experiment. Under rainfed conditions, soil water potential was used as input for variety parameterization.

Table 2. Experiments and associated varieties implemented in Los Baños (LB) and Nueva Ecija (NE), Philippines, during the dry seasons of 2011 (2011-DS), 2012 (2012-DS), 2013 (2013-DS) and wet season of 2013 (2013-WS). RF indicates switching from continuously flooded to rainfed condition after panicle initiation while CF indicates continuously flooded throughout the season. The datasets were marked for calibration (C) and evaluation (E).

Variety	Experiments					
	2011-DS-RF	2011-DS-CF	2012-DS-RF	2012-DS-CF	2013-DS-CF	2013-WS-CF
FFZ	C	C	E	E		
GSR IR1-1-Y4-Y1			C	E		
GSR IR1-5-S8-D3-SUB1			C	E		
GSR IR1-5-S10-D1-D1	E	C	C	E		
GSR IR1-5-S14-SI2-Y2			C	E		
GSR IR1-8-S6-S3-Y2			C	E		
GSR IR1-8-S12-Y2-D1	E	C	C	E		
GSR IR1-12-D10-S1-D1	E	C	C	E		
IR-74371-70-1-1	E	C	C	E		
NSICRc158	C	C			C(LB), E(NE)	E(LB, NE)
PSBRc82	E	C	C	E	C(LB), E(NE)	E(LB, NE)

The parameters controlling and affecting leaf development, light interception, assimilate partitioning, leaf death, drought tolerance, rooting depth, and nitrogen uptake were calibrated using an auto-calibration tool to minimize the differences between the measured and simulated values on LAI, total above-ground biomass (AGB), and biomass of green leaves, dead leaves, stems, and panicles.

The auto-calibration program was developed based on differential evolution algorithms of global optimization [20] and applied on the earlier studies [21, 22] for parameterization of ORYZA2000. The documentation of varietal parameterization using the auto-calibration program is available online (https://sites.google.com/a/irri.org/oryza2000/downloads/ new-release/download-new-version). The varietal parameters used in this study were fixed for ensuring that the differences between simulated and measured values are similar to the coefficients of variation (CV) of field measurements [23, 24], the desired target differences were set to 5% of the measured values for storage organ biomass, 10% for green leaf and total AGB, and 15% for LAI and dead leaf biomass. The initial values of varietal parameters were from a popular variety, IR72, and their variation ranges were assumed to be 25 to 50% to ensure that the calibrated parameter values are physiologically reasonable. The values of key varietal parameters for each GSR variety developed from the calibration process were presented in S1 Table (Hereafter, 'S' after table/figure numbers indicates supporting information).

Validation of model prediction. Model predictions on rice growth and yield must be validated using independent experiments before the model can be used for further simulations over different environments. To ensure the predictability of ORYZA (v3) for the growth and development of different varieties under different environments, the model was initially evaluated against a standard set of measurement data from actual field experiments as indicated by 'E' in Table 2.

The experiment-specific weather and soil data, as well as crop management information, were fed to the model for validation. Measurements on the actual and simulated sequential AGB, panicle biomass (PB), and commercial grain yield (GY) from the experiments for all GSR and check varieties were integrated into X and Y data pairs for each sampling date. With the data pairs, the statistical analyses were conducted to quantify the differences between measured actual (X) and simulated (Y) values for AGB, PB, and GY. The calculated statistical parameters were the linear regression parameters for slope (α), intercept (β), and correlation coefficient (R^2), and the Student's t-test with unequal means assumption ($P(t)$). In addition, the normalized root mean square errors ($RMSE_n$) and index of modeling agreement (M_{eff}) [25] were calculated using Eqs 1 and 2, where n is the number of measurements, *i* is the data pair index, and x̄ is the average of all observations for a variable.

$$RMSE_n = \frac{\left(n \sum_{i=1}^{n} (Y_i - X_i)^2\right)^{0.5}}{\sum_{i=1}^{n} X_i} \times 100 \quad (1)$$

$$M_{eff} = 1 - \frac{\sum_{i=1}^{n} (Y_i - X_i)^2}{\sum_{i=1}^{n} (|Y_i - \overline{X}| + |X_i - \overline{X}|)^2} \quad (2)$$

Model reproduces experimental data best when α, R^2 and M_{eff} are close to 1, β is close to 0, *P(t)* is larger than 0.05, and $RMSE_n$ is similar to the *CV* of measured values.

Simulations with numerous environments. GEI needs to be quantified using thousands of simulations in various environments in order to achieve the study targets. Consequently, the simulations were designed to have the same crop management across environments for all varieties to ensure that the variation in rice growth and development will come from the GEI. In this study, crop management was excluded as a factor of environment because it is an integration of human activities and could be modified easily.

All GSR varieties used in this study were bred for drought-prone environments, thus, the simulations were designed for rainfed conditions. The current rice cultivation regions in Southern Asia were gridded at 15 arc-minute geo-resolution, with each grid cell assigned as one environment, thereby producing 3796 environments. The corresponding historical weather (across 15 years) and soil information of each grid cell was extracted from IRRI database and from the World Inventory of Soil Emission Potential (WISE), respectively [26], to determine the temporal and spatial variations of the target environments. Temporally, the 15 years of weather information from 1998 to 2012 explained the dynamics and variation of the environment season by season. Spatially, the environments of adjacent grid cells may be similar to or significantly different from each other due to varying weather and soil information.

For each GSR variety and each environment, two groups of simulations were designed to differentiate the performance of the crop under the two types of water management practices—continuously flooded (group 1) and rainfed (group 2) in Table 2. For both groups, fertilizer was fully supplied, and rice was assumed to grow under conditions without any biotic-stress. Therefore, radiation and temperature were the only limiting factors for rice growth in group 1, while soil moisture was an additional limiting factor in group 2. In each group, 24 sowing dates per year were included, starting from 1 January at 15-day intervals, over 14 years (from 1998 to 2011). The rice crop sown in the later seasons of 2011 grew through early 2012 and required the weather information for 2012 for simulation. In total, 672 simulations (24 sowing dates y^{-1} × 14 years × 2 groups) were implemented for each variety in each environment. Other crop management practices, except for water management, were completely the same for both groups. The field bund height was set to 10 cm and the seedlings were transplanted at 14 days after sowing with a density of 66 seedling m^{-2}.

The model predicted commercial GY (grain biomass with 14% moisture). The average drought index (varying from 0 to 1 for severe to no drought stresses) in the vegetative and reproductive stages was used for varietal performance evaluation over a large number of environments and for the identification of TPEs.

2.3 Analysis of data

Simulation data cleaning. Three variables were extracted from the outputs of simulations for each variety in each environment and each group, namely GY, average daily drought stress index in the vegetative stage (*DIV*) and reproductive stage (*DIR*). The outputs of each variable in groups 1 and 2 (i.e. irrigated and rainfed water management, respectively) were organized into matrices *P* and *A*, respectively (Eqs 3 and 4), where *p* and *a* stand for the simulated values of a variable among *GY*, *DIV*, and *DIR* from groups 1 and 2. The subscripts *i* and *j* represent the indices of sowing dates in a year and years of simulation, respectively. Subscript *i* changes from 1 (first sowing date on 1 January) to *n* (last sowing date), while *j* varied from 1 (start year 1989) to *m* (last year of simulation). In this study, the *n* and *m* were 24 and 14, respectively.

$$P = |p_{ij}| \ (i = 1, 2, \cdots, n; j = 1, 2, \cdots, m) \tag{3}$$

$$A = |a_{ij}| \ (i = 1, 2, \cdots, n; j = 1, 2, \cdots, m) \tag{4}$$

The matrix *P* was only organized for *GY*, assigned as *PY*, but *A* was organized for variables *GY*, *DIV*, and *DIR*, indicated as *AY*, *AV*, and *AR*, respectively. In *PY*, any zero value of p_{ij}implied that the ith sowing date was not suitable for rice growth. Therefore, the corresponding $p_{i.}$ and a_i in all four matrices were removed and excluded in the succeeding analyses.

Classification of TPE. To define the characteristics of each environment, the matrices *PY*, *AY*, *AV*, and *AR* of all varieties were combined into four big matrices of *PY′*, *AY′*, *AV′*, and *AR′*, respectively. The TPE for rainfed rice was classified based on information on the best rainfed season in each environment.

To identify the best rainfed season, the average rainfed yield (*RY*) and *CV* over the years and all varieties under rainfed condition were calculated for each sowing date from AY' (Eqs 5 and 6), where *c* is the number of varieties in this study, subscript *v* denotes varieties, and *i* stands for sowing seasons. The 24 sowing dates were tested to derive *RY* and a maximum of 24 values were derived for *CV*. The best rainfed season was selected from the top three *RY* for the smallest *CV*.

$$RY_i = \frac{\sum_{v=1}^{c} \sum_{j=1}^{m} a_{ijv}}{m \times c} \quad (5)$$

$$CV_i = \frac{\sqrt{\frac{1}{m \times c} \sum_{v=1}^{c} \sum_{j=1}^{m} (a_{ijv} - a_i')^2}}{AY_i} \quad (6)$$

For each environment, the data of the best rainfed season was extracted from PY', AY', AV', and AR' to data arrays py_{jv}, ay_{jv}, av_{jv}, and ar_{jv} for irrigated and rainfed grain yields, and the drought indices at the vegetative and reproductive stages, in which *j* and *v* are the year and variety indices among all tested varieties and years. These four arrays were used to determine the TPE class of a given environment.

For each environment, the TPE class for rainfed rice was defined by the severity and type of drought stress it was subjected to. The severity of drought stress was presented by the frequency (f_{75}) at which ay_{jv}/py_{jv} is lower than 0.75. Here, we assumed that the yield reduction by 25% caused by drought is acceptable for a single season, but a higher frequency of such reduction among seasons was considered a significant drought impact. Severity of drought stress was classified into three groups: 1) severe drought stress for $f_{75} \geq 50\%$; 2) moderate drought stress for $25\% \leq f_{75} < 50\%$; and 3) none to mild drought stress for $f_{75} < 25\%$ (Table 3).

The impact of drought stress is highly related to the timing of its occurrence in the different rice growth stages. Drought stress is relatively more harmful when it occurs in the reproductive rather than the vegetative stage [27]. In this study, drought stress was classified into three types according to the growth stage at which it occurred: 1) vegetative; 2) reproductive; and 3) mixed, when drought stress occurred in both vegetative and reproductive growth stages.

To determine the drought type of an environment, $\overline{av}$ and $\overline{ar}$ were derived from av_{jv} and ar_{jv} as average values of all their elements. Drought was vegetative type if $\overline{av}$ is smaller than $\overline{ar}$, or reproductive type if $\overline{ar}$ is smaller than 90% of $\overline{av}$, otherwise, the drought was mixed type (Table 3). Finally, all environments were classified into 9 TPE groups based on the GEI generated by simulation results of ORYZA (v3).

Table 3. TPE classification of environment in the study area based on the drought severity and drought timing. The f75 is simply the frequency of the rainfed yields occurring less than 75% of irrigated yield.

TPE	Drought Timing	Drought Severity
L1	Vegetative	None or Mild Drought (f_{75}<25%)
L2	Reproductive	
L3	Vegetative + Reproductive	
M1	Vegetative	Moderate Drought (25%<f_{75}<50%)
M2	Reproductive	
M3	Vegetative + Reproductive	
S1	Vegetative	Severe Drought (f_{75}>50%)
S2	Reproductive	
S3	Vegetative + Reproductive	

Yield stability and adaptability of variety. Yield stability of a given variety across environments was evaluated with simulated grain yields under irrigated and rainfed conditions among the best rainfed rice seasons. For a variety, the irrigated and rainfed yields in the best rainfed season were extracted from each environment to form two new data matrices YP and YA (Eqs 7 and 8), where *j* stands for the simulation years changing from 1 to *m* (Eqs 1 and 2), *k* stands for the number of environments, and *q* is the total number of environments.

$$YP = |yp_{jk}| \ (j = 1, 2, \cdots, m; k = 1, 2, \cdots, q) \tag{7}$$

$$YA = |ya_{jk}| \ (j = 1, 2, \cdots, m; k = 1, 2, \cdots, q) \tag{8}$$

Five indicators, γ, δ, λ, φ and ψ, were derived from the data of YP and YA to quantify the yield stabilities of a variety in a class of TPE or all tested environments where parameter γ is the average rainfed yield derived from *YA*; δ presents how far the rainfed rice yield is from the irrigated rice yield (Eq 9); λ is the coefficient of variation of rainfed yield in *YA* (Eq 10), representing the yield variation cross over both spatial and temporal scales; φ is the spatial variability of yields among environments; and ψ is the temporal variability of yield among different growth seasons (Eqs 11 and 12). The parameters λ, φ, and ψ, represent the yield stability over all environments, or on a given TPE, depending on the yield dataset used in the analysis. Variety has higher yield stability when the values of these parameters are lower.

$$\delta = \frac{\sqrt{m \times q \times (\sum_{j=1}^{m} \sum_{k=1}^{q} (yp_{jk} - ya_{jk})^2)}}{\sum_{j=1}^{m} \sum_{k=1}^{q} yp_{jk}} \tag{9}$$

$$\lambda = \frac{\sqrt{\frac{1}{m \times q} \sum_{j=1}^{m} \sum_{k=1}^{q} (ay_{jk} - \gamma)^2}}{\gamma} \tag{10}$$

$$\varphi = \frac{\sqrt{\frac{1}{q} \sum_{k=1}^{q} \left(\left(\frac{1}{m} \sum_{j=1}^{m} ay_{jk} \right) - \gamma \right)^2}}{\gamma} \tag{11}$$

$$\psi = \frac{\sqrt{\frac{1}{m} \sum_{j=1}^{m} \left(\left(\frac{1}{q} \sum_{k=1}^{q} ay_{jk} \right) - \gamma \right)^2}}{\gamma} \tag{12}$$

The analyses above were implemented for all environments and 9 TPE classes. To derive the five indicators for all environments as an entity, the data in PY and AY were used. For each TPE class in Table 3, the five indicators were derived from the YP and YA, which were categorized into 9 groups based on TPE classes (Table 4).

A better variety should have a large value of γ to indicate high productivity in drought-prone environments, and a small value of δ to indicate strong tolerance to drought stress or low risk in drought-prone environments. A small value of λ is attributed to good yield stability in spatial and temporal dimensions, or even a good spatial or temporal yield stability with the small value of φ or ψ.

Moreover, the values of λ, φ and ψ derived from the whole matrices of PY and AY also represented the properties for varietal adaptability among TPEs which may vary from severe to no drought stress conditions. The good adaptability for many different kinds of TPEs does not necessarily mean a good productivity and lower risk to drought stress. For a rice breeder, a variety with low values of δ, λ, φ and ψ, and a high value of γ over different kinds of environments is a good set of selection criteria. The selection would be very difficult for all TPE classes but is achievable for a few TPE classes. For a local farmer, a variety with lower values of λ and ψ and a higher value of γ is desirable, while a low value of δ implies easy crop management in rainfed rice production systems.

Table 4. The datasets used to determine TPE Classes and yield stability in the TPE. YP and YA are the irrigated and rainfed rice grain yields in the best rainfed season. The λ is the coefficient of variation of rainfed yield in YA, φ is the spatial variability of yields among environments, and the ψ is the temporal variability of yield among different growth seasons.

TPE Classes	Yield datasets to define TPE	Yield stability (= 1.0—value of the parameter)
L1	YP_{l1} and YA_{l1}	λ, φ and ψ for domain of TPE: L1
L2	YP_{l2} and YA_{l2}	λ, φ and ψ for domain of TPE: L2
L3	YP_{l2} and YA_{l}	λ, φ and ψ for domain of TPE: L3
M1	YP_{m1} and YA_{m1}	λ, φ and ψ for domain of TPE: M1
M2	YP_{m2} and YA_{m2}	λ, φ and ψ for domain of TPE: M2
M3	YP_{m3} and YA_{m3}	λ, φ and ψ for domain of TPE: M3
S1	YP_{s1} and YA_{s1}	λ, φ and ψ for domain of TPE: S1
S2	YP_{s2} and YA_{s2}	λ, φ and ψ for domain of TPE: S2
S3	YP_{s3} and YA_{s3}	λ, φ and ψ for domain of TPE: S3

The cluster analysis. Hierarchical clustering analysis was conducted to directly show the difference between the GSR and the check varieties in terms of plant physiological parameters using the ORYZA (v3) model and varietal adaptability. The varietal performance in a given TPE or across all TPEs was clustered into five groups based on the five parameters γ, δ, λ, φ and ψ to rank the performances of the varieties from 1 (best) to 5 (worst).

3. RESULTS

3.1 The reliability of ORYZA (v3) for production prediction

Following the calibration and validation processes, each variety used in this study was parameterized and evaluated in independent experiments (Table 2). In comparing the simulated to measured values, ORYZA (v3) performed well in representing the three key plant growth variables–AGB, PB, and GY–in both calibration and validation datasets for all varieties (Table 5). The estimations on AGB, PB, and GY were reliable for individual varieties because all statistical indicators were close to the desirable values (Table 5), despite different values among varieties (S2 and S3 Tables). The graphical analysis shows that more than 90% of the simulated plant biomass and GY values were within the range of confidence defined by CV of measurements (S1 Fig).

Table 5. Statistical analysis for the calibration and validation datasets of all tested varieties in this study. AGB is above-ground biomass, PB is panicle biomass, and GY is grain yield.

Variable	Data pairs	R^2	α	β	$p(t)$	$RMSEn$ (%)	M_{eff}
	Calibration dataset						
AGB	102	0.969	284.959	0.919	0.470	8.592	0.998
PB	50	0.973	120.395	0.955	0.492	4.566	0.999
GY	26	0.898	567.020	0.849	0.493	8.987	0.993
	Validation dataset						
AGB	116	0.955	191.209	0.929	0.425	7.851	0.998
PB	55	0.971	-93.475	0.971	0.357	6.636	0.999
GY	27	0.885	314.758	0.871	0.305	9.282	0.990

Particularly, the similar accuracies in estimating biomass accumulation and grain yields in rainfed conditions as those in full irrigated conditions implied that drought stress in rainfed was a reliable estimate.

In summary, ORYZA (v3) was able to represent the dynamics of AGB and PB during the growing season, and also well predicted the end-season GY for all tested varieties and environments. The simulated values on yield, biomass accumulation, and drought stress were reliable inputs for the classification of TPE and evaluation of varietal performance in a large number of different environments.

3.2 Classification of TPE based on GEI

Using the calibrated and validated crop parameters, ORYZA (v3) predicted the yields of all varieties based on GEI (Table 1) in 3796 environments (i.e. soil and climate combinations) in Southern Asia. Using the predicted yields under fully irrigated and rainfed conditions, these environments were classified into 9 TPE classes depending on drought severities and time of drought occurrence relative to the crop growth stage in the best local rainfed season (Fig 1). The major TPE was identified with mild drought stress at the vegetative stage, with 15.4% of the total environments under mild drought throughout the growth season. Moderate drought stress was mainly affecting the whole growth season or reproductive stage while severe drought mainly occurred in the reproductive stage for 5.4% of the environments. There was no severe drought stress for the best rainfed season in Southeastern Asia. In this region, there was no significant yield penalty for rainfed rice if it is grown in the best local rainfed season. Rainfed rice would suffer severe drought stress in western parts of South Asia, which implies that rice yield could not reach 75% of the irrigated rice yield in more than 50% of the best rainfed seasons.

3.3 Yield stability of GSR varieties

The TPE type for severe drought stress in the vegetative stage (S1) did not exist for the best rainfed rice season in Southern Asia (Fig 1). Consequently, S1 was excluded in all succeeding analyses (Fig 2). The TPE types, S3, M1, and L2 only occurred in a few environments, hence, were also excluded in the next analysis for yield spatial variation (Fig 2D).

It was not surprising that the rainfed rice yields increased for all varieties as the drought stress changed from severe to mild stress among environments (Fig 2A), but the yield penalty caused by drought and the yield under severe drought stress were significantly different (Fig 2A and 2B). The drought stress at the reproductive stage had a stronger impact on rainfed rice production than the other two types of drought stresses (vegetative only, and combined vegetative and reproductive drought stresses) (Fig 2B). Yield variability followed a similar trend of yield penalty among TPEs (Fig 2C). Similar to yield penalty and variation (Fig 2B and 2C), the seasonal variation of yield also decreased as drought stress decreased (Fig 2E). However, this does not hold true for yield spatial variation (Fig 2D), where yield spatial variation was higher in mild drought stress environments than in moderate drought stress environments. As expected, the yield spatial and seasonal variations were much higher in severe drought stress than in moderate and mild drought stress environments because the impact was normally magnified in severe drought stress environments. This implies that the little changes on drought severity and/or occurring time would result in large changes in rice production.

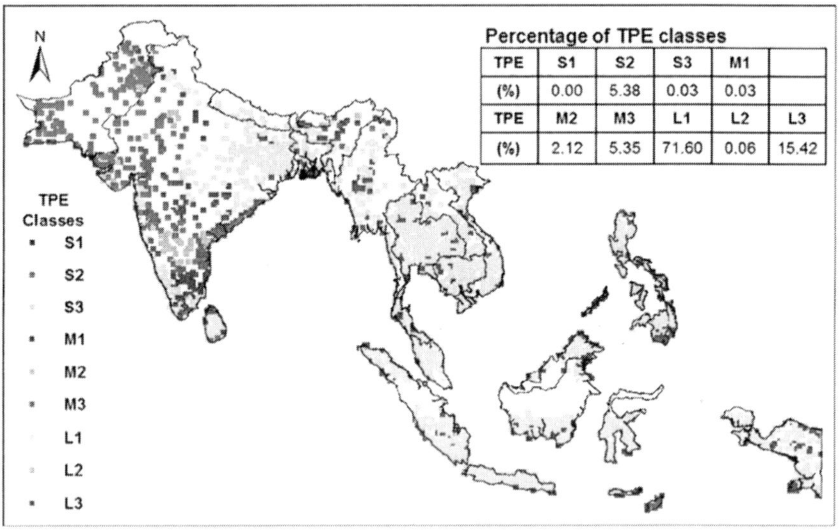

Figure 1. The TPE classes of the tested environments in South Asia. The TPE classes of drought stress to rainfed rice at possible areas and the best rainfed season.

Considering the yields and their stability over different TPEs for 3796 environments, the variety GSR-IR1-1-Y4-Y1 was outstanding among all other varieties due to its significantly higher GY, relatively lower yield penalty of drought, and relatively good grain yield stability over locations and seasons (Fig 2C–2E). Varying crop characteristics resulted in the different performance of all other varieties (Fig 2F). GSR-IR1-8-S6-S3-Y2 was also demonstrated to perform better than the check varieties PSBRC82,

NSICRc158, and IR74371-70-1-1 and was found to have good adaptability to all environments. The varieties GSR-IR1-5-S8-D3-SUB1 and GSR-IR1-8-S12-Y2-D1 were similar to the check varieties PSBRC82 and NSICRc158 (Table 6). In severe drought environments, where the yield penalty of drought may reach 85%, only GSR-IR1-1-Y4-Y1 and GSR-IR1-8-S6-S3-Y2 had better adaptability than or equal to the check variety PSBRc82 (Table 6). GSR-IR1-1-Y4-Y1, GSR-IR1-5-S8-D3-SUB1, GSR-IR1-5-S10-D1-D1, GSR-IR1-8-S6-S3-Y2, and GSR-IR1-8-S12-Y2-D1 were relatively more competitive with the check varieties under moderate and mild drought stress environments. Except for GSR-IR1-1-Y4-Y1 and GSR-IR1-5-S10-D1-D1, the other GSR varieties did not have an advantage to adapt to drought stresses at reproductive or combined vegetative and reproductive stages, while most GSR varieties adapted well to drought stress at the vegetative stage.

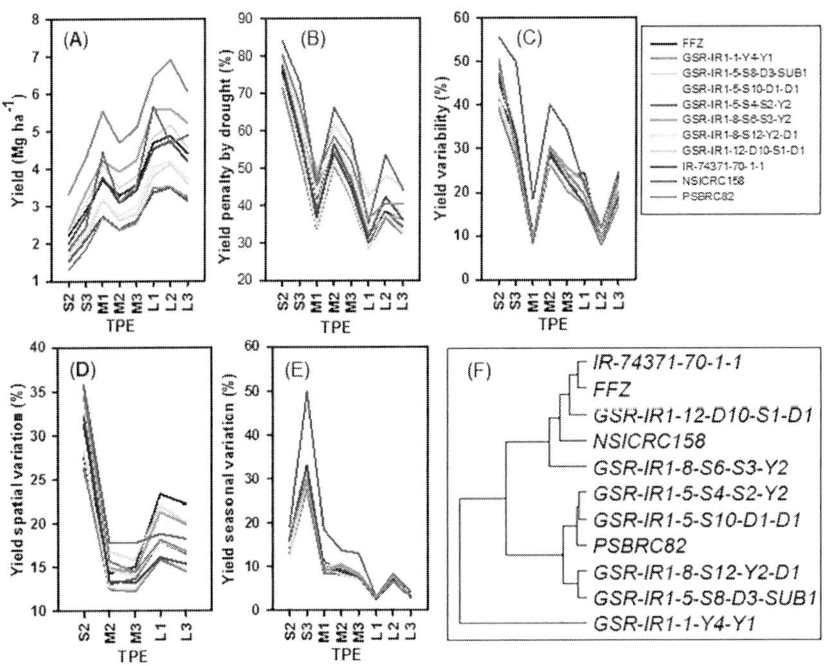

Figure 2. The rainfed rice yields, yield variation and stability of tested varieties over different TPEs. Yield and yield spatial variation.

Table 6. Adaptability ranking of the varieties to different levels of drought stress. Severity can be severe (S), moderate (M), and none to mild (L) which can occur at vegetative (V), reproductive (R), and combined vegetative and reproductive (V+R) timing.

VARIETY	Drought severity			Drought timing			Sum
	S	M	L	V	R	V+R	
FFZ	4	5	4	5	3	5	5
GSR-IR1-1-Y4-Y1	1	1	1	1	1	1	1
GSR-IR1-5-S8-D3-SUB1	5	3	2	2	3	2	3
GSR-IR1-5-S10-D1-D1	5	3	2	2	2	3	4
GSR-IR1-5-S14-S2-Y2	5	4	3	3	3	3	4
GSR-IR1-8-S6-S3-Y2	2	2	3	4	2	2	2
GSR-IR1-8-S12-Y2-D1	5	3	2	2	3	2	3
GSR-IR1-12-D10-S1-D1	4	4	3	4	3	4	4
IR-74371-70-1-1	3	5	4	5	3	5	5
NSIC Rc158	3	3	3	4	3	3	3
PSBRc82	2	4	3	3	3	2	3

3.4 Identification of potential adaptation regions for outstanding varieties

The performance of GSR-IR1-1-Y4-Y1 was outstanding among all other GSR and check varieties in all types of TPEs, hence, having a high potential for wide dissemination in Southern Asia. The suitability of this variety in different regions in Southern Asia should be explored (Fig 3), with a high suitability found in Southeast Asia and a low suitability in some western parts of South Asia (i.e. Pakistan and West India).

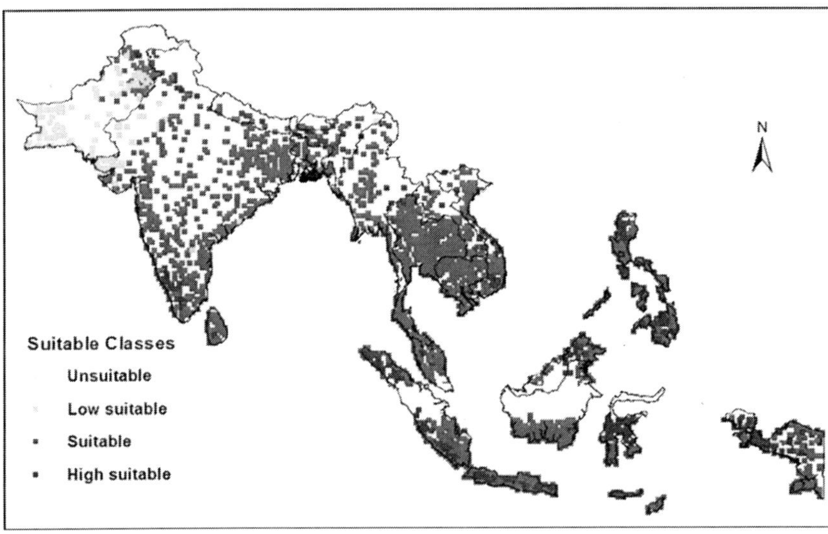

Figure 3. The potential dissemination areas of GSR-IR1-1-Y4-Y1. The potential dissemination regions of the identified outstanding variety GSR-IR1-1-Y4-Y1 in Southern Asia.

4. DISCUSSION

4.1 The consistency between field experiment and simulation

ORYZA (v3) well represented the biomass accumulation and final GY at site-specific conditions (Table 5, S2 and S3 Tables, and S1 Fig). The difference in rice yield was not statistically significant between the model predictions and the actual observations in the field experiments, which confirmed that

ORYZA (v3) is able to provide a reliable estimation of rice yield in various environments.

Using these reliable rice yield predictions, the performance of GSR varieties was evaluated across a large number of environments in comparison with check varieties. The same evaluation was also conducted using data from field experiments under limiting environments. The evaluation results from the simulated data were inconsistent with the results from the measured GYs (Table 7). The key parameters were not comparable between the results from the field experiments in limited environments and the simulation in a large number of environments in terms of average yield, variation, and penalty of yield in drought-prone environments, resulting in a different performance ranking of the tested varieties (Table 7, MET_{order} vs. P_{order}). This result suggests that the sampled environments of the field experiments did not cover the highly diverse environments across the region in simulation. While the varietal performance ranking was constrained only for experimental sites and 14 best rainfed seasons (S4 Table), the MET_{order} and S_{order} still did not completely agree with each other, but the ranking difference between them was insignificant (Table 7). However, S_{order} of the top performing varieties corresponds significantly to MET_{order} ($R^2 = 0.86$, $p<0.0001$) in this case. In other words, with site specific soil and weather information, the ORYZA (V3) model was able to correctly evaluate the varietal performances in the MET experiments for specific sites. However, the accuracy in evaluating varietal performances using the model approach was still constrained by the quality of input information on weather and soil. Further study should be conducted to assess how these kinds of information cause inconsistency of varietal performances, particularly under extreme (good and bad) environments. In practice, any variety should be definitely considered promising, provided it had a top ranking in grain yield in both field experiments under a limited number of environments and simulation results from a larger number of environments. In this study, the GSR IR1-1-Y4-Y1-Y1 and GSR IR1-8-S6-S3-Y2 were confirmed to be the most promising varieties for the rainfed areas using this approach, as they showed 10–20% and 20–40% grain yield advantages over the advanced check NSICRC158 and drought tolerant check IR74371-70-1-1, respectively, across large number of environments.

4.2 Evaluation from GEI to TPE

In the ORYZA (V3) model, the genotype is characterized by a series of quantitative parameters, including the weather, soil, and crop management information. These parameters are inputs for simulations, in which the genetic parameters for a rice variety are constant and the environment parameters constantly change.

Table 7. Results derived from field experiments in limited environments and simulations under a large number of environments.

ID	Field experiment						Large simulation				
	N	Yield	CV	Yield penalty	MET_{order}	S_{order}	N	Yield	CV	Yield penalty	P_{order}
FFZ	2x2	4.16	0.83	81.8	5	5	3796x14	4.42	0.29	39.0	5
GSR IR1-12-D10-S1-D1	2x3	3.84	0.84	82.0	8	6	3796x14	4.58	0.23	35.6	4
GSR IR1-1-Y4-Y1-Y1	2x1	4.23	0.45	61.9	4	3	3796x14	6.11	0.22	35.3	1
GSR IR1-5-S10-D1-D1	2x2	4.12	0.66	70.9	6	8	3796x14	3.31	0.27	39.1	9
GSR IR1-5-S14-S2-Y2	2x1	5.07	0.47	64.3	2	2	3796x14	3.17	0.25	37.4	11
GSR IR1-5-S8-D3-Sub1	2x1	6.29	0.59	74.5	1	1	3796x14	3.58	0.28	47.9	8
GSR IR1-8-S12-Y2-D1	2x2	3.16	0.91	86.8	11	9	3796x14	3.78	0.26	40.6	7
GSR IR1-8-S6-S3-Y2	2x1	4.68	0.49	66.0	3	4	3796x14	5.25	0.28	39.2	2
IR74371-70-1-1	2x2	3.87	0.65	71.4	7	7	3796x14	4.25	0.24	38.8	6
NSIC Rc158	3x1	3.67	0.38	91.6	10	11	3796x14	5.16	0.30	43.9	3
PSBRc82	2x1+3x1	3.73	0.68	69.0	9	10	3796x14	3.26	0.26	42.6	10

MET_{order} is the varietal performance rank of all 11 varieties from the best (1) to the worst (11) ranked by the data from MET field experiment, S_{order} is the varietal performance rank using simulation data under MET site-specific condition, and P_{order} is the varietal performance rank using simulation data under numerous environments. The N is number of environments while CV presents the coefficient of variation of yields among environments.

As a result, GEI, which was simply quantified by the simulated varietal yields through a series of eco-physiological processes in the model, may vary considerably due mainly to various environments. Consequently, environments can be classified using the GEI results represented by the simulation yields in this case. Similar approach has been proven as an effective and widely used method to classify environments of crop growth [28, 29, 30, 31]. It was used as classification of TPE for rice growth [32, 33]. In this study, the quantification of GEI by the ORYZA (V3) model used a set of variables for production environments as well as the number of environments and seasons [34], for classifying the TPEs of the rainfed ecosystems. However, the number and types of genotypes could significantly affect the classification of TPE (Fig 4). Thus, the number of different genotypes to be used in the model should not be less than ten in order to accurately define the types of TPE from GEI.

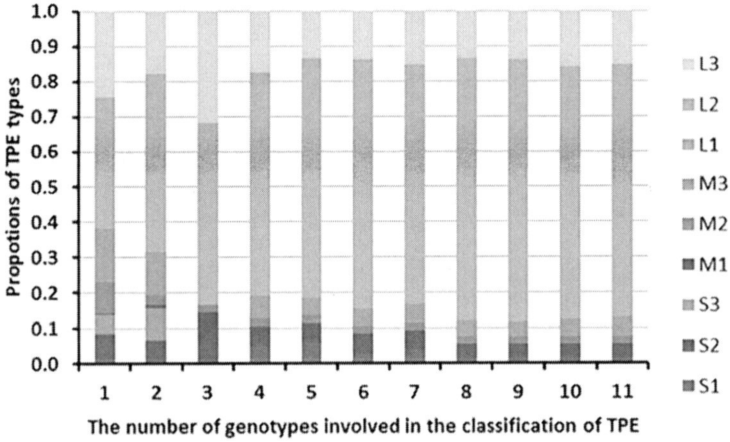

Figure 4. Proportional change in TPE types with increase in number of genotypes used for TPE classification. The TPE types in this study were defined by the drought stress severity and types.

It was evident that the variation of genotypes also affected the confidence levels pertaining to the definition of TPE types. In this study, all genotypes were assumed to have some degree of drought tolerance. The proportion of TPE with severe drought would increase when more drought-sensitive genotypes were used. Moreover, the best rainfed seasons used in the classification of TPE may differ to local rice farming seasons, introducing bias in the classification, which should be addressed in future studies. On the other hand, the defined TPE, as in severe drought, would be very useful in the selection of drought tolerant varieties. The drought tolerance of a new line would be better than the varieties used in this study if it performed better in a typical drought TPE.

5. CONCLUSIONS

ORYZA (v3) was able to reliably predict the total above-ground and panicle biomass, as well as the final grain yield of rice. The uncertainty of around 10% on grain yields was comparable to the coefficient of variation of the field measurements on these variables. In this study, calibration and validation were carried out in multiple environments. The reliable predictions on grain yield ensure the capacity of ORYZA (v3) to evaluate the performances of these calibrated varieties in a large number of environments. The crop modeling approach was demonstrated to effectively evaluate varietal performances over large numbers of environments with minimal costs and time. High quality evaluation on varietal performance using the modeling approach can be achieved with improved soil and weather information. The developed approach can also be used for other rice growing regions whenever the crop, soil, and weather data are available. Logically, this approach can also be applied on other crops by replacing the ORYZA (v3) rice model with the corresponding crop models and following the stepwise process.

Upon integration of the predicted grain yields of all the tested varieties, the environments involved in this study were classified into different drought stress TPEs. In Southern Asia, 86% of the current rice cultivated areas experience mild drought stress, of which 75% occurs at the vegetative stage and only 15% during the whole season if the cropping season was adjusted according to the local soil-climatic condition. Severe drought stress at the reproductive stage could possibly occur in 5.4% of the total area, located mainly in the North Western part (i.e. Pakistan and North West India) of Southern Asia. The TPE classification would be reliable only if the number of tested genotypes was ten or larger. Two GSR varieties, GSR-IR1-1-Y4-Y1 and GSR-IR1-8-S6-S3-Y2 consistently performed better than others across all types of TPE, particularly in TPEs with severe drought stress or drought stress at reproductive stage.

ACKNOWLEDGMENTS

This study was a collaboration between the Green Super Rice (GSR) and the Stress-Tolerant

REFERENCES

1. Perkins J. Green Revolution. 2011. Available: http://www.eoearth.org/view/ article/ 153125
2. Ali J, Santiaguel JF. Greener Rice. Rice Today. 2011. Available: http://ricetoday. irri.org/greener-rice/
3. Ali J, Xu JL, Gao Y, Fontanilla M, Li Z. Breeding for yield potential and enhanced productivity across different rice ecologies through green super rice (GSR) breeding strategy. In: Muralidharan K and Siddiq EA, eds. International Dialogue on Perception and Prospects of Designer Rice. Hyderabad, India: Society for the Advancement of Rice Research, Directorate of Rice Research; 2013b. pp. 60–68.
4. Ali J, Xu JL, Gao YM, Fontanilla MA, Li Z. Innovative Green Super Rice (GSR) molecular breeding strategy: achievements and advances. In: Proc 12th SABRAO Congress on Plant breeding towards 2025: Challenges in a rapidly changing world. Chiang Mai, Thailand: The Society for the Advancement of Breeding Research in Asia and Oceania; 2012. pp.16–17.
5. Marcaida M III, Li T, Angeles O, Evangelista GK, Fontanilla MA, Xu JL, et al. Biomass accumulation and partitioning of newly developed Green Super Rice (GSR) cultivars under drought stress during the reproductive stage. Field Crops Res. 2014; 162: 30–38.
6. Yorobe JM Jr, Ali J, Pede VO, Rejesus RM, Velarde OP, Wang H. Yield and income effects of rice varieties with tolerance of multiple abiotic stresses: the case of green super rice (GSR) and flooding in the Philippines. Agr Econ. 2016; 47: 1–11.
7. Li T, Raman AK, Marcaida M III, Kumar A, Angeles O, Radanielson A. Simulation of genotype performances across a larger number of environments for rice breeding using ORYZA2000. Field Crops Res. 2013a; 149: 312–321.
8. Poulton PL, Vesna T, Dalgliesh NP, Seng V. Applying simulation to improve rice varieties in reducing the on-farm yield gap in Cambodian lowland rice ecosystems. Experimental Agriculture. 2015; 51: 264–284.
9. Li T, Hasegawa T, Yin XY, Zhu Y, Boote K, Adam M, et al. Uncertainties in predicting rice yield by current crop models under a wide range of climatic conditions. Glob Chang Biol. 2015a; 21: 1328–1341. pmid:25294087
10. Li T, Angeles O, Radanielson A, Marcaida M III, Manalo E. Drought stress impacts of climate change on rainfed rice in South Asia. Clim Change. 2015b; 133: 709–720.
11. Anantha MS, Patel D, Quintana M, Swain P, Jawaharlal L, Dwivedi JL, et al. Trait Combinations That Improve Rice Yield under Drought: Sahbhagi Dhan and New Drought-Tolerant Varieties in South Asia. Crop Science 2015; 56: 408–421.
12. Peng SB, Khush GS, Virk P, Tang QY, Zou YB. Progress in ideotype breeding to increase rice yield potential. Field Crops Res. 2008; 108: 32–38
13. Bandillo N, Raghavan C, Muyco PA, Sevilla MAL, Lobina IT, Dilla-Ermita CJ, et al. Multi-parent advanced generation inter-cross (MAGIC) populations in rice: progress and potential for genetics research and breeding. Rice (2013); 6: 11. pmid:24280183
14. Li T, Bouman BAM, Boling A. The calibration and validation of ORYZA2000. IRRI. 2009. Available: https://sites.google.com/a/irri.org/oryza2000/calibration-and-validation.
15. Yadav S, Li T, Humphreys E, Gill G, Kukal SS. Evaluation and application of ORYZA2000 for irrigation scheduling of puddled transplanted rice in north west India. Field Crops Res. 2011; 122: 104–117.
16. Tayefe M. Calibration and evaluation of ORYZA2000 model under different N options in irrigated rice in Iran. J Biodivers Environ Sci. 2013; 3: 77–82.
17. Wang WG, Sun FC, Luo YF, Xu JZ. Changes of Rice Water Demand and Irrigation Water Requirement in Southeast China under Future Climate change. Procedia Eng. 2012; 28: 341–345.
18. International Rice Research Institute. Annual Report. 2013.

19. Bouman BAM, Kropff MJ, Tuong TP, Wopereis MCS, Ten Berge HFM, Van Laar HH. ORYZA2000: modeling lowland rice. Philippines: International Rice Research Institute, Netherlands: Wageningen University and Research Centre; 2001.
20. Storn R, Price K. Differential evolution–a simple and efficient heuristic for global optimization over continuous spaces. J Glob Optim. 1997; 11: 341–359.
21. Soundharajan B, Sudheer KP. Sensitivity analysis and auto-calibration of ORYZA2000 using simulation-optimization framework. Paddy and Water Environment. 2013; 11: 59–71.
22. Zhang T, Li T, Yang X, Simelton E. Model biases in rice phenology under warmer climates. Scientific Reports. 2016; 6:27355. pmid:27273847
23. Belder P, Bouman BAM, Spiertz JHJ. Exploring options for water savings in lowland rice using a modelling approach. Agr Syst. 2007; 92: 91–114.
24. Jing Q, Bouman BAM, van Keulen H, Hengsdjik H, Cao WX, Dai TB. Disentangling the effect of environmental factors on yield and nitrogen uptake of irrigated rice in Asia. Agr Syst. 2008; 98: 177–188.
25. Krause P, Boyle DP, Bäse F. Comparison of different efficiency criteria for hydrological model assessment. Advances in Geosciences. 2005; 5: 89–97.
26. Batjes NH. ISRIC-WISE derived soil properties on 5 by 5 arc-minutes global grid (ver. 1.1); 2006. Report 2006/02, ISRIC–World Soil Information, Wageningen.
27. Guan YS, Serraj R, Liu SH, Xu JL, Ali J, Wang WS, et al. Simultaneously improving yield under drought stress and non-stress conditions: A case study of rice (Oryza sativa L.). J Exp Bot. 2010; 61: 4145–4156. pmid:20660496
28. Chapman SC, Cooper M, Hammer GI. Using crop simulation to generate genotype by environment interaction effects for sorghum in water-limited environments. Aust J Agr Res. 2002; 53: 379–389.
29. Chapman SC, Cooper M, Hammer GI, Butler D. Genotype by environment interactions affecting grain sorghum. II. Frequencies of different seasonal patterns of drought stress are related to location effects on hybrid yield. Aust J Agr Res. 2000; 50: 209–222.
30. Chapman SC, Cooper M, Podlich D, Hammer GI. Evaluating plant breeding strategy by simulating gene action in dryland environment effects. Agron J. 2003; 95: 99–113.
31. Löffler CM, Wei J, Fast T, Geogerty J, Langton S, Bergman M, et al. Classification of maize environments using crop simulation and geographic information system. Crop Sci. 2005; 45: 1708–1716.
32. Heinemann AB, Camilo BP, Julian RV, Arango-Londono D, Bonilla-Findji O, Medeiros JC, et al. Variation and impact of drought-stress patterns across upland rice target population of environments in Brazil. J Exp Bot. 2015; 66: 3625–3638. pmid:25873681
33. Heinemann AB, Dingkuhn M, Luquet D, Combres JC, Chapman SC. Characterization of drought stress environments for upland rice and maize in Central Brazil. Euphytica. 2008; 162: 395–410.
34. Cooper M, Byth DE. Understanding plant adaptation to achieve systematic applied crop improvement: A fundamental challenge. In: Cooper M, Hammer GL, editors. Plant Adaptation and Crop Improvement. CAB International, IRRI, UK: 1996; pp. 5–23.

CHAPTER 18

SMART – Sunflower Mutant population And Reverse genetic Tool for crop improvement

Anish PK Kumar[1†], Adnane Boualem[2,3,4†], Anjanabha Bhattacharya[1], Seema Parikh[1], Nirali Desai[1], Andres Zambelli[5], Alberto Leon[5], Manash Chatterjee[1,6*] and Abdelhafid Bendahmane[2,3,4]

[1]*Bench Bio Pvt Ltd., c/o Jai Research Foundation, Vapi, Gujarat 396195, India*

[2]*INRA, UMR1165 Unité de Recherche en Génomique Végétale URGV, Evry F-91057, France.*

[3]*UEVE, UMR Unité de Recherche en Génomique Végétale URGV, Evry F-91057, France.*

[4]*CNRS, ERL8196 UMR Unité de Recherche en Génomique Végétale URGV, Evry F-91057, France.*

[5]*Biotechnology Research Centre, Nutrisun Business Unit- Advanta Semillas SAIC, Belcarce, Argentina.*

[6]*National University of Ireland Galway (NUIG), Galway, Ireland.*

AUTHORS' CONTRIBUTIONS

PKAK initiated EMS mutagenesis & standardization experiments, TILLING & maintaining TILLING population (MTP); ABo coordinated experiments, initiated forward genetic screens, participated in the bioinformatic analysis and carefully edited the manuscript; ABh drafted the MS, and participated in the mutant collection phenotyping; SP, ND carried the TILLING and bioinformatics analysis; AZ and AL provided critical inputs on genes to be

TILLed; MC, AB conceived the study, coordinated experiments and carefully edited the manuscript. All authors read and approved the final manuscript.

ABSTRACT

Background: Sunflower (*Helianthus annuus L.*) is an important oilseed crop grown widely in various areas of the world. Classical genetic studies have been extensively undertaken for the improvement of this particular oilseed crop. Pertaining to this endeavor, we developed a "chemically induced mutated genetic resource for detecting SNP by TILLING" in sunflower to create new traits.

Results: To optimize the EMS mutagenesis, we first conducted a "kill curve" analysis with a range of EMS dose from 0.5% to 3%. Based on the observed germination rate, a 50% survival rate i.e. LD_{50}, treatment with 0.6% EMS for 8 hours was chosen to generate 5,000 M2 populations, out of which, 4,763 M3 plants with fertile seed set. Phenotypic characterization of the 5,000 M2 mutagenised lines were undertaken to assess the mutagenesis quality and to identify traits of interest. In the M2 population, about 1.1% of the plants showed phenotypic variations. The sunflower TILLING platform was setup using Endo-1-nuclease as mismatch detection system coupled with an eight fold DNA pooling strategy. As proof-of-concept, we screened the M2 population for induced mutations in two genes related to fatty acid biosynthesis, FatA an acyl-ACP thioesterase and SAD the stearoyl-ACP desaturase and identified a total of 26 mutations.

Conclusion: Based on the TILLING of *FatA* and *SAD* genes, we calculated the overall mutation rate to one mutation every 480 kb, similar to other report for this crop so far. As sunflower is a plant model for seed oil biosynthesis, we anticipate that the developed genetic resource will be a useful tool to identify novel traits for sunflower crop improvement.

Keywords: Sunflower TILLING Functional genomics Oil quality Fatty acids Crop improvement

1. BACKGROUND

Sunflower (*Helianthus annuus* L.), 2n =34, belongs to family Asteraceae, with an estimated genome size of 3000 Mbp, and is the fourth most important oilseed crop [1]. It is a native of North America [2] and widely cultivated in the world with an annual production of about 24 million tonnes (http://www.fao.org). The genome size of this crop is large compared to model species like Arabidopsis (125 Mbp), Rice (430 Mbp), Sorghum (750 Mbp), Soybean (1100 Mbp) or Tomato (950 Mbp) (reviewed by [3]). The major drivers of this large genome size are mainly the recent polyploidization event and the important amplification of transposable elements [2, 4]. Many wild sunflower species are increasingly used for improvement of cultivated varieties by conventional breeding [5]. With continued biased selection of the cultivated sunflower for traits such as yield, many alleles conferring useful traits got lost. Therefore a method to create variability for breeding this crop, in this era of increasing global food crisis and changing climatic regimes, is highly desirable. For many decades, plant breeders have concentrated their efforts on improvement of sunflower through traditional breeding and recently, molecular mapping has been successfully undertaken for marker assisted breeding (MAS) [6]. Furthermore, chemical mutagenesis has been used by breeders to create variability in crops including sunflower but these have been mostly restricted to dominant traits. Therefore, many desirable mutations that are recessive have been missed during selection [7]. Finally, traits such as oil quality cannot be selected by visual inspection of plants.

Sunflower is one of the major oil seed crops grown all over the world and considerable research efforts have been put to understand lipids biosynthesis and modification. This research mainly focused on (i) the production of more stable sunflower oils for biolubrication (high oleic acid content) and (ii) the increase of healthy substitutes for food industry [8]. The *de novo* synthesis of fatty acids in plant storage tissues is an intraplastidial process in which the multienzyme fatty acid synthase (FAS) complex catalyses a series of enzymatic reactions. In sunflower, the main products of intraplastidial fatty acid synthesis are, first, palmitoyl-ACP (16:0-ACP), which is further elongated by the FAS II complex to produce stearoyl-ACP (18:0-ACP). In turn, this molecule is the substrate for stearoyl-ACP desaturase (SAD) that introduces a double bond in the carbon chain to produce oleoyl-ACP (18:1-ACP). All of these products can be exported from the plastid after the hydrolysis between the acyl molecule and ACP by the acyl-ACP thioesterase. Two types of acyl-ACPs thioesterases have been identified in higher plants such as sunflowers: FatA and FatB. FatA thioesterases preferentially act on long chain fatty acids, and have particularly high specificity for 18:1-ACP and a lower affinity for 16:0-ACP and 18:0-ACP [9]. To address the needs of the confectionery industry for saturated fatty acids, high stearic acid content oils have been developed mainly by genetic modification of the FatA stearoyl-ACP thioesterase and the SAD stearoyl-ACP desaturase [10, 11]. Stearic fatty acid is considered cardiovascular neutral and do not modify the plasmatic cholesterol levels in humans [12].

TILLING (Targeting Induced Local Lesion IN Genome) is a technology to detect induced and natural polymorphisms (SNP) in plants [13]. The critical steps in TILLING procedure include the purity of seed material, the quality of the mutagenesis and the size of the mutant collection. TILLING relies on the ability of a group of enzymes, the endonucleases, to detect mismatches in a specific gene ([14]). With the introduction of high throughput TILLING technology ([15]), detection of even rare recessive mutation is possible [16, 17]. The significant advantages of TILLING method in creating and discovering new traits is that it significantly costs less for product development compared to transgenic crop plants. This makes it an attractive tool for crop improvement. As sequencing technology is getting advanced, deciphering small changes, like SNPs, which plays an important role in adaptive response and evolution of species is possible [3].

Sunflower, cultivated as a source of vegetable oil and protein is an attractive model for investigating seed oil quality. To investigate such trait, we have developed a reference EMS mutant population under controlled conditions and established a TILLING platform. As proof-of-concept, we screened for mutations in two genes, *FatA* and *SAD*, involved in accumulation of short to medium chain fatty acids and identified 26 induced mutations.

2. RESULTS

2.1 EMS mutagenesis

A relatively large TILLING resource comprising 5,000 M2 families was developed for sunflower crop improvement using variety BBS-1. To find the optimum dose of EMS that produce a maximum mutation density without causing extensive sterility and lethality; a 'kill curve' analysis was conducted on batches of 100 seeds treated with different EMS doses, form 0% to 3%, during 8 hours (Table 1). At 0.5% and 0.6% EMS, seed germination rate was 54% and 50%, respectively. Drastic plant lethality was observed for seeds treated with an EMS dose greater than 1% and total lethality was observed at 3% EMS. Finally, 0.6% EMS treatment during 8 hours, corresponding to the LD50 was retained and tested on large batches of seeds.

A total of 25,000 seeds were mutagenised and sown in pots to raise the nursery for creating the TILLING sunflower genetic resource. About 12,500 M1 plants were obtained, showing 50% lethality at seedling stage. After transplanting, further mortality occurred and a total of 6,500 M1 plants produced seeds, among which 5,000 M1 plants produced a good seed set.

M2 seeds were harvested from the individual M1 plants. In total, 5,000 M2 seed stocks were harvested and corresponding M3 seeds were produced.

Table 1 Impact of EMS concentration on seed germination

EMS dose (%)	Treated seeds	Seed germination (%)
0	100	78
0,5	100	54
0,6	100	50
0,75	100	44
1	100	38
2	100	3
3	100	0

2.2 M2 plant phenotyping

Ten seeds per M2 family were sown in pots to raise the nursery and to screen for visual developmental phenotypes. After 4 weeks, plants were transferred to the fields and about 95% of the M2 plants were recovered. The phenotypic scoring was based on the observation of each plant, using the untreated sunflower phenotype as a reference. The phenotypic screen included dwarfism, plant height, flowering pattern, the leaf architecture and sterility. The data collected are shown in Table 2. Photographs were taken at regular intervals for record purposes (Figure 1). A range of phenotype was observed and some are described below. Compared to the untreated sunflower, 10 and 4 plants showed dwarf and tall phenotype, respectively (Table 2). Some of these dwarf plants can be used in creating hybrids with enhanced yield and dwarf stature. For the flower phenotype, we identified 12 plants with large flower heads (potential to increase yield), 13 plants with twin flowers, 8 plants with fused and/or closed flower and 3 plants with upright flower (Table 2 and Figure 1B-D). Three plants harbored variation in flower color, 1 plant produced small seeds and 3 plants were sterile as both disc and ray florets were missing. In total 57 M2 plants, representing 1.1% of the population, showed at least one altered trait.

2.3 Mutation discovery by TILLING

Screening for mutations in the two genes, *FatA* and *SAD*, was used as proof-of-concept to determine the quality of the sunflower mutant collection and to estimate the mutation density. Both of these genes control production of short to medium chain fatty acids [18, 19]. SAD (Stearoyl-ACP desaturase) controls the production of stearic acid over oleic acid and inhibition of this enzyme leads to an increase stearic acid synthesis and a reduction of oleic acid production [18]. There are two isoforms of acyl-ACP thioesterase, FatA and FatB which dictate the length of medium chain fatty acids production. *FatA* controlling essential role in chain termination of fatty acid synthesized [19] was used in the present study.

Table 2 List of mutant phenotype classes

Category	Sub-category	Nb of plants	%
Plant height	Tall	4	0,08
	Dwarf	10	0,2
Flowering	Large flower head	12	0,24
	Twin flower	13	0,26
	Fused flower	2	0,04
	Closed flower	6	0,12
	Sterile Flower	3	0,06
	Colour	3	0,06
	Upright flower	3	0,06
Leaf architecture	Aberrant	1	0,02
Total		57	1,1

We screened for mutations in the whole exonic and promoter regions of *FatA* and *SAD* genes. Mutations were detected in the amplified targets using the mismatch-specific endonuclease ENDO1 as previously described (Figure 2, [20, 21]). TILLING of FatA yielded 12 independent point mutations, which correspond to 6 silent, 3 missense, 1 stop codon and 2 promoter region mutations (Table 3). TILLING of SAD yielded 14 independent point mutations, which correspond to 8 silent and 6 missense mutations (Table 3). All the identified mutations were confirmed by sequencing (Figure 3) and, as expected for EMS mutagenesis, single nucleotide substitutions were identified both in coding and non-coding regions and classical EMS mutations G to A and C to T were induced (data not shown).

Figure 1 Phenotypes identified in SMART genetic resource. **A**). albino seedling, **B**) sterile flower head, **C**) fused flowers and **D**) twin flowers.

We calculated the mutation frequency in the *FatA* and *SAD* genes (Table 3) according to Dalmais et al.[20]. We estimated the SMART average mutation density to one mutation every 480 kb. This mutation rate is similar to those reported in the other diploid crops (reviewed by [3]).

The large number of mutations identified in the TILLING screens, i.e. 26 independent mutations, confirms that the genetic resource developed is of high quality. Background mutations in the interesting mutant line can be removed by backcrossing with the parental non-mutated line and cleaned lines can be released after trials as a variety. As the mutation is known, the released varieties can be followed in the field by Marker Assisted Selection (MAS). Mutations resulting in STOP codon and in splicing junction of exon-intron boundary are of great potential as found in the present study. Currently bioinformatic tools are available to predict the severity of the non-synomonous mutations [22]. Synonymous substitutions have the potential to alter the production and function of a protein through folding changes, although the exact mechanism and frequency of such event is still a mystery [23].

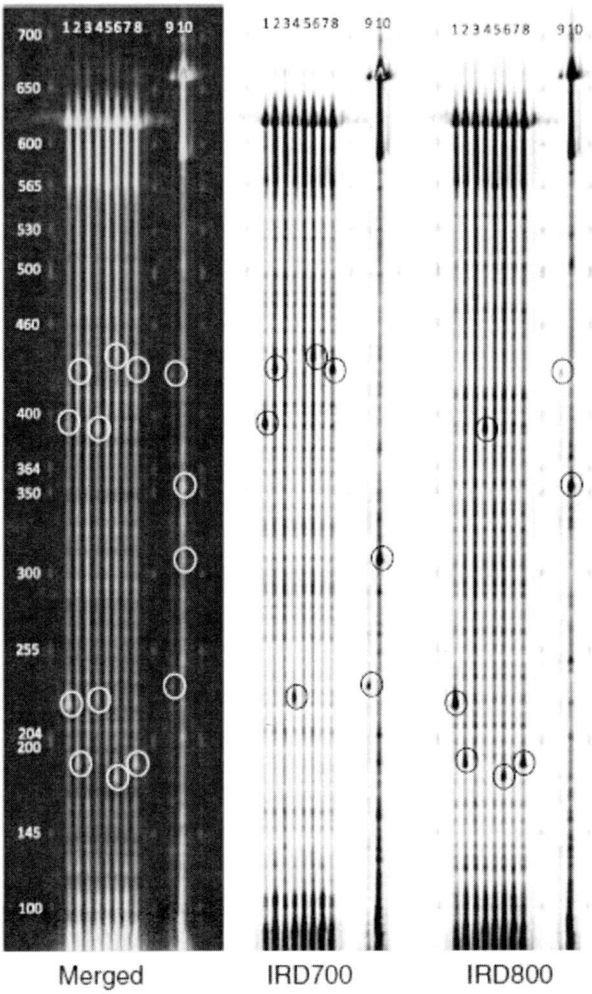

Figure 2 Li-CoR Gel electrophoresis and identification of mutants.

3. DISCUSSION

As the world population continues to grow at a rapid pace, crop varieties with novel traits such as improved growth and yield potential are needed to feed in the future. Plant breeding is an old science which played a significant role during green revolution for its contribution in increasing crop yields.

Table 3 Types of induced mutation in FatA and SAD and mutation frequency

Target genes	Function	Fragment Size (bp)	Identified mutations						Mutation frequency (1/kb)
			M ssense	Stop codon	Silent	Intronic	Promoter region	Total	
FatA	Fatty acid biosynthesis	1,881	3	1	6	1	2	12	1/783
SAD	Fatty acid biosynthesis	619	6	0	8	0	0	14	1/221
Total		2,500	9	1	13	1	2	26	1/480

The mutation frequency for each amplicon is calculated as follows:
[(size of the amplicon) × (total number of samples screened)] / (total number of identified mutants). The average mutation frequency was estimated to be one mutation per 480 kb.

At present, there is skepticism against growing transgenic crops and transgenic technology, mainly because of the high cost incurred in creating a variety due to lengthy regulatory hurdles, which is almost absent with mutation breeding. TILLING provides a way to overcome problems associated with other techniques used in plant improvement like RNAi, over expression of genes, antisense technique, site directed mutagenesis by zinc fingers, T-DNA knockouts and transposon tagging [3, 24].

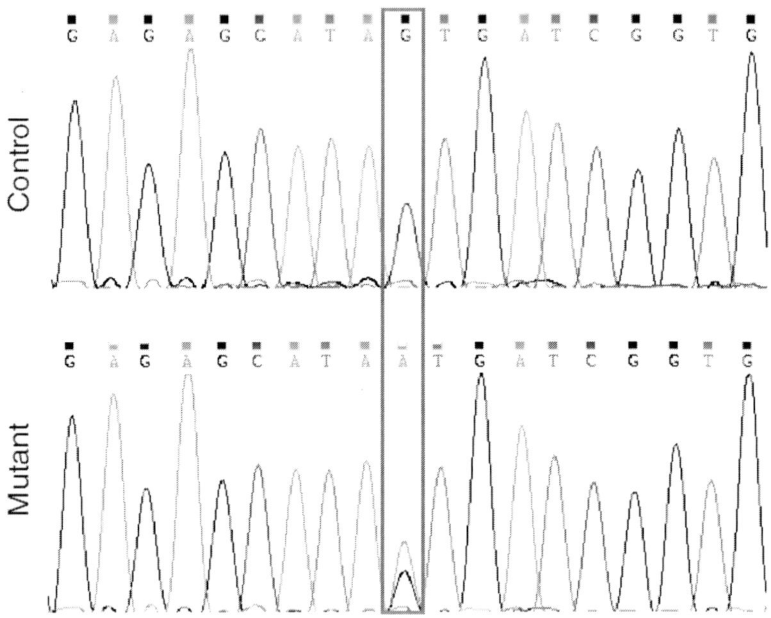

Figure 3 Mutant detection by sequencing. Individual plants from the same M2 family have been sequenced to indentify the nucleotide substitution. Sequences of a control plant (top) and a heterozygous mutant (bottom) are shown.

Many useful alleles involved in qualitative and quantitative parameters that exist in natural crop populations have been eliminated in the numerous rounds of selection for targeted traits. Therefore, creation of novel genetic resource by mutagenesis is a useful tool in crop improvement. In this study, we present a sunflower TILLING genetic resource produced by EMS chemical mutagenesis. To set up the sunflower TILLING platform, we developed an EMS mutant collection under controlled condition. The 5,000 M2 mutant collection was phenotyped and genomic DNA was prepared from the mutant plants and organized in pools for bulked screens. The mutant collection phenotyping showed that 1.1% of the mutant plant present at least on altered trait and confirm the quality of the mutagenesis. The sunflower mutant collection could be used for both forward and reverse genetics.

Sunflower has not been sequenced until now, owing to its large genome size (about 3,000Mbp). However, 443,858 ESTs are available at NCBI as of today. Studies by Chapman et al. [25] conducted genome scan in cultivated sunflower to predict genes involved in evolution of modern day cultivated variety. Lai et al. [26] published SNPs identified from submitted EST information in gene banks. Therefore, comparative genomic analysis can successfully be performed to pinpoint key gene information relating to various traits in this species by comparing known genetic information from other sequenced species.

To validate the quality of mutagenised population, we screened for mutants in FatA and SAD genes involved in oil biosynthesis. Many other published reports are available today pertaining to changes in seed oil composition in crop varieties. For example, mutations induced in the gene encoding Δ12 fatty acid desaturase (FAD2). Screening for FAD2 variants which increase oleic acid has been confirmed in soybean [27] and novel FAD2 alleles have been pursued in peanut [28] and sunflower [24]. In the SMART mutant population, the mutation frequency was estimated to be 1/480 kb similar to the recently published reports in sunflower (1/475, [24]) and other crops (Arabidopsis 1/300 kb, Soybean 1/250 kb, Rice 1/265 kb, Tomato 1/322 kb; reviewed by [3]).

Chemical mutagenesis and TILLING approach together can be used to develop plants not only for the purpose of increasing productivity and quality parameters, but also in phytoremediation [29]. TILLING has been successfully used in many crops to obtain useful alleles such as long shelf-life in melon and tomato [13, 30, 31, 32, 33], improved starch quality in potato [34], virus resistance in tomato and pepper [35, 36, 37]. A number of crops are being sequenced at present, sequence information of more than 30 crops are available now and more than hundred crop plants are being sequenced (NCBI Genome Project database;). Many key value-added traits can be obtained in sunflower with candidate gene polymorphism.

4. CONCLUSION

A TILLING genomic resource was developed for cultivated sunflower. Novel mutants having useful mutation in genes can be found by screening the population through high throughput ADP (Allele Discovery Platform) technique as mentioned in this article. Results obtained by screening two proof-of-concept genes are highly promising and confirms that the population is well mutagenized. Additional candidate genes can be screened from the resource developed. Such approach can also be used to develop markers for MAS (Marker assisted selection) strategy in sunflower breeding programs.

5. METHODS

5.1 Standardization of EMS dose and determination of LD_{50}value

Standardization of EMS (Ethyl Methane Sulfonate) mutagenesis treatment was done to determine LD_{50} value (lethal dose causing 50% reduction in seed germination) for the sunflower variety BBS-1 (BenchBio Sunflower-1). Kill curve analysis was done by treating 100 seeds with different EMS concentration (from 0 to 3% in deionized water) during 8 hours. Seeds were then thoroughly washed and transferred to wet tissue papers and kept in a growth chamber at 22°C with 12 hours photoperiod. This analysis was done in triplicate.

Finally, the optimal condition (0.6% EMS for 8 hours) was selected to mutagenise 25,000 M1 seeds with EMS. The seeds were sown in BenchBio Field 1–3, with a spacing of 60 cm × 45 cm. Standard cultivation practices were followed to grow the M1 plants. Self pollination was encouraged by covering inflorescence with a cloth bag before anthesis. Mature seed head were carefully tagged and collected from individual plants. The M2 seeds were harvested, packed and stored in cold room (8°C with 45-50% RH).

5.2 Genomic DNA extraction

Ten seeds per M2 families were sown in small pots inside the glasshouse and leaf material was collected from 3rd and 4th leaf of eight individual plants per M2 family, 3 weeks post germination, in 96-well plates containing 2 steel beads (4 mm) per well. Tissues were ground using a custom built bead mill [38]. Approximately 100 mg was used for genomic DNA extraction using the Dneasy 96 Plant Kit (Qiagen, Hilden, Germany). The extracted genomic DNAs were run on a 1% agarose gel (HiMedia, India) to check quality. The quantity and purity of the genomic DNA samples were confirmed by UV spectrophotometer 1770 (Shimadzu Co., Japan). Individual DNA concentrations were normalized to 40 ng/ul and then used to prepare 8 fold pooling plates for TILLING screens.

5.3 PCR amplification and mutation detection

PCR amplification is based on nested-PCR. The first PCR amplification is a standard PCR reaction using target-specific primers (Table 4) and 4 ng of sunflower genomic DNA. One µl of the first PCR served as a template for the second nested PCR amplification, using combination of specific primers labelled at the 5′end and 3′end with infra-red dyes IRD700 and IRD800 (Table 4), respectively. Mutation detection was carried out as described previously [31]. The identity of the mutations was determined by sequencing. TILLING request should be addressed to the corresponding author.

Table 4 Primers used

Primers	Gene name	Sequence (5′..3′)
BBPL0078F	FATA	CACTATACATAACCCACTCCGTACC
BBPL0136R	FATA	TTAACAAAAACCTAACCTGCAACAG
BBPL0138F	FATA	TGATGCATTTAGATGATGAATTTTG
BBPL0078R	FATA	AAAAGAAAGAACCGTGTTATCAGAG
BBPL0080F	FATA	TGAAATTTACAGATATCCTGCTTGG
BBPL0080R	FATA	GGATAAAAATGGCAAGATTTCAAAC
BBPL0082F	FATA	AAAGACTGAATTGCCCAAGTTTAC
BBPL0082R	FATA	CTATATCCCAGAATAGGCGAAATG
BBPL0146F_IRD700	FATA	CAAAACCTTCACACCTAACCTTTTC
BBPL0146R_IRD800	FATA	GTCTTATTAATCCCGACCTCATAAC
BBPL0147F_IRD700	FATA	CGTATTATAGGAGGTAGGAGGAAATC
BBPL0147R_IRD800	FATA	ATATCGTCATTGACTTTCTGGAGTC
BBPL0107F_IRD700	FATA	ATCAAAGATCATTCCAATGGTGAG
BBPL0107R_IRD800	FATA	ACGAATGGCATTTTACTCTTCTTG
BBPL0090F_IRD700	FATA	AAATTCGATGGAGTTAACGAGCTG
BBPL0090R_IRD800	FATA	AAGGATAAATACGAAACACAGATGC
BBPL0039F	SAD	AGACGTTGTTGACGTGAAACAC
BBPL0039R	SAD	CCATGGAAAGCAACAGAACC
BBPL0069F_IRD700	SAD	TTTGGGCGACTTATTTAATGC
BBPL0069R_IRD800	SAD	CTTGTTTTTTCAGAGCTTCAC

5.4 Forward Genetic Screen (FGS)

Five seeds per M2 family were grown to maturity and phenotyped, at 4 months post-germination, in the field for traits related to yield and plant architecture. From each M2 plant, M3 seeds were collected and packed. Number of fertile M2 plants was also evaluated.

ACKNOWLEDGEMENTS

We would like to thank the Shroff family, and Shroff family fund for financial support and encouragement. The authors wish to thank JRF Directors and Mr. Sudhansu Gupta and Mr. Bipin Parmar for administrative support. Mr. Lalman Paul and Mr. Sanjay Parmar for their support in growing mutant population and caring of plants.

REFERENCES

1. Kolkman JM, Berry ST, Leon AJ, Slabaugh MB, Tang S, Gao W, Shintani DK, Burke JM, Knapp SJ: Single nucleotide polymorphisms and linkage disequilibrium in sunflower. Genetics. 2007, 177 (1): 457-468. 10.1534/genetics.107.074054.
2. Harter AV, Gardner KA, Falush D, Lentz DL, Bye RA, Rieseberg LH: Origin of extant domesticated sunflowers in eastern North America. Nature. 2004, 430: 201-205. 10.1038/nature02710.
3. Rashid M, He G, Guanxiao Y, Khurram Z: Relevance of tilling in plant genomics. AJCS. 2011, 5 (4): 411-420.
4. Staton SE, Bakken BH, Blackman BK, Chapman MA, Kane NC, Tang S, Ungerer MC, Knapp SJ, Rieseberg LH, Burke JM: The sunflower (Helianthus annuus L.) genome reflects a recent history of biased accumulation of transposable elements. Plant J. 2012, 10.1111/j.1365-313X.2012.05072.
5. Seiler GJ: Utilization of wild sunflower species for the improvement of cultivated sunflower. Field Crop Res. 1992, 30 (Suppl 3–4): 195-230.
6. Micic Z, Hahn V, Bauer E, Schon C, Melchinger A: QTL mapping of resistance to Sclerotinia midstalk rot in RIL of sunflower population NDBLOSsel × CM625. Theor Appl Genet. 2005, 110 (Suppl 8): 1490-1498.
7. Barkley NA, Wang ML: Application of TILLING and EcoTILLING as reverse genetic approaches to elucidate the function of genes in plants and animals. Curr Gen. 2008, 9: 212-226. 10.2174/138920208784533656.
8. Garcés R, Martínez-Force E, Salas JJ, Venegas-Calerón M: Current advances in sunflower oil and its applications. Lipid Technology. 2009, 21: 79-82. 10.1002/lite.200900016.
9. Pleite R, Martínez-Force E, Garcés R: Inhibitors of fatty acid biosynthesis in sunflower seeds. J Plant Physiol. 2006, 163 (9): 885-b94. 10.1016/j.jplph.2005.11.017.
10. Facciotti MT, Bertain PB, Yuan L: Improved stearate phenotype in transgenic canola expressing a modified acyl-acyl carrier protein thioesterase. Nat Biotechnol. 1999, 17: 593-597. 10.1038/9909.
11. Zhang P, Burton JW, Upchurch RG, Whittle E, Shanklin J, Dewey RE: Mutations in a Δ9-stearoyl-ACP-desaturase gene are associated with enhanced stearic acid levels in soybean seeds. Crop Sci. 2008, 48: 2305-2313. 10.2135/cropsci2008.02.0084.

12. Kris-Etherton PM, Griel AE, Psota TL, Gebauer SK, Zhang J, Etherton TD: Dietary stearic acid and risk of cardiovascular disease: intake, sources, digestion, and adsorption. Lipids. 2005, 40: 1193-1200. 10.1007/s11745-005-1485-y.
13. Colbert T, Till BJ, Tompa R, Reynolds S, Steine MN, Yeung AT, McCallum CM, Comai L, Henikoff S: High-throughput screening for induced point mutations. Plant Physiol. 2001, 126: 480-484. 10.1104/pp.126.2.480.
14. Triques K, Sturbois B, Gallais S, Dalmais M, Chauvin S, Clepet C, Aubourg S, Rameau C, Caboche M, Bendahmane A: Characterization ofArabidopsis thalianamismatch specific endonucleases: application to mutation discovery by TILLING in pea. Plant J. 2007, 51: 1116-1125. 10.1111/j.1365-313X.2007.03201.x.
15. Till BJ, Reynolds SH, Greene EA, Codomo CA, Enns LC, Johnson JE, Burter C, Odden AR, Young K, Taylor NE, Henikoff JG, Comai L, Henikoff S: Largescale discovery of induced point mutations with high-throughput TILLING. Genome Res. 2003, 13: 524-530. 10.1101/gr.977903.
16. McCallum CM, Comai L, Greene EA, Henikoff S: Targeting Induced Local Lesions IN Genomes (TILLING) for plant functional genomics. Plant Physiol. 2000, 123: 439-442. 10.1104/pp.123.2.439.
17. Colbert TG, Hurst SR, Slade AJ: Tomatoes that soften more slowly post-harvest due to non-transgenic alterations in an expansin gene. US: US Patent Application; 2001. 20110113507
18. Zarhloul KM, Stoll C, Luhs W, Syring-Ehemann A, Hausmann L, Toper R, Friedt W: Breeding high-stearic oilseed rape (Brassica napus) with high and low-erucic background using optimised promoter-gene constructs. Mol Breed. 2007, 18: 241-251.
19. Jones A, Davles HM, Voelker TA: Palmitoyl-Acyl carrier protein (ACP) thioesterase and the evolutionary origin of plant Acyl-ACP Thioesterases. Plant Cell. 1995, 7: 359-371.
20. Dalmais M, Schmidt J, Signor CL, Moussy F, Burstin J, Savois V, Aubert G, Brunaud V, Oliveira Y, Guichard C, Thompson R, Bendahmane A: UTILLdb, a pisum sativum in silico forward and reverse genetics tool. Genome Biol. 2008, 9: 43R. 10.1186/gb-2008-9-2-r43.
21. Triques K, Piednoir E, Dalmais M, Schmidt J, Le Signor C, Sharkey M, Caboche M, Sturbois B, Bendahmane A: Mutation detection using ENDO 1: application to disease diagnostics in humans and TILLING and Eco-TILLING in plants. BMC Mol Biol. 2008, 9: 42. 10.1186/1471-2199-9-42.
22. Taylor NE, Greene EA: PARSESNP: a tool for the analysis of nucleotide polymorphisms. Nucleic Acids Res. 2003, 31: 3808-3811. 10.1093/nar/gkg574.
23. Tsai CJ, Sauna ZE, Kimchi-Sarfaty C, Ambudkar SV, Gottesman MM, Nussinov R: Synonymous mutations and ribosome stalling can lead to altered folding pathways and distinct minima. J Mol Biol. 2008, 383: 281-291. 10.1016/j.jmb.2008.08.012.
24. Sabetta W, Alba V, Blanco A, Montemurro C: sunTILL: a TILLING resource for gene function analysis in sunflower. Plant Methods. 2011, 7: 20. 10.1186/1746-4811-7-20.
25. Chapman MA, Pashley CH, Wenzler J, Hvala J, Tang S, Knapp SJ, Burke JM: A genomic scan for selection reveals candidates for genes involved in the evolution of cultivated sunflower (Helianthus annuus). Plant Cell. 2008, 20 (Suppl 11): 2931-45.
26. Lai Z, Livingstone K, Zou Y, Church SA, Knapp SJ, Andrews J, Rieseberg LH: Identification and mapping of SNPs from ESTs in sunflower. Theor Appl Genet. 2005, 111: 1532-44. 10.1007/s00122-005-0082-4.
27. Dierking EC, Bilyeu KD: New sources of soybean seed meal and oil composition traits identified through TILLING. BMC Plant Biol. 2009, 9: 89. 10.1186/1471-2229-9-89.
28. Knoll J, Ramos ML, Zeng Y, Holbrook CC, Chow M, Chen S, Maleki S, Bhattacharya A, Ozias-Akins P: TILLING for allergen reduction and improvement of quality traits in Peanut. BMC Plant Biol. 2011, 11: 81. 10.1186/1471-2229-11-81.
29. Nehnevajova E, Herzig R, Federer G, Erismann KH, Schwitzguébel JP: Chemical mutagenesis - a promising technique to increase metal concentration and extraction in sunflowers. Int J Phytoremediation. 2007, 9 (Suppl 2): 149 65.
30. McCallum CM, Slade AJ, Colbert TG, Knauf VC, Hurst S: Tomatoes having reduced polygalacturonase activity caused by non-transgenic mutations in the polygalacturonase gene. US: US Patent; 2008. 7393996

31. Dahmani-Mardas F, Troadec C, Boualem A, Leveque S, Alsadon AA, Aldoss AA, Dogimont C, Bendahmane A: Engineering melon plants with improved fruit shelf life using the TILLING approach. PLoS One. 2010, 5: e15776. 10.1371/journal.pone.0015776.
32. Hurst SR, Loeffler DL, Steine MN, Amen A, Vafeados A: Non-transgenic tomato varieties having increased shelf life post-harvest due to alterations in beta-galactosidase 4. 2011, US: US Patent Application, 20110159168
33. Gonzalez M, Xu M, Esteras C, Roig C, Monforte AJ, Troadec C, Pujol M, Nuez F, Bendahmane A, Garcia-Mas J, Pico B: Towards a TILLING platform for functional genomics in Piel de sapo melons. BMC Research Notes. 2011, 4: 289. 10.1186/1756-0500-4-289.
34. Muth J, Hartje S, Twyman RM, Hofferbert HR, Tacke E, Prüfer D: Precision breeding for novel starch variants in potato. Plant Biotech. Journal. 2011, 6: 576-584.
35. Piron F, Nicolai M, Minoia S, Piednoir E, Moretti A, Salgues A, Zamir D, Caranta C, Bendahmane A: An induced mutation in tomato eIF4E leads to immunity to two potyviruses. PLoS One. 2010, 5: e11313. 10.1371/journal.pone.0011313.
36. Ibiza VP, Cañizares J, Nuez F: EcoTILLING in Capsicum species: searching for new virus resistances. BMC Genomics. 2010, 11: 631. 10.1186/1471-2164-11-631.
37. Jeong HJ, Kwon JK, Pandeya D, Hwang J, Hoang NH, Bae JH, Kang BC: A survey of natural and ethyl methane sulfonate-induced variations of eIF4E using high-resolution melting analysis in Capsicum. Mol Breeding. 2011. 10.1007/s11032-011-9550-5.
38. Abraham T, Kumar PK, Chatterjee M, Shroff J, Shroff V: Method and apparatus for gravity assisted high throughput disruption of cellular material. India Patent Application; 2009: 2308/MUM/09.http://ipindiaservices.gov.in/patentsearch/search/index.aspx.

INDEX